Advanced Composite Biomaterials

Advanced Composite Biomaterials

Editors

Stefan Ioan Voicu
Marian Miculescu

MDPI • Basel • Beijing • Wuhan • Barcelona • Belgrade • Manchester • Tokyo • Cluj • Tianjin

Editors
Stefan Ioan Voicu
University Politehnica of
Bucharest
Romania

Marian Miculescu
University Politehnica of
Bucharest
Romania

Editorial Office
MDPI
St. Alban-Anlage 66
4052 Basel, Switzerland

This is a reprint of articles from the Special Issue published online in the open access journal *Materials* (ISSN 1996-1944) (available at: https://www.mdpi.com/journal/materials/special_issues/advanced_composite_biomaterials).

For citation purposes, cite each article independently as indicated on the article page online and as indicated below:

LastName, A.A.; LastName, B.B.; LastName, C.C. Article Title. *Journal Name* **Year**, *Volume Number*, Page Range.

ISBN 978-3-0365-0764-4 (Hbk)
ISBN 978-3-0365-0765-1 (PDF)

© 2021 by the authors. Articles in this book are Open Access and distributed under the Creative Commons Attribution (CC BY) license, which allows users to download, copy and build upon published articles, as long as the author and publisher are properly credited, which ensures maximum dissemination and a wider impact of our publications.

The book as a whole is distributed by MDPI under the terms and conditions of the Creative Commons license CC BY-NC-ND.

Contents

About the Editors vii

Preface to "Advanced Composite Biomaterials" ix

Stefan Ioan Voicu and Marian Miculescu
Advanced Composite Biomaterials
Reprinted from: *Materials* 2021, 14, 625, doi:10.3390/ma14030625 1

Koichiro Hayashi, Atsuto Tokuda, Jin Nakamura, Ayae Sugawara-Narutaki and Chikara Ohtsuki
Tearable and Fillable Composite Sponges Capable of Heat Generation and Drug Release in Response to Alternating Magnetic Field
Reprinted from: *Materials* 2020, 13, 3637, doi:10.3390/ma13163637 5

Armin Thumm, Regis Risani, Alan Dickson and Mathias Sorieul
Ligno-Cellulosic Fibre Sized with Nucleating Agents Promoting Transcrystallinity in Isotactic Polypropylene Composites
Reprinted from: *Materials* 2020, 13, 1259, doi:10.3390/ma13051259 17

Hanxiao Huang, Yunshui Yu, Yan Qing, Xiaofeng Zhang, Jia Cui and Hankun Wang
Ultralight Industrial Bamboo Residue-Derived Holocellulose Thermal Insulation Aerogels with Hydrophobic and Fire Resistant Properties
Reprinted from: *Materials* 2020, 13, 477, doi:10.3390/ma13020477 35

Andreea Madalina Pandele, Andreea Constantinescu, Ionut Cristian Radu, Florin Miculescu, Stefan Ioan Voicu and Lucian Toma Ciocan
Synthesis and Characterization of PLA-Micro-structured Hydroxyapatite Composite Films
Reprinted from: *Materials* 2020, 13, 274, doi:10.3390/ma13020274 51

Li Xu, Yushu Zhang, Haiqing Pan, Nan Xu, Changtong Mei, Haiyan Mao, Wenqing Zhang, Jiabin Cai and Changyan Xu
Preparation and Performance of Radiata-Pine-Derived Polyvinyl Alcohol/Carbon Quantum Dots Fluorescent Films
Reprinted from: *Materials* 2020, 13, 67, doi:10.3390/ma13010067 65

Yanchen Li, Beibei Wang, Yingni Yang, Yi Liu and Hongwu Guo
Preparation and Characterization of Dyed Corn Straw by Acid Red GR and Active Brilliant X-3B Dyes
Reprinted from: *Materials* 2019, 12, 3483, doi:10.3390/ma12213483 85

Huichao Jin, Wei Bing, Limei Tian, Peng Wang and Jie Zhao
Combined Effects of Color and Elastic Modulus on Antifouling Performance: A Study of Graphene Oxide/Silicone Rubber Composite Membranes
Reprinted from: *Materials* 2019, 12, 2608, doi:10.3390/ma12162608 99

Saleh Zidan, Nikolaos Silikas, Abdulaziz Alhotan, Julfikar Haider and Julian Yates
Investigating the Mechanical Properties of ZrO_2-Impregnated PMMA Nanocomposite for Denture-Based Applications
Reprinted from: *Materials* 2019, 12, , doi:10.3390/ma12081344 111

Madalina Oprea and Stefan Ioan Voicu
Cellulose Composites with Graphene for Tissue Engineering Applications
Reprinted from: *Materials* **2020**, *13*, 5347, doi:10.3390/ma13235347 . . . . **125**

Madalina Oprea and Stefan Ioan Voicu
Recent Advances in Applications of Cellulose Derivatives-Based Composite Membranes with Hydroxyapatite
Reprinted from: *Materials* **2020**, *13*, 2481, doi:10.3390/ma13112481 . . . . **149**

About the Editors

Stefan Ioan Voicu is Professor at the Faculty of Applied Chemistry and Materials Science, the University Politehnica of Bucharest. His work is primarily in the field of polymeric membrane materials and processes in the Department of Analytical Chemistry and Environmental Engineering. Previously, he worked for Honeywell Automation and Controlled Solutions–Sensors and Wireless Laboratory Bucharest in the field of chemical matrixes for sensors. He has a BSc in Organic Chemistry, an MSc in Environmental Engineering, a Ph.D. in Polymeric Membranes, and a Habilitation in Chemical Engineering, all from the University Politehnica of Bucharest, Romania. He has 60 SCI journal articles with an H index of 23, three granted US patents, and 10 book chapters in the field of polymers, polymer composites, and polymeric membranes (in applications from water purification to sensors, fuel cells to biomedical applications).

Marian Miculescu, Ph.D., is specialized in biomaterials, phase constitution, and material properties. He is Full Professor at the Faculty of Materials Science and Engineering at the University Politehnica of Bucharest, Romania. He earned his Ph.D. in Materials Science in 2010, and in 2019, he presented the Habilitation Thesis "Contributions on Synthesis of Natural Origin Ceramics and Analysis on Biomaterials Surface and Interfaces". He works in the field of materials science (material properties, material synthesis and characterization, thermal treatments, and advanced materials) and has over 15 years experience in the domain, in which time he has participated in more than 25 national research projects in the field of materials science, engineering, and technology. He has received more than 20 international and national awards for his contributions to science and is a member of several professional associations throughout Europe. He constantly supervises a team of Ph.D., MSc, and BSc students.

Preface to "Advanced Composite Biomaterials"

'Biomaterials' is one of the most important fields of study in terms of development in the 21st century. This is due to the progress in medical science, material science, chemistry, and physics, and the large number of materials with practical uses in osteointegration, controlled drug release systems, and tissue engineering, etc. This book contributes to the broad field of biomaterials by presenting both review articles in the area of polymeric biomaterials based on cellulose, e.g., hydroxyapatite or graphene composites for tissue engineering, and original research articles focused on various applications, such as controlled drug release and composites based on polylactic acid or poly (methyl methacrylate). The papers herein discuss, for example, the controlled release of active pharmaceutical substances under the influence of stimuli in the magnetic field through heat generation, which can be achieved by the synthesis of implantable polymeric biomaterials. Moreover, improving the mechanical properties of propylene-based biomaterials, which are widely used in abdominal surgery, by obtaining ligno-cellulose fiber composites is examined. These lines of research have huge practical implications in the field of surgery. Furthermore, aerogels have multiple uses in the field of biomaterials, and the synthesis of new such systems and the search for easily accessible materials with improved properties are a continuous challenge. Thus, improving the mechanical and thermal properties of aerogels, which is crucial for potential applications in the biomedical field, is discussed. Furthermore, polylactic acid, one of the most widely used biocompatible polymers, is investigated. Hydroxyapatite compounds of this polymer can be successfully used as precursors in 3D printing, as well as for obtaining polymeric biomaterial membranes with potential applications in osteointegration and various other techniques. The synthesis and characterization of composite films based on polyvinyl alcohol, cellulate nanofibrils, and carbon quantum dots are explored, with a focus on potential applications in the manufacture of packaging with special properties, e.g., antibacterial, transparent, and resistant to ultraviolet radiation. In the field of optical properties, the capacity and staining mechanism of horn claw (a biocompatible material obtained from biomass) is investigated. In addition, the clogging of polymer membranes is a major problem when separating viruses and bacteria. A large study is presented regarding the synthesis of silicone rubber composite membranes and graphene oxide with remarkable anticlogging properties. Moreover, the study demonstrated the influence of the color of the surface of the membrane under conditions of hydrodynamic separation (instead of the conditions of static separation), which opened a new field of research. In the field of precursors for implantable dental materials, a new composite based on poly (methyl methacrylate) and ZrO_2 is reported, the composite material demonstrating remarkable mechanical properties compared to classical resin. Cellulose derivatives are among the most widely used biocompatible polymers due to the fact that their degradation exclusively releases glucose. For this reason, cellulose-based composites are currently the most studied and developed form of precursor. Of the many fillers that can be used, graphene (with potential applications in tissue engineering) and hydroxyapatite (for osteointegration) are explored in detail in two reviews.

Stefan Ioan Voicu, Marian Miculescu
Editors

Editorial

Advanced Composite Biomaterials

Stefan Ioan Voicu [1,2,*] and Marian Miculescu [3]

1. Advanced Polymer Materials Group, Faculty of Applied Chemistry and Material Science, University Polytehnica of Bucharest, str. Gheorghe Polizu 1-7, 011061 Bucharest, Romania
2. Faculty of Applied Chemistry and Materials Science, University Politehnica of Bucharest, Gheorghe Polizu 1-7, 011061 Bucharest, Romania
3. Faculty of Materials Science, University Politehnica of Bucharest, Splaiul Independentei 313, 060042 Bucharest, Romania; marian.miculescu@upb.ro
* Correspondence: svoicu@gmail.com

Received: 25 January 2021; Accepted: 27 January 2021; Published: 1 February 2021

"Biomaterials" is one of the most important fields of study in terms of its development in the 21st century. This is due to both the progress of medical science, material science, chemistry, and physics, and the large number of practical needs its scope can address—materials for favoring osteointegration, systems for controlled release of drugs, materials and sites for tissue engineering, etc. From Figure 1, we can see a steady increase in interest in research in the field of biomaterials in the last 10 years, following a simple search of the keyword "biomaterial" across two main databases—Thomson Reuters ISI and Scopus, respectively. The graph shows that, in the evaluated period, approximately 45,000 articles were published, which is well above other areas.

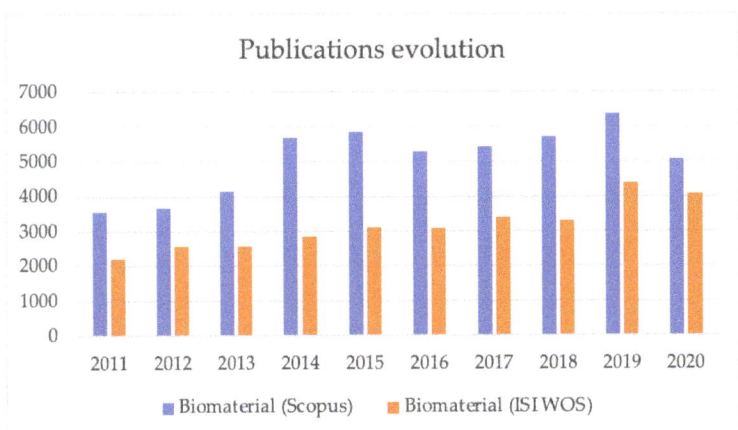

Figure 1. Evolution of the number of publications on Scopus and ISI Web of Knowledge containing "biomaterial" during the period 2011–2020.

This special number contributes to the vast field of biomaterials presenting both review articles in the field of polymeric biomaterials based on cellulose (hydroxyapatite or graphene composites for tissue engineering), as well as original research articles with various applications, such as controlled release of drugs, composites based on polylactic acid or poly methyl methacrylate. Thus, the controlled release of active pharmaceutical substances by heat generation, under the influence of the stimuli in the magnetic field, can be obtained by the synthesis of implantable polymeric biomaterials, having a practical and crucial contribution in the field of surgery [1]. Propylene-based biomaterials are widely used in abdominal surgery; their mechanical properties and biocompatibility can be improved by obtaining lignocellulose fiber composites [2]. Aerogels have multiple uses in the field of biomaterials; the synthesis of new such systems or the search for materials that are increasingly accessible or with improved properties, being a continuous challenge. The properties of aerogels are crucial for potential applications in the biomedical field, and the improvement of their mechanical and thermal characteristics are reported, including for medical applications that require these properties [3]. One of the most widely used biocompatible polymers is polylactic acid. Hydroxyapatite compounds of this polymer can be successfully used as precursors for 3D printing, as well as for obtaining polymeric biomaterial membranes with potential applications in osteointegration or various separations [4]. The synthesis and characterization of composite films based on polyvinyl alcohol, cellulate nanofibrils and carbon quantum dots have been reported, alongside the potential applications of films obtained in package manufacturing that have special properties (antibacterial, transparent and resistant to ultraviolet [5]). In the field of the optical properties of biomaterials, the capacity and staining mechanism for horn claw (a biocompatible material obtained from biomass) were investigated [6]. The clogging of polymer membranes, especially of separations of viruses or bacteria, is a major problem in the field of the separation of these species. A large study has been reported on the synthesis of silicone rubber composite membranes and graphene oxide with remarkable anti-clogging properties [7]. Moreover, the study showed, in the first place, the influence of the color of the surface of the membrane under conditions of hydrodynamic separation (instead of the conditions of static separation)—the reported data opening a new scientific field. In the field of precursors for implantable dental materials, a new composite based on poly-methyl methacrylate and ZrO_2 was reported—the composite material proving to have remarkable mechanical properties compared to classical resin [8]. Cellulose derivatives are among the most widely used biocompatible polymers due to the fact that their degradation exclusively releases glucose. For this reason, cellulose-based composites are the most studied and developed precursors at present. Of the many fillers that can be used, graphene (with potential applications in tissue engineering) [9] and hydroxyapatite (especially for osteointegration) [10] are presented in detail in two reviews.

Author Contributions: Conceptualization, S.I.V. and M.M.; writing—review and editing, S.I.V. All authors have read and agreed to the published version of the manuscript.

Funding: This research received no external funding.

Acknowledgments: The Guest Editors kindly acknowledge the administrative help provided by Clark Xu for this Special Issue.

Conflicts of Interest: The authors declare no conflict of interest.

References

1. Hayashi, K.; Tokuda, A.; Nakamura, J.; Sugawara-Narutaki, A.; Ohtsuki, C. Tearable and Fillable Composite Sponges Capable of Heat Generation and Drug Release in Response to Alternating Magnetic Field. *Materials* **2020**, *13*, 3637. [CrossRef] [PubMed]
2. Thumm, A.; Risani, R.; Dickson, A.; Sorieul, M. Ligno-Cellulosic Fibre Sized with Nucleating Agents Promoting Transcrystallinity in Isotactic Polypropylene Composites. *Materials* **2020**, *13*, 1259. [CrossRef] [PubMed]
3. Huang, H.; Yu, Y.; Qing, Y.; Zhang, X.; Cui, J.; Wang, H. Ultralight Industrial Bamboo Residue-Derived Holocellulose Thermal Insulation Aerogels with Hydrophobic and Fire Resistant Properties. *Materials* **2020**, *13*, 477. [CrossRef] [PubMed]

4. Pandele, A.M.; Constantinescu, A.; Radu, I.C.; Miculescu, F.; Ioan Voicu, S.; Hammer, L.T. Synthesis and Characterization of PLA-Micro-structured Hydroxyapatite Composite Films. *Materials* **2020**, *13*, 274. [CrossRef] [PubMed]
5. Xu, L.; Zhang, Y.; Pan, H.; Xu, N.; Mei, C.; Mao, H.; Zhang, W.; Horses, J.; Xu, C. Preparation and Performance of Radiata-Pine-Derived Polyvinyl Alcohol/Carbon Quantum Dots Fluorescent Films. *Materials* **2020**, *13*, 67. [CrossRef] [PubMed]
6. Li, Y.; Wang, B.; Yang, Y.; Liu, Y.; Guo, H. Preparation and Characterization of Dyed Corn Straw by Acid Red GR and Active Brilliant X-3B Dyes. *Materials* **2019**, *12*, 3483. [CrossRef] [PubMed]
7. Jin, H.; Bing, W.; Tian, L.; Wang, P.; Zhao, J. Combined Effects of Color and Elastic Modulus on Antifouling Performance: A Study of Graphene Oxide/Silicone Rubber Composite Membranes. *Materials* **2019**, *12*, 2608. [CrossRef] [PubMed]
8. Zidan, S.; Silikas, N.; Alhotan, A.; Haider, J.; Yates, J. Investigating the Mechanical Properties of ZrO_2-Impregnated PMMA Nanocomposite for Denture-Based Applications. *Materials* **2019**, *12*, 1344. [CrossRef] [PubMed]
9. Oprea, M.; Voicu, S.I. Cellulose Composites with Graphene for Tissue Engineering Applications. *Materials* **2020**, *13*, 5347. [CrossRef] [PubMed]
10. Oprea, M.; Voicu, S.I. Recent Advances in Applications of Cellulose Derivatives-Based Composite Membranes with Hydroxyapatite. *Materials* **2020**, *13*, 2481.

© 2021 by the authors. Licensee MDPI, Basel, Switzerland. This article is an open access article distributed under the terms and conditions of the Creative Commons Attribution (CC BY) license (http://creativecommons.org/licenses/by/4.0/).

Article

Tearable and Fillable Composite Sponges Capable of Heat Generation and Drug Release in Response to Alternating Magnetic Field

Koichiro Hayashi [1,*], Atsuto Tokuda [2], Jin Nakamura [2], Ayae Sugawara-Narutaki [2] and Chikara Ohtsuki [2]

[1] Department of Biomaterials, Faculty of Dental Science, Kyushu University3-1-1, Maidashi, Higashi-ku, Fukuoka 812-8582, Japan
[2] Department of Materials Chemistry, Graduate School of Engineering, Nagoya University, Furo-cho, Chikusa-ku, Nagoya 464-8603, Japan; tokuda.atsuto@f.mbox.nagoya-u.ac.jp (A.T.); nakamura@chembio.nagoya-u.ac.jp (J.N.); ayae@energy.nagoya-u.ac.jp (A.S.-N.); ohtsuki@chembio.nagoya-u.ac.jp (C.O.)
* Correspondence: khayashi@dent.kyushu-u.ac.jp; Tel.: +81-92-842-6345

Received: 16 July 2020; Accepted: 14 August 2020; Published: 17 August 2020

Abstract: Tearable and fillable implants are used to facilitate surgery. The use of implants that can generate heat and release a drug in response to an exogenous trigger, such as an alternating magnetic field (AMF), can facilitate on-demand combined thermal treatment and chemotherapy via remote operation. In this study, we fabricated tearable sponges composed of collagen, magnetite nanoparticles, and anticancer drugs. Crosslinking of the sponges by heating for 6 h completely suppressed undesirable drug release in saline at 37 °C but allowed drug release at 45 °C. The sponges generated heat immediately after AMF application and raised the cell culture medium temperature from 37 to 45 °C within 15 min. Heat generation was controlled by switching the AMF on and off. Furthermore, in response to heat generation, drug release from the sponges could be induced and moderated. Thus, remote-controlled heat generation and drug release were achieved by switching the AMF on and off. The sponges destroyed tumor cells when AMF was applied for 15 min but not when AMF was absent. The tearing and filling properties of the sponges may be useful for the surgical repair of bone and tissue defects. Moreover, these sponges, along with AMF application, can facilitate combined thermal therapy and chemotherapy.

Keywords: magnetic nanoparticles; composite; DDS; hyperthermia; collagen

---

1. Introduction

Surgery is a standard treatment for removing bone and soft tissue tumors [1,2]. To reconstruct the defects formed by surgery, restorative materials are frequently implanted into the defect site [3–5]. The use of restorative materials that can release anticancer drugs may play a role in the prevention of the recurrence of tumors as well as the reconstruction of the defects [6–9].

To date, porous materials have been developed as implantable carriers for drug delivery systems (DDS) [10–12]. Notably, sponge-like materials are promising implantable carrier materials for DDS because they easily fill defects owing to their tearable and flexible properties [13–17]. To date, various sponge-like materials composed of chitosan, silk, and gelatin have been fabricated by freeze-drying and electrospinning methods [13–17]. However, in many cases, drug release from implantable carriers is difficult to control because they are spontaneously dissolved in the body [17]. Therefore, researchers have tried to actively control drug release by developing implantable carriers with responsiveness to stimuli such as light [18] and electric fields [19,20]. Although these stimuli have the advantage of being switchable, they cannot penetrate deep into the body. Therefore, the use of these stimuli is considered challenging for the active control of drug release.

Although chemotherapy controlled by DDS is a promising tumor treatment, the effectiveness of anticancer drugs depends on the cancer type and stage [20]. In contrast, treatments such as thermal therapy show therapeutic effects regardless of cancer type and stage [21–23]. It has been reported that the combination of chemotherapy and thermal treatment, enhances treatment efficacy [24,25]. Notably, magnetic nanoparticles and anticancer drug-loaded materials can be used in a way to exploit the heat generated by magnetic nanoparticles in response to alternating magnetic field (AMF) exposure [26–29], which can subsequently act as a trigger for releasing drugs from the materials, achieving a remotely controllable on-demand administration of combined chemotherapy and thermal treatment [30–34].

In this study, we synthesized magnetic nanoparticles and anticancer drug-loaded collagen sponges and evaluated their ability to generate heat and drug release behavior. Furthermore, the therapeutic efficacy of combined chemotherapy and thermal treatment through the use of the composite sponge and AMF application was evaluated using in vitro assays.

2. Experimental Section

2.1. Synthesis of Magnetic Nanoparticles (MNPs)

The MNPs were prepared using a previously reported method [35,36]. Briefly, iron (III) acetylacetonate, Fe(acac)$_3$, (Nihon Kagaku Sangyo, Tokyo, Japan) was dissolved in ethanol. Subsequently, hydrazine monohydrate (Kishida Chemical, Osaka, Japan) and distilled water were added to the Fe(acac)$_3$ solution, and then the mixture was stirred at 78 °C for 24 h. The magnetic properties of iron oxide (magnetite and/or maghemite) nanoparticles were controlled by adjusting the Fe(acac)$_3$ concentration and the amounts of hydrazine monohydrate and distilled water added to the mixture. The Fe(acac)$_3$ concentration and the amounts of hydrazine monohydrate and distilled water used are shown in Table S1. The MNPs were collected by centrifugation of the solution at 10,000 rpm for 10 min. The obtained MNPs were washed with ethanol and distilled water three times, respectively.

2.2. Fabrication of the MNPs and Anticancer Drug-Loaded Collagen Sponge

The anticancer drug we used was doxorubicin hydrochloride (DOX; Tokyo Chemical Industry, Tokyo, Japan). For the fabrication of the MNPs and DOX-loaded collagen sponge (MDC sponge), MNPs with the highest saturation magnetization (MNPs-8) were used. The MNPs (3.5 mg) and DOX (85 µg) were mixed with a type-I collagen solution (5 mg/mL, Nitta Gelatin, Osaka, Japan), and the MDC sponge was fabricated by freeze-drying the mixture at −80 °C for 48 h. To control DOX release from the MDC sponge, the MDC sponge was crosslinked by heat treatment at 140 °C for 1.5, 6, or 24 h under vacuum. To confirm that the MDC sponge contained DOX using Fourier-transform infrared (FTIR) spectroscopy, a collagen sponge loaded with MNPs (without DOX; MC sponge) was fabricated.

2.3. Characterization of the MNPs and MDC sponge

The microstructures of the MNPs and MDC sponge were observed using a transmission electron microscope (TEM; JEM-2100Plus, JEOL, Tokyo, Japan) and a scanning electron microscope (SEM; JSM-5600, JEOL, Tokyo, Japan). The crystal phases of the MNPs and MDC sponge were confirmed using X-ray diffraction (XRD; RINT-2100/PC, Rigaku, Tokyo, Japan). The crystallite size of the MNPs was calculated using Scherrer's equation and the 311-diffraction peak. The FTIR spectra were obtained using an FTIR spectrometer (FT/IR-6100, JASCO, Tokyo, Japan). The inorganic and organic percentages of the MDC sponge were measured using thermogravimetric and differential thermal analysis (TG-DTA; DTG-60AH, Shimadzu, Kyoto, Japan). The magnetic properties of the MNPs and MDC sponge were measured at room temperature using a vibrating sample magnetometer (VSM; BHV55, Riken Denshi, Tokyo, Japan).

2.4. Heat Generation Properties of the MNPs

The MNPs with the highest saturation magnetization (MNPs-8) were uniformly suspended in distilled water by sonication (50 W, 20 kHz, 30 s) using an ultrasonic oscillator (VCX-50PB, Ieda Trading, Tokyo, Japan) at a concentration of 1 mg/mL. The MNPs were suspended for a few hours at least and no precipitation was observed. The suspension was placed inside the coil of an induction heater (Easy Heat, Alonics, Tokyo, Japan). Subsequently, the suspension was exposed to a magnetic field of 74 Oe and a frequency of 216 kHz for 10 min. The temperature of the suspension was measured every 30 s using an infrared thermal imaging camera (InfReC G100EX, Nippon Avionics, Tokyo, Japan). The heat generation properties of MNPs were assessed based on the specific absorption rate (SAR) of the particles. SAR was calculated using the following equation:

$$\text{SAR} = \frac{mC}{m_{Fe_3O_4}} \times \frac{dT}{dt}, \tag{1}$$

where, C is the specific heat of water (4.2 J/(g·K)); m is the mass of the sample; $m_{Fe_3O_4}$ is the mass of the MNPs in the sample; T is the temperature; t is the application time of the AMF; and dT/dt is the slope of the curve of temperature vs. application time of the AMF in the first 30 s [37].

2.5. DOX Release from the MDC Sponge Without the AMF Application

An untreated MDC sponge (4.4 mg) and an MDC sponge (4.4 mg) that was crosslinked for 1.5, 6, or 24 h, were immersed in phosphate-buffered saline (PBS, 1.0 mL) at 37 °C and 45 °C. The UV-vis spectra of the supernatant (400-800 nm of wavelength range) were measured using UV-vis spectroscopy (V-670 spectrophotometer, JASCO, Tokyo, Japan). The amount of released DOX was estimated using the Beer–Lambert law based on the absorbance at 480 nm.

2.6. Heat Generation Properties of MDC Sponge

An MDC sponge crosslinked for 6 h (11 mg) was immersed in a cell culture medium (1.3 mL) and exposed to AMF (magnetic field of 74 Oe and frequency of 216 kHz) for 20 min using an induction heater. The temperature of the MDC sponge in the cell culture medium was measured every 30 s using an infrared thermal imaging camera.

2.7. Control of Heat Generation and DOX Release by Switching the AMF on and off

An MDC sponge crosslinked for 6 h was immersed in a cell culture medium, and the system temperature was kept at 37 °C. The MDC sponge was exposed to AMF (magnetic field of 74 Oe of magnetic field and frequency of 216 kHz) at 15-min intervals by switching the AMF on and off. The solution temperature during AMF application was measured using an infrared thermal camera. The amount of DOX released from the MDC sponge every 15 min was estimated using the Beer–Lambert law based on the absorbance at 480 nm, which was measured using UV-vis spectroscopy.

2.8. Destructive Ability of the MDC Sponge on HeLa Cells in the Presence of an AMF

HeLa cells (Riken, Tsukuba, Japan) were cultured in Dulbecco's modified Eagle's medium (DMEM; Fujifilm Wako Pure Chemical, Osaka, Japan) supplemented with fetal bovine serum (FBS; final concentration 10%, Sigma Aldrich, MO), MEM non-essential amino acids solution (final concentration 1%, Fujifilm Wako Pure Chemical), and a penicillin–streptomycin solution (final concentration 1%, Fujifilm Wako Pure Chemical). Cells were seeded at a density of 2.5×10^4 cells per well in a 24-well plate and cultured under 5% CO_2 at 37 °C for 24 h. Cells were enumerated using a cell counter (Cell Counting Kit-8, Dojindo Laboratories, Kumamoto, Japan). The MDC sponge crosslinked for 6 h (3.5 mg) was placed in a well and exposed to AMF (magnetic field of 74 Oe and frequency of 216 kHz) for 15 min using the induction heater. Cell viability was measured using cytotoxicity assays and a tetrazolium salt (CCK-8 assay system, Takara Bio, Shiga, Japan) at days 3 and 5 after AMF application.

In the CCK-8 assay, the absorbance at 460 nm was measured using a microplate reader (Epoch 2, BioTek Instrument, VT). As a control, we measured the cell viability of the non-treated cells and cells cultured with the MDC sponge in the absence of the AMF application. Significant differences were estimated by multiple comparisons between groups using a general multiple comparison method, the Tukey–Kramer method. $p < 0.05$ was considered statistically significant.

3. Results and Discussion

3.1. The Structure, Magnetic Properties, and Heat Generation Ability of the MNPs

The XRD patterns showed that all the MNPs were composed of magnetite and/or maghemite (Figure S1). The crystallite size of the MNPs was increased with increasing $Fe(acac)_3$ concentration and the additive amounts of hydrazine monohydrate and distilled water (Figure S2). We have previously demonstrated that the crystallite size increased by increasing the amount of the iron source. Furthermore, hydrolysis of the iron complex was promoted by increasing the amounts of hydrazine and water, resulting in an increase in crystallite size. Thus, the results of this study are consistent with those of our previous reports [31].

The magnetization curves of all the MNPs showed neither coercivity nor remnant magnetization (Figure S3), indicating that the MNPs were superparamagnetic. It has been reported that MNPs less than 10 nm in diameter exhibit superparamagnetic properties [38]. As the crystallite sizes of all the MNPs in our study were less than 10 nm in diameter, all the MNPs exhibited superparamagnetic properties. Furthermore, the magnetization of the MNPs increased when the $Fe(acac)_3$ concentration (Figure S3A) and the additive amounts of hydrazine monohydrate (Figure S3B) and distilled water (Figure S3C) increased. Thus, the magnetization of the MNPs increased as the crystallite size increased. As the MNPs-8 had the highest magnetization (76.8 emu/g) at 15 kOe, they were used to fabricate the MDC sponges.

The MNPs-8 were uniformly suspended in distilled water and generated heat in response to AMF exposure (74 Oe and 216 kHz), raising the water temperature from 28.5 to 56.8 °C for 10 min (Figure S4). The SAR was 70.6 W/g. Hergt et al. reported that the SAR of Endorem, a magnetic resonance imaging contrast agent consisting of 6-nm-MNPs, was < 0.1 W/g at 300 kHz and 82 Oe [39]. Timko et al. reported that the SAR of MNPs with a diameter of 10–140 nm enveloped by a biological membrane consisting of phospholipids and specific proteins was 171 W/g at 750 kHz and 63 Oe [40]. Drake et al. reported that the SAR of Gd-doped iron oxide was 36 W/g at 52 kHz and 246 Oe [41]. Generally, SAR is proportional to the frequency and the square of the amplitude of the magnetic field [42]. Thus, the MNPs-8 had higher heat generation abilities than the reported materials. We have previously demonstrated that the dead layer of MNPs synthesized using the same method used in this study was thin, providing high heat generation abilities [43].

3.2. The Structure of the MDC Sponges

The MDC sponges were flexible and tearable (Figure 1A,B), characteristics that facilitate the filling of the defects formed by surgery. The MDC sponges were primarily composed of collagen fibers (Figure 1C), which contained DOX and MNPs (Figure 1D,E). The crosslinking of collagen reportedly impacts mechanical properties, such as the elastic modulus and elongation [44,45]. Owing to this crosslinking effect, the handleability for filling the MDC sponges into the defects seems to be improved.

The diffraction peaks of magnetite and/or maghemite were detected in the XRD patterns (Figure 2A). In the FTIR spectra of the MC and MDC sponges (Figure 2B), absorption bands attributable to amide groups in collagen were detected at 1650 cm^{-1} (amide I band), 1560 cm^{-1} (amide II band), and 1235 cm^{-1} (amide III band) [46]. Furthermore, in the spectra of the MDC sponges, bands attributable to DOX were detected at 1283 cm^{-1} (v C-O-C), 1114 cm^{-1} (primary alcohol, v C-O), 1070 cm^{-1} (secondary alcohol, v C-O), and 988 cm^{-1} (tertiary alcohol, v C-O) [47]. The XRD and FTIR results demonstrated that the MDC sponges contained MNPs and DOX in the collagen matrix. The TG-DTA curves showed weight losses due to dehydration in the range of 30–100 °C and due to the burnout of organics in the range of 200–800 °C (Figure 2C). The TG result revealed that the magnetite percentage was 38.7 wt %.

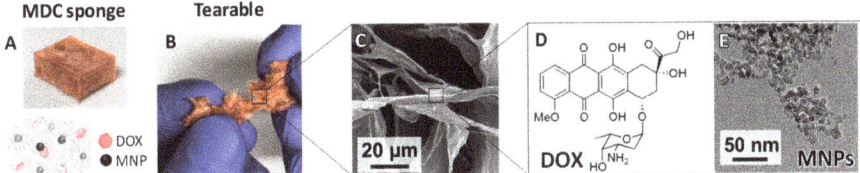

Figure 1. Characteristics of the MNPs and DOX-loaded collagen (MDC) sponges. (**A**) Photograph of the MDC sponge and illustration showing its structure. (**B**) Photograph showing that the MDC sponge is tearable. (**C**) SEM images of the MDC sponge. (**D**) Chemical structure of DOX. (**E**) TEM images of MNPs-8.

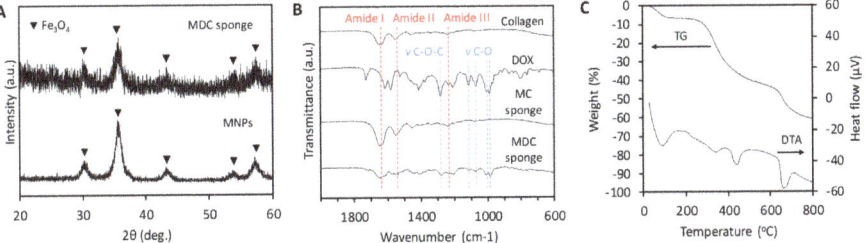

Figure 2. The diffraction peaks of magnetite were detected in the X-ray diffraction (XRD) patterns (**A**) XRD patterns of the magnetite nanoparticles (MNPs) and DOX-loaded collagen (MDC) sponge and MNPs. (**B**) FTIR spectra of collagen, DOX, MC sponge, and MDC sponge. (**C**) Thermogravimetric and differential thermal analysis (TG-DTA) curves of MDC sponge.

3.3. Control of DOX Release from the MDC Sponges in the Absence of AMF Application

To control the DOX release from the MDC sponge in the absence of AMF exposure, the MDC sponge was crosslinked by heat treatment at 140 °C for 1.5, 6, or 24 h under vacuum. The crosslink density was calculated by the area ratio between the P1 band due to the free amino group (1541 cm^{-1}) and the invariant P2 band, ascribed to –CH$_2$ in-plane bending vibration (1456 cm^{-1}) in the FTIR spectrum [48]. The P2/P1 values were calculated (Figure S5B) based on the FTIR results of the MDC sponges before and after crosslinking for 1.5, 6, and 24 h (Figure S5A). The P2/P1 values increased as the crosslinking time increased (Figure S5B). Thus, the crosslink density increased as a result of the increase in crosslinking time.

In the MDC sponges that were not crosslinked (Figure 3A), almost all the DOX was released from the sponges within 10 min after immersion in PBS both at 37 °C (body temperature) and 45 °C (thermal treatment temperature). In contrast, in the sponges that were crosslinked for 1.5 h, rapid DOX release immediately after immersion was suppressed (Figure 3B). However, a gradual DOX release still occurred at 37 °C, while heating at 45 °C prompted DOX release. Furthermore, a 6 h crosslinking treatment of the MDC sponges led to the complete suppression of DOX release at 37 °C and a gradual DOX release at 45 °C (Figure 3C). DOX was not released from the MDC sponges that were crosslinked for 24 h either at 37 °C or 45 °C (Figure 3D). The crosslinking of collagen reportedly reduces the swelling [44,45]. Thus, this crosslinking effect may allow for the prevention of DOX release from the MDC sponges that underwent crosslinking treatment at 37 °C. The above results demonstrated that an on-demand release of DOX using an exogenous trigger such as AMF exposure could be achieved by crosslinking the MDC sponges for 6 h.

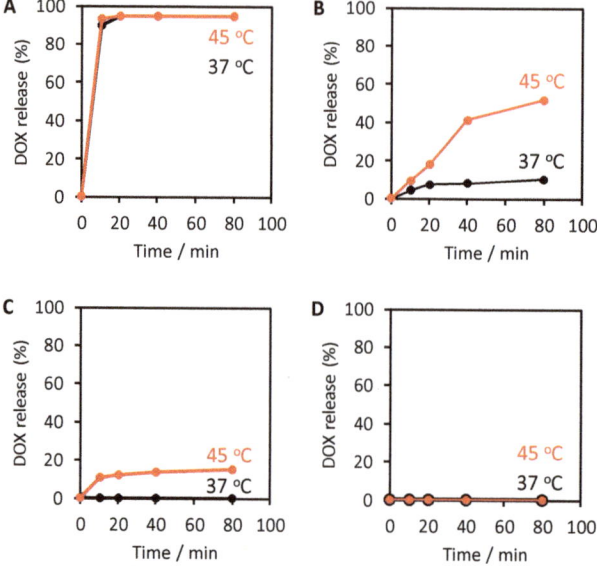

Figure 3. DOX release from the MDC sponge without crosslinking treatment (**A**) and with 1.5 h (**B**), 6 h (**C**), and 24 h (**D**) crosslinking treatment in phosphate-buffered saline (PBS) and the absence of alternating magnetic field (AMF) at 37 °C and 45 °C.

3.4. The Magnetic Properties and Heat Generation Ability of the MDC Sponges

Consistent with the MNPs findings, the MDC sponges had neither coercivity nor remnant magnetization at room temperature and exhibited superparamagnetic properties (Figure 4). The saturation magnetization of the MDC sponges was 44.8 emu/g, which was very close to the value (44.4 emu/g) that is required to obtain thermal treatment efficacy [49].

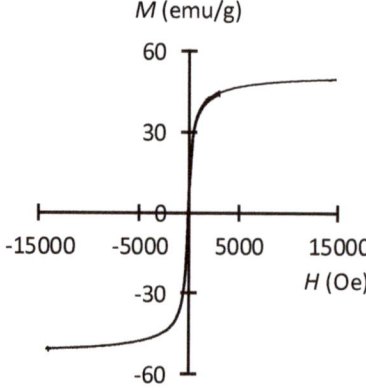

Figure 4. Magnetization curve of the MDC sponge at room temperature.

The MDC sponges generated heat in the cell culture medium in response to AMF exposure (Figure 5A) and heated the cell culture medium from 27.5 to 44.9 °C for 20 min (Figure 5B).

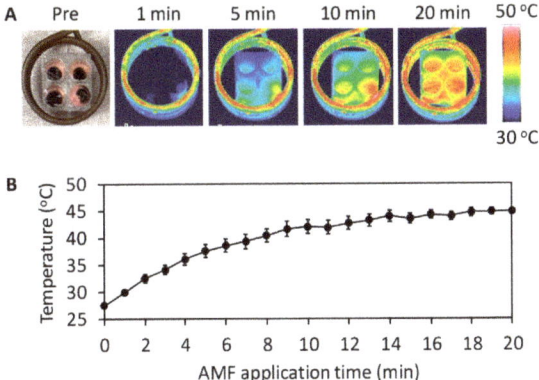

Figure 5. (**A**) Photograph and thermal images of MDC sponges in the cell culture medium before and during AMF application. (**B**) Change in medium temperature with AMF application time.

3.5. Control of DOX Release from the MDC Sponges by Switching the AMF on and off

The MDC sponges were exposed to AMF at 15-min intervals by switching the AMF on and off in the cell culture medium. The MDC sponges responded immediately to AMF exposure and generated heat, increasing the medium temperature to 45 °C within 15 min (Figure 6A). The MDC sponges stopped generating heat immediately after the AMF was switched off, resulting in a rapid fall in the medium temperature to 37 °C. Thus, heat generation by the MDC sponges was controlled by switching the AMF on and off.

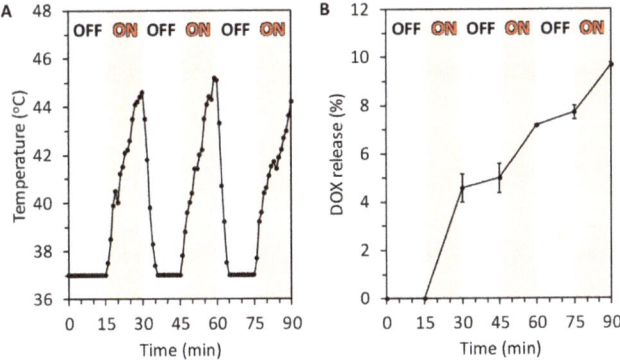

Figure 6. (**A**) Temperature change in MDC sponge-immersed in the cell culture medium in response to switching the AMF on and off. (**B**) DOX release from the MDC sponge when AMF was repeatedly applied and removed.

Although the MDC sponges released no DOX before the first AMF application, they started releasing DOX immediately after the AMF was switched on (Figure 6B). When AMF application was stopped, DOX release from the MDC sponges became slow. When AMF was applied again, a fast DOX release from the MDC sponges was again observed. A similar DOX release response was observed as the AMF was subsequently turned on and off. The above results demonstrated that both the heat generation ability of the MDC sponges and DOX release from them were remotely and actively controlled by switching the AMF on and off.

3.6. The Tumor Cell Killing Ability of the MDC Sponges

The viability of the tumor cells (HeLa cells) that were incubated with the MDC sponge in the absence of AMF exposure for 3 and 5 days was approximately 100% (Figure 7A,B), suggesting that there was no DOX release from the MDC sponge. The viabilities of cells that were subjected to a 15-min AMF application in the presence of the MC sponge and subsequently incubated with the MC sponge in the absence of AMF at days 3 and 5 were 27.6% ± 2.8% and 8.1% ± 4.9%, respectively (Figure 7A,B). Thus, thermal treatment was effective in killing tumor cells. To evaluate the effects of combined thermal treatment and chemotherapy, the cells were subjected to a 15-min AMF application in the presence of the MDC sponge. Cell viabilities at days 3 and 5 were 7.5% ± 0.8% and 2.1% ± 0.3%, respectively (Figure 7A,B). These results demonstrated that the combination of thermal treatment and DOX action led to higher tumor cell killing effects than thermal treatment alone. Furthermore, gradual DOX release from the MDC sponges after switching the AMF off continued to promote tumor cell killing and cell growth inhibition. Thus, with a simple surgical procedure, the MDC sponges may be utilized to achieve on-demand treatment and subsequent remotely controlled administration of efficient treatments.

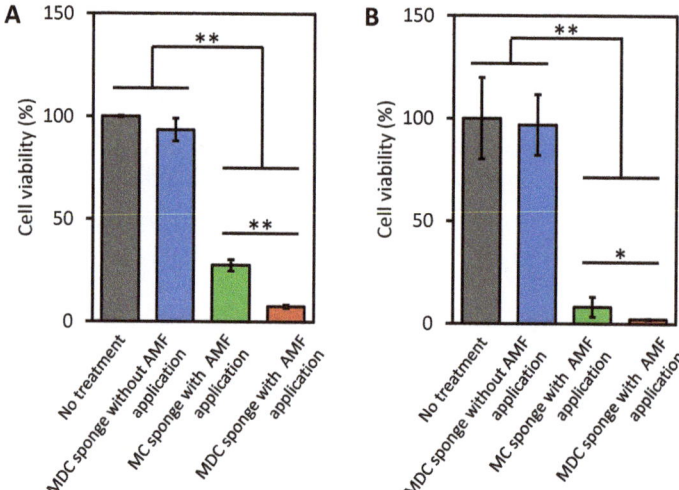

Figure 7. Viability of untreated HeLa cells incubated without sponges and AMF application, with the MDC sponge and no AMF application, with MC sponge (without DOX) and 15-min AMF application, and with MDC sponge and 15-min AMF application at 3 (**A**) and 5 (**B**) days, respectively (* $p < 0.05$ and ** $p < 0.01$).

Currently, collagen sponges are used in clinical practice due to their handleability. For example, Teruplug® is filled into tooth extraction wounds for hemostasis, the protection of the wound surface, and the promotion of tissue formation [50]. This fact suggests that MDC sponges, which have favorable handleability, are considered applicable in clinical practice.

4. Conclusions

This study successfully produced tearable MDC sponges. The degree of crosslinking of the MDC sponges was controlled by adjusting the crosslinking time. A 6-h crosslinking prevented undesirable DOX release from the MDC sponges in the absence of AMF exposure, but it allowed DOX release after AMF exposure. Heat generation and DOX release were controlled by switching the AMF on and off. Furthermore, AMF application in the presence of the MDC sponges had profound effects on destroying tumor cells, and the effect continued after termination of AMF exposure. Although the present study evaluated the handleability and cell killing abilities of MDC sponges through in vitro experiments, in vivo experiments are necessary for the evaluation of practical usefulness. Therefore, in future studies, we will evaluate the usefulness of MDC sponges through in vivo experiments.

Supplementary Materials: The following are available online at http://www.mdpi.com/1996-1944/13/16/3637/s1, Figure S1: XRD patterns of MNPs, Figure S2: Relationship between crystallite size and Fe(acac)$_3$ concentration, Figure S3: Magnetization curves of MNPs at room temperature, Figure S4: Photograph and thermal images of MNPs-8 dispersed in distilled water before and during AMF application, Figure S5: FTIR spectra of MDC sponges before and after crosslinking for 1.5, 6, and 24 h, Table S1: Synthesis conditions of MNPs.

Author Contributions: Conceptualization, K.H.; methodology, K.H., A.T., J.N., A.S.-N., and C.O.; validation, K.H., J.N., A.S.-N., and C.O.; formal analysis, A.T.; investigation, A.T.; data curation, A.T.; writing—original draft preparation, K.H.; writing—review and editing, J.N., A.S.-N., and C.O.; supervision, K.H., J.N., A.S.-N., and C.O.; project administration, K.H.; funding acquisition, K.H. All authors have read and agreed to the published version of the manuscript.

Funding: This study was supported partially by JSPS KAKENHI grant no. JP19K22970.

Acknowledgments: This study was supported partially by JSPS KAKENHI grant no. JP19K22970. TEM observations were conducted in the High Voltage Electron Microscope Laboratory at Nagoya University. VSM measurements were conducted at Nagoya University, supported by "Nanotechnology Platform Program" of the Ministry of Education, Culture, Sports, Science, and Technology (MEXT), Japan, Grant Number JPMXP09F19NU0072.

Conflicts of Interest: The authors declare no conflict of interest.

Abbreviations

AMF	alternating magnetic field
MNPs	magnetite nanoparticles
DOX	doxorubicin hydrochloride
FTIR	Fourier-transform infrared
MDC	MNPs and DOX-loaded collagen sponge
SEM	scanning electron microscope
TEM	transmission electron microscope

References

1. Weller, M.; van den Bent, M.; Tonn, J.C.; Stupp, R.; Preusser, M.; Cohen-Jonathan-Moyal, E.; Henriksson, R.; Le Rhun, E.; Balana, C.; Chinot, O.; et al. European Association for Neuro-Oncology (EANO) guideline on the diagnosis and treatment of adult astrocytic and oligodendroglial gliomas. *Lancet Oncol.* **2017**, *18*, E315–E329. [CrossRef]
2. Harbeck, N.; Gnant, M. Breast cancer. *Lancet* **2017**, *389*, 1134–1150. [CrossRef]
3. Tetsworth, K.; Block, S.; Glatt, V. Putting 3D modelling and 3D printing into practice: Virtual surgery and preoperative planning to reconstruct complex post-traumatic skeletal deformities and defects. *J. Soc. Int. Chir. Orthop. Traumatol.* **2017**, *3*, 16, (SICOT). [CrossRef] [PubMed]
4. Brennan, T.; Tham, T.M.; Costantino, P. The Temporalis Muscle Flap for Palate Reconstruction: Case Series and Review of the Literature. *Int. Arch. Otorhinolaryngology* **2017**, *21*, 259–264. [CrossRef]
5. Chan, D.S.; Fnais, N.; Ibrahim, I.; Daniel, S.; Manoukian, J. Exploring polycaprolactone in tracheal surgery: A scoping review of in-vivo studies. *Int. J. Pediatr. Otorhinolaryngol.* **2019**, *123*, 38–42. [CrossRef]
6. Palamà, I.E.; Arcadio, V.; D'Amone, S.; Biasiucci, M.; Gigli, G.; Cortese, B. Therapeutic PCL scafold for reparation of resected osteosarcoma defect. *Sci. Rep.* **2017**, *7*, 12672. [CrossRef]
7. Sarkar, N.; Bose, S. Liposome-Encapsulated Curcumin-Loaded 3D Printed Scaffold for Bone Tissue Engineering. *ACS Appl. Mater. Interfaces* **2019**, *11*, 17184–17192. [CrossRef]
8. Chew, S.A.; Danti, S. Biomaterial-Based Implantable Devices for Cancer Therapy. *Adv. Healthc. Mater.* **2017**, *6*, 1600766. [CrossRef]
9. Huang, W.W.; Ling, S.J.; Li, C.M.; Omenetto, F.G.; Kaplan, D.L. Silkworm silk-based materials and devices generated using bio-nanotechnology. *Chem. Soc. Rev.* **2018**, *47*, 6486–6504. [CrossRef]

10. Farid-ul-Haq, M.; Haseeb, M.T.; Hussain, M.A.; Ashraf, M.U.; Naeem-ul-Hassan, M.; Hussain, S.Z.; Hussain, I. A smart drug delivery system based on Artemisia vulgaris hydrogel: Design, on-off switching, and real-time swelling, transit detection, and mechanistic studies. *J. Drug Deliv. Sci. Technol.* **2020**, *58*, 101795. [CrossRef]
11. Yazdi, M.K.; Zarrintaj, P.; Hosseiniamoli, H.; Mashhadzadeh, A.H.; Saeb, M.R.; Ramsey, J.D.; Ganjali, M.R.; Mozafari, M. Zeolites for theranostic applications. *J. Mater. Chem. B* **2020**, *8*, 5992–6012. [CrossRef] [PubMed]
12. Lu, H.; Zhang, N.; Ma, M.M. Electroconductive hydrogels for biomedical applications. Wiley Interdiscip. Rev.-Nanomed. *Nanobiotechnology* **2019**, *11*, e1568.
13. Robert, M.C.; Frenette, M.; Zhou, C.; Yan, Y.; Chodosh, J.; Jakobiec, F.A.; Stagner, A.M.; Vavvas, D.; Dohlman, C.H.; Paschalis, E.I. A Drug Delivery System for Administration of Anti-TNF-alpha Antibody. *Transl. Vis. Sci. Technol.* **2016**, *5*, 11. [CrossRef] [PubMed]
14. Pawar, V.; Bulbake, U.; Khan, W.; Srivastava, R. Chitosan sponges as a sustained release carrier system for the prophylaxis of orthopedic implant-associated infections. *Int. J. Biol. Macromol.* **2019**, *134*, 100–112. [CrossRef]
15. Pritchard, E.M.; Valentin, T.; Panilaitis, B.; Omenetto, F.; Kaplan, D.L. Antibiotic-Releasing Silk Biomaterials for Infection Prevention and Treatment. *Adv. Funct. Mater.* **2013**, *23*, 854–861. [CrossRef] [PubMed]
16. Zhang, Z.Y.; Kuang, G.Z.; Zong, S.; Liu, S.; Xiao, H.H.; Chen, X.S.; Zhou, D.F.; Huang, Y.B. Sandwich-Like Fibers/Sponge Composite Combining Chemotherapy and Hemostasis for Efficient Postoperative Prevention of Tumor Recurrence and Metastasis. *Adv. Mater.* **2018**, *30*, 1803217. [CrossRef]
17. Ibrahim, H.K.; Fahmy, R.H. Localized rosuvastatin via implantable bioerodible sponge and its potential role in augmenting bone healing and regeneration. *Drug. Deliv.* **2016**, *23*, 3181–3192. [CrossRef]
18. Cen, D.; Wan, Z.; Fu, Y.K.; Pan, H.Q.; Xu, J.J.; Wang, Y.F.; Wu, Y.J.; Li, X.; Cai, X.J. Implantable fibrous 'patch' enabling preclinical chemo-photothermal tumor therapy. *Colloid Surf. B-Biointerfaces* **2020**, *192*, 111005. [CrossRef]
19. Perez-Martinez, C.J.; Chavez, S.D.M.; del Castillo-Castro, T.; Ceniceros, T.E.L.; Castillo-Ortega, M.M.; Rodriguez-Felix, D.E.; Ruiz, J.C.G. Electroconductive nanocomposite hydrogel for pulsatile drug release. *React. Funct. Polym.* **2016**, *100*, 12–17. [CrossRef]
20. Shah, S.A.A.; Firlak, M.; Berrow, S.R.; Halcovitch, N.R.; Baldock, S.J.; Yousafzai, B.M.; Hathout, R.M.; Hardy, J.G. Electrochemically Enhanced Drug Delivery Using Polypyrrole Films. *Materials* **2018**, *11*, 1123. [CrossRef]
21. Leary, M.; Heerboth, S.; Lapinska, K.; Sarkar, S. Sensitization of Drug Resistant Cancer Cells: A Matter of Combination Therapy. *Cancers* **2018**, *10*, 483. [CrossRef] [PubMed]
22. Sanai, N.; Berger, M.S. Surgical oncology for gliomas: The state of the art. *Nat. Rev. Clin. Oncol.* **2018**, *15*, 112–115. [CrossRef] [PubMed]
23. Nault, J.C.; Sutter, O.; Nahon, P.; Ganne-Carrie, N.; Seror, O. Percutaneous treatment of hepatocellular carcinoma: State of the art and innovations. *J. Hepatol.* **2018**, *68*, 783–797. [CrossRef] [PubMed]
24. Chen, J.J.; Wang, Y.T.; Ma, B.Y.; Guan, L.; Tian, Z.F.; Lin, K.L.; Zhu, Y.F. Biodegradable hollow mesoporous organosilica-based nanosystems with dual stimuli-responsive drug delivery for efficient tumor inhibition by synergistic chemo- and photothermal therapy. *Appl. Mater. Today* **2020**, *19*, 100655. [CrossRef]
25. Fearon, K.; Strasser, F.; Anker, S.D.; Bosaeus, I.; Bruera, E.; Fainsinger, R.L.; Jatoi, A.; Loprinzi, C.; MacDonald, N.; Mantovani, G.; et al. Definition and classification of cancer cachexia: An international consensus. *Lancet Oncol.* **2011**, *12*, 489–495. [CrossRef]
26. Lin, Y.; Zhang, K.; Zhang, R.; She, Z.; Tan, R.; Fan, Y.; Li, X. Magnetic nanoparticles applied in targeted therapy and magnetic resonance imaging: Crucial preparation parameters, indispensable pre-treatments, updated research advancements and future perspectives. *J. Mater. Chem. B* **2020**, *8*, 5973–5991. [CrossRef] [PubMed]
27. Shasha, C.; Krishnan, K.M. Nonequilibrium Dynamics of Magnetic Nanoparticles with Applications in Biomedicine. *Adv. Mater.* **2020**, 1904131. [CrossRef] [PubMed]
28. Wang, X.; Law, J.; Luo, M.; Gong, Z.; Yu, J.; Tang, W.; Zhang, Z.; Mei, X.; Huang, Z.; You, L.; et al. Magnetic Measurement and Stimulation of Cellular and Intracellular Structures. *ACS Nano* **2020**, *14*, 3805–3821. [CrossRef] [PubMed]
29. Israel, L.L.; Galstyan, A.; Holler, E.; Ljubimova, J.Y. Magnetic iron oxide nanoparticles for imaging, targeting and treatment of primary and metastatic tumors of the brain. *J. Control. Release* **2020**, *320*, 45–62. [CrossRef]
30. Xiao, Y.; Du, J. Superparamagnetic nanoparticles for biomedical applications. *J. Mater. Chem. B* **2020**, *8*, 354–367. [CrossRef]

31. Hayashi, K.; Sato, Y.; Sakamoto, W.; Yogo, T. Theranostic Nanoparticles for MRI-Guided Thermochemotherapy: "Tight" Clustering of Magnetic Nanoparticles Boosts Relaxivity and Heat-Generation Power. *ACS Biomater. Sci. Eng.* **2017**, *3*, 95–105. [CrossRef]
32. Hayashi, K.; Sakamoto, W.; Yogo, T. Smart Ferrofluid with Quick Gel Transformation in Tumors for MRI-Guided Local Magnetic Thermochemotherapy. *Adv. Funct. Mater.* **2016**, *26*, 1708–1718. [CrossRef]
33. Pucci, C.; De Pasquale, D.; Marino, A.; Martinelli, C.; Lauciello, S.; Ciofani, G. Hybrid Magnetic Nanovectors Promote Selective Glioblastoma Cell Death through a Combined Effect of Lysosomal Membrane Permeabilization and Chemotherapy. *ACS Appl. Mater. Interfaces* **2020**, *12*, 29037–29055. [CrossRef] [PubMed]
34. Chen, W.; Cheng, C.A.; Zink, J.I. Spatial, temporal, and dose control of drug delivery using noninvasive magnetic stimulation. *ACS Nano* **2019**, *13*, 1292–1308. [CrossRef] [PubMed]
35. Hayashi, K.; Sakamoto, W.; Yogo, T. Magnetic and rheological properties of monodisperse Fe_3O_4 nanoparticle/organic hybrid. *J. Magn. Magn. Mater.* **2009**, *321*, 450–457. [CrossRef]
36. Hayashi, K.; Sakamoto, W.; Yogo, T. One-pot synthesis of magnetic nanoparticles assembled on polysiloxane rod and their response to magnetic field. *Colloid Polym. Sci.* **2013**, *291*, 2837–2842. [CrossRef]
37. Latham, A.H.; Williams, M.E. Controlling Transport and Chemical Functionality of Magnetic Nanoparticles. *Acc. Chem. Res.* **2008**, *41*, 411–420. [CrossRef]
38. Jeong, U.; Teng, X.W.; Wang, Y.; Yang, H.; Xia, Y.N. Superparamagnetic colloids: Controlled synthesis and niche applications. *Adv. Mater.* **2007**, *19*, 33–60. [CrossRef]
39. Hergt, R.; Andrä, W.; d'Ambly, C.G.; Hilger, I.; Kaiser, W.A.; Richter, U.; Schmidt, H.-G. Physical Limits of Hyperthermia Using Magnetite Fine Particles. *IEEE Trans. Magn.* **1998**, *34*, 3745–3754. [CrossRef]
40. Timko, M.; Dzarova, A.; Kovac, J.; Skumiel, A.; Józefczak, A.; Hornowski, T.; Gojżewski, H.; Zavisova, V.; Koneracka, M.; Sprincova, A.; et al. Magnetic properties and heating effect in bacterial magnetic nanoparticles. *J. Magn. Magn. Mater.* **2009**, *321*, 1521–1524. [CrossRef]
41. Drake, P.; Cho, H.J.; Shih, P.S.; Kao, C.H.; Lee, K.F.; Kuo, C.H.; Lin, X.Z.; Lin, Y.J. Gd-doped iron-oxide nanoparticles for tumour therapy via magnetic field hyperthermia. *J. Mater. Chem.* **2007**, *17*, 4914–4918. [CrossRef]
42. Hiergeist, R.; Andrä, W.; Buske, N.; Hergt, R.; Hilger, I.; Richter, U.; Kaiser, W. Magnetic properties and heating effect in bacterial magnetic nanoparticles. *J. Magn. Magn. Mater.* **1999**, *201*, 420. [CrossRef]
43. Hayashi, K.; Moriya, M.; Sakamoto, W.; Yogo, T. Chemoselective Synthesis of Folic Acid-Functionalized Magnetite Nanoparticles via Click Chemistry for Magnetic Hyperthermia. *Chem. Mater.* **2009**, *21*, 1318–1325. [CrossRef]
44. Ruszczak, Z.; Friess, W. Collagen as a carrier for on-site delivery of antibacterial drugs. *Adv. Drug Deliv. Rev.* **2003**, *55*, 1679–1698. [CrossRef] [PubMed]
45. Angele, P.; Abke, J.; Kujat, R.; Faltermeier, H.; Schumann, D.; Nerlich, M.; Kinner, B.; Englert, C.; Ruszczak, Z.; Mehrl, R.; et al. Influence of different collagen species on physico-chemical properties of crosslinked collagen matrices. *Biomaterials* **2004**, *25*, 2831–2841. [CrossRef]
46. Fernandes, L.L.; Resende, C.X.; Tavares, D.S.; Soares, G.A.; Castro, L.O.; Granjeiro, J.M. Cytocompatibility of Chitosan and Collagen-Chitosan Scaffolds for Tissue Engineering. *Polimeros* **2011**, *21*, 1–6. [CrossRef]
47. Kanwal, U.; Bukhari, N.I.; Rana, N.F.; Rehman, M.; Hussain, K.; Abbas, N.; Mehmood, A.; Raza, A. Doxorubicin-loaded quaternary ammonium palmitoyl glycol chitosan polymeric nanoformulation: Uptake by cells and organs. *Int. J. Nanomed.* **2019**, *14*, 1–15. [CrossRef]
48. Madaghiele, M.; Calo, E.; Salvatore, L.; Bonfrate, V.; Pedone, D.; Frigione, M.; Sannino, A. Assessment of collagen crosslinking and denaturation for the design of regenerative scaffolds. *J. Biomed. Mater. Res. Part A* **2016**, *104*, 186–194. [CrossRef]

49. Hayashi, K.; Nakamura, M.; Sakamoto, W.; Yogo, T.; Miki, H.; Ozaki, S.; Abe, M.; Matsumoto, T.; Ishimura, K. Superparamagnetic Nanoparticle Clusters for Cancer Theranostics Combining Magnetic Resonance Imaging and Hyperthermia Treatment. *Theranostics* **2013**, *3*, 366–376. [CrossRef]
50. Hur, J.-W.; Yoon, S.-J.; Ryu, S.-Y. Comparison of the bone healing capacity of autogenous bone, demineralized freeze dried bone allograft, and collagen sponge in repairing rabbit cranial defects. *J. Korean Assoc. Oral Maxillofac. Surg.* **2012**, *38*, 221–230. [CrossRef]

 © 2020 by the authors. Licensee MDPI, Basel, Switzerland. This article is an open access article distributed under the terms and conditions of the Creative Commons Attribution (CC BY) license (http://creativecommons.org/licenses/by/4.0/).

Article

Ligno-Cellulosic Fibre Sized with Nucleating Agents Promoting Transcrystallinity in Isotactic Polypropylene Composites

Armin Thumm, Regis Risani, Alan Dickson and Mathias Sorieul *

Scion, Forest Research Institute Ltd., 49 Sala street, 3020 Rotorua, New Zealand;
Armin.thumm@scionresearch.com (A.T.); Regis.risani@scionresearch.com (R.R.);
Alan.dickson@scionresearch.com (A.D.)
* Correspondence: Mathias.sorieul@scionresearch.com; Tel.: +64-7343-5514

Received: 20 February 2020; Accepted: 9 March 2020; Published: 10 March 2020

Abstract: The mechanical performance of composites made from isotactic polypropylene reinforced with natural fibres depends on the interface between fibre and matrix, as well as matrix crystallinity. Sizing the fibre surface with nucleating agents to promote transcrystallinity is a potential route to improve the mechanical properties. The sizing of thermo-mechanical pulp and regenerated cellulose (Tencel™) fibres with α- and β-nucleating agents, to improve tensile strength and impact strength respectively, was assessed in this study. Polarised microscopy, electron microscopy and differential scanning calorimetry (DSC) showed that transcrystallinity was achieved and that the bulk crystallinity of the matrix was affected during processing (compounding and injection moulding). However, despite substantial changes in crystal structure in the final composite, the sizing method used did not lead to significant changes regarding the overall composite mechanical performance.

Keywords: nucleating agent; isotactic polypropylene; transcrystallinity; natural fibres; Tencel™

1. Introduction

Isotactic polypropylene (iPP) is an important engineering thermoplastic polyolefin used in many different commercial applications such as packaging and automotive parts [1]. The main advantages of iPP are its easy processing and low manufacturing cost [2]. Its mechanical behaviour, thermal properties, and chemical resistance depend on its semi-crystalline structure and fraction (typically 50%–70%) [3]. Isotactic polypropylene is a non-polar, polymorphic polymer. The three basic crystal forms of iPP are: the monoclinic alpha (α) form, the trigonal beta (β) form and the orthorhombic gamma (γ) form [4,5]. The α-form crystals are obtained under common processing conditions, between 60 °C and 188 °C, with a maximum crystallisation rate around 80 °C [6,7]. In composites, the presence of α-crystals improves thermodynamic stability and mechanical performance but also reduces the impact strength, especially at low temperature. The β-form crystals are metastable thermodynamically, and have several advantages over the α-form, such as improved elongation at break and improved impact strength [8,9]. The β-crystals fan-shaped morphology has the ability to dissipate the impact energy, therefore improving toughness [10]. The β-form can be obtained using a temperature gradient method [11], shear flow-induced crystallization [12,13] or specific nucleating agents (NA) [14]. A temperature of crystallisation around 130 °C is best to promote nucleation and growth of β-iPP crystals [15,16].

It is a common strategy to improve the mechanical properties of polyolefin-based materials via the introduction of reinforcing agents (e.g., carbon fibres, glass fibres, clays, lignocellulosic materials). The strength of composite materials is influenced by: matrix properties, intrinsic properties of the reinforcement, dispersion of the reinforcement in the matrix, degree of orientation, quality of the interface and volume fraction of reinforcement [17,18]. In the resulting composite, the stress concentration develops at the interface between the matrix and reinforcement agent and is mainly influenced by the following: the volume fraction; the thermal expansion coefficient difference between each material; the interphase; and the crystallisation of the matrix. To achieve high mechanical performance, good interfacial adhesion allowing stress transfer between the matrix and reinforcement agent is crucial [19,20]. Several methods have been developed to improve interfacial adhesion between highly polar fibres and nonpolar polymers. The most common strategies used are sizing, compatibilisation, polymer grafting and interfacial crystallisation [10]. Transcrystallinity (TC) or transcrystalline layers describes a crystalline structure with limited thickness located at the interphase region, it originates from a high density of heterogeneous nuclei with a crystal growth orientation mainly perpendicular to the fibre surface until the growing front impinges with spherulites nucleated in the bulk [21,22]. The development of a TC structure is a promising route for improving the load transfer between the semi-crystalline matrix and the fibre [23].

TC enhances fibre–matrix adhesion, reduces stress concentration, creates a mechanical interlock and a protective layer around the fibres leading to an efficient stress transfer from the matrix across the interphase [24]. Good TC is obtained when the crystallisation parameters are more favourable to the TC formation compared to the crystallisation of spherulites in the bulk [25]. In composites, the development of the TC layer is influenced by factors such as matrix type, thermal history, temperature of polymer crystallisation, chain mobility, rate of cooling, occurrence of shearing forces during crystallisation and thermal expansion coefficients of individual components [24,26]. The surface topology, composition and sizing of the fibre itself are major parameters influencing the TC formation [27]. To achieve good TC, high nucleation site density along the fibre surface is crucial [28]. The proximity of nucleation sites restricts the crystal growth to the lateral direction, leading to the development of a columnar layer around fibres [22,29].

Lignocellulosic fibres have many beneficial features such as biodegradability, renewability, availability, low cost, and ease of preparation [30]. Polymer composites incorporating lignocellulosic fibres offer several advantages including low density, good mechanical properties and high damping capabilities [31,32]. The lignocellulosic fibres chosen for this study were High Temperature Thermo-Mechanical Pulp fibres (HT TMP) and Tencel™. The HT TMP fibres are naturally coated with a thin, relatively non-polar, lignin layer [33–35]. Softwood HT TMP fibres are of industrial relevance due to their low cost of production and low weight advantage when compounded into plastic composites [36,37]. Tencel™ is a regenerated cellulose fibre which exhibits high tensile properties (strength of 1.4 GPa and modulus 36 GPa) [38] and was included in this study as a reference fibre, as its high ductility improves the impact characteristics of brittle polymer matrices [39].

This study investigates the potential of α- or β-nucleating agent (NA) as a sizing additive for improving the mechanical properties of an iPP-natural fibre composite. The hypothesis is that the NA will generate a high nucleation density on the fibre surface leading to an α or β specific TC, improving both tensile properties and/or impact resistance. The addition of the NA usually occurs at the compounding stage of composite manufacture. In this study, we developed an original approach by directly adding the NA onto the fibres that could be adopted to tailor the lignocellulosic fibre composite characteristics eliminating the need for the addition of NA at the compounding stage. Our novel approach was to coat the fibres with the NA prior to compounding with the iPP to create a TC structure with improved matrix-fibre stress transfer.

2. Materials and Methods

2.1. Materials

2.1.1. Fibres

Wood chips (*Pinus radiata*), were thermomechanically pulped at 180 °C at Scion's fibre and pulp processing pilot plant (Rotorua, New Zealand). The resulting HT TMP fibres were dried in a tube drier to approximately 12% moisture content. An earlier study has found that chips processed in this way lead to fibres with a cellulose content of 39%–42%, lignin content of 28%–31% and 22%–25% hemicelluloses [40]. Prior to processing the fibres were approximately 1.25 mm long and 0.03 mm wide. The length and aspect ratio were significantly reduced during compounding [41]. Tencel™ fibres were purchased from Lenzing Fibers GmbH (Heiligenkreuz, Austria).

2.1.2. Nucleating Agents

The α-NA was Hyperform® HPN-68L (Milliken Chemical, Blacksburg, SC, USA) and the β-NA was NU-100 (NJStar), (New Japan Chemical Co. Ltd., Osaka, Japan). Hyperform® HPN-68L is a powdered α-NA comprising a dicarboxylate sodium-based compound known as HPN-68L. It is used to reduce cycle time and improve the tensile and flexural mechanical properties of the iPP [42]. Nu-100 (NJSTAR NU-100) is a β-NA. Its active ingredient, N,N′-dicyclohexyl-2,6-naphthalenedicarboxamide (DCNDCA), has been found to be an effective β-NA [15]. However, DCNDCA is dual selective and can induce both α-phase and β-phase, depending on the thermal condition applied [15].

2.1.3. Polypropylene and Compatibiliser

Isotactic homo-polypropylene SJ-170, Hopelen was sourced from Lotte Chemical Co., Seosan, South Korea. The compatibiliser was 3% w/w maleic anhydride grafted polypropylene (MAPP) Epolene G3015 (Eastman Chemical Co., Kingsport, TN, USA). A 3% MAPP loading was used according to the manufacturer recommendation for 30% wt lignocellulosic fibre content.

2.2. Methods

2.2.1. Sizing of the Fibre and Fibre Pellet Production

Fibres were blended with an adhesive thermoplastic acrylated emulsion [36] and NA in a dry blender consisting of a 12 m steel loop (Ø: 0.15 m) with a fan that creates a turbulent flow. The adhesive was administered onto the fibre in the loop via an inserted spray gun. The NA powder was then slowly added into the loop and the materials were circulated in the loop for a further 2 min. The adhesive was added at a 4% wt loading with respect to oven-dried fibre. The NA was added at a 1.7% loading with respect to fibre (0.5% wt loading with respect to the finished composite). Coated fibres were hot-pressed into 3 mm thin sheets and subsequently cut into small dice (4 × 4 mm) with a pneumatic chopper. The process is described in detail in Warnes et al., [36].

2.2.2. Nitrogen Analysis

Between 0.25 and 0.5 g of fibre dice were heated in a stream of high purity oxygen in a Leco furnace (Laboratory Equipment Corporation, St Joseph, MI, USA) to produce CO_2, N_2 and NO_x. A subsample of the combustion gases was passed through a heated copper catalyst that further reduced the NO_x to N_2, which was then measured by thermal conductivity. This results in the percentage of total nitrogen. This percentage was converted to NA content of the final sample by subtracting the N content of a NA free reference from the sample containing NA and then dividing the resulting difference by the nitrogen content of the pure NA. This method could only be applied to the β-NA as the α-NA does not contain nitrogen.

2.2.3. Compounding

The HT TMP fibre dice and the Tencel™ fibre dice were compounded at 30% wt loading into iPP with 3% wt maleic anhydride PP (MAPP) as a coupling agent. The LabTech™ extruder (LTE26-40, LabTech Engineering Co. Ltd., Samut Prakan, Thailand) had a 26 mm screw diameter, with co-rotating twin screws, with a 40 L/D (length/diameter) ratio. The PP and MAPP were dry blended and fed into the main feed throat using a Weighbatch™ DS20 (Weighbatch, Hamilton, New Zealand) gravimetric feeder. The fibre dice, which were oven dried overnight at 105 °C, were side-fed into the extruder using a K-Tron twin-screw gravimetric feeder (Coperion K-Tron, Sewell, NJ, USA). For the "no fibre" reference, the NAs were dry-blended with the iPP at the same time as the MAPP. Two atmospheric venting ports and one crammer vacuum (0.7 bar) degassing port were used to remove the entrapped air in the melt along with volatile organic compounds (VOCs). The melt was extruded through a 2-strand die and pulled into a water bath before being granulated into 3 mm long pellets. Each formulation was compounded in a single extrusion run at 200 rpm screw speed and 8 kg h^{-1} total extrusion throughput to ensure proper fibre mixing with a gentle screw design and reverse barrel temperature profile (220 °C to 190 °C).

2.2.4. Injection Moulding

Prior to injection moulding, the compounded pellets were oven dried at 105 °C, to obtain a pellet with residual moisture content below 0.3% wt. The compounded pellets were injection moulded with a BOY 35 machine (BOY Spritzgiessautomaten, Neustadt-Fernthal, Neustadt-Fernthal, Germany) into ISO multipurpose injection moulded test specimens (dogbone) type A (ISO 3167). The barrel temperature of the injection moulder was 190 °C and the mould was kept at 30 °C. The injection speed was 100 mm s^{-1}. The mould was filled with a screw speed of 100 rpm with 15 bar back pressure. Cooling time was 20 s. The injection moulding parameters were the same for all composites. Injection moulded parts were collected once their weight was constant. Thirty test specimens were collected for each treatment with the first ten disregarded for analysis.

2.2.5. Polarised Light Optical Microscopy (PLOM)

Cross-polarized optical microscopy was performed with a Leica DMRB microscope (Leica Mikroskopie & Systeme GmbH, Wetzlar, Germany), using a Leica EC3 camera (Leica Microsystems Ltd., Singapore) and a 10× magnification lens (Leica PL Fluotar). The temperature was controlled during the experiments with a programmable Mettler-Toledo hot stage HS82 (Mettler-Toledo GmbH, Greifensee, Switzerland).

For each experiment, a piece of iPP thin film was placed on a glass microscopy slide. Individual fibres are then positioned onto the film. The slide was placed into the temperature-controlled stage [43]. The stage temperature was raised to 200 °C and once the iPP was molten, the sample was covered with a cover slip. After a 5-min holding period at 200 °C, to erase the thermal history of the sample, the temperature was decreased with a 10 °C/min ramp and stabilized at 133 °C to monitor the isothermal crystallisation (Scheme 1). The samples were imaged after 5 min of isothermal crystallisation at 133 °C. The temperature of 133 °C was chosen as it is the optimal temperature to promote β over α crystallisation [16,25].

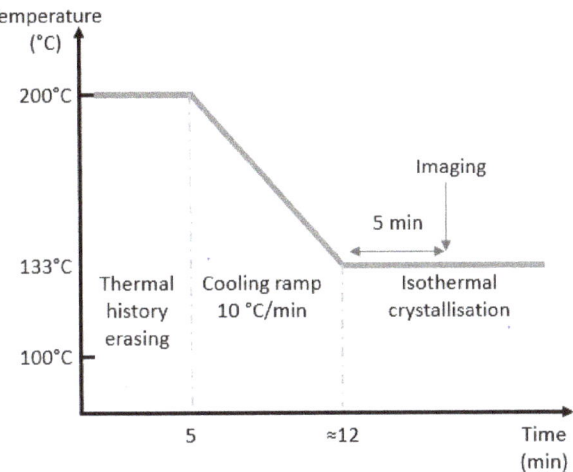

Scheme 1. Schematic representation of thermal protocol applied on the Isotactic polypropylene (iPP)/fibre samples during polarised light optical microscopy (PLOM) measurements.

2.2.6. Etching

Injection moulded samples were etched based on the method of Olley [44]. This method removes any amorphous material so that the remaining crystalline structure could be observed by electron microscopy. A 35 mm section was cut from the middle part of a tensile specimen. The section was gently stirred for 6 h in a solution of 1.3% wt potassium permanganate ($KMnO_4$), 32.9% wt concentrated sulfuric acid (H_2SO_4) and 65.8% wt concentrated phosphoric acid (H_3PO_4) at room temperature. It was then cleaned with hydrogen peroxide for 5 min, rinsed with water, and dried overnight in an oven at 105 °C.

2.2.7. Scanning Electron Microscopy (SEM)

All SEM samples were coated with chromium using an Emitech K575X sputter coater (Quorum Technologies Ltd., Kent, UK) and imaged using a JEOL JSM 6700F (JEOL Ltd., Tokyo, Japan) at 3 kV accelerating voltage. For publication purposes the images were contrast enhanced using the enhanced local contrast (CLAHE) plugin in the 'Fiji' distribution of ImageJ (V1.5h) [45].

2.2.8. Crystal Size Measurement

The SEM images generated from the etched samples were analysed using V++ software (Digital Optics, Version 5.0, Wellington, New Zealand) to determine the diameter of the crystals. Approximately 50 crystals were measured for each treatment.

2.2.9. Differential Scanning Calorimetry (DSC)

For DSC analysis, all measurements were performed in triplicate. A transverse section (0.35 mm) was cut from the middle part of a tensile specimen. To avoid measuring the crystallinity of the skin of the test specimens, only the core of the transverse section (~5 mg) was used. All the calorimetric experiments were performed with a Discovery (TA instruments, USA) differential scanning calorimeter under nitrogen atmosphere (50 mL min^{-1}). The temperature scale was calibrated using indium, lead and tin as standards to ensure reliability of the data obtained. Melting temperatures, enthalpies, crystallisation and fusion peaks were determined by TRIOS software (TA Instruments, USA). The degree of crystallinity (X_c) was estimated using Equation (1) where ΔH_m is the measured melting enthalpy of the polymeric part of the sample, Wf the fibre weight fraction in the composite, and ΔH 100% the equilibrium melting enthalpy of 100% crystalline PP assumed to be 207 J g^{-1} [46].

$$X_c = 100 \times \frac{\Delta Hm}{\Delta H\ 100\%\ (1-Wf)} \quad (1)$$

The thermal gradient of the DSC measurements is described in Scheme 2.

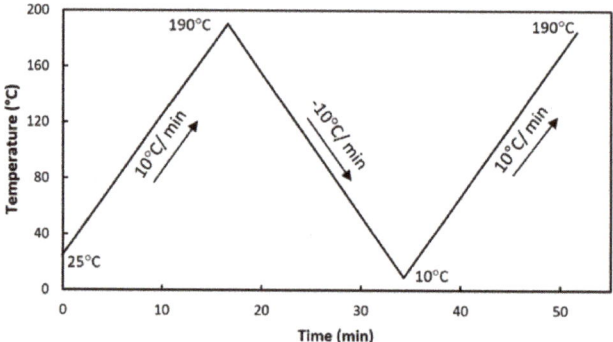

Scheme 2. Schematic illustration of the thermal protocol of the differential scanning calorimetry (DSC) measurements.

The enthalpy variations of the polymer during the initial heating ramp (Scheme 2), gives an indication of the thermal history related to the injection moulding conditions. A maximum temperature of 190 °C was used to limit the natural fibre degradation but, as a consequence, the thermal history might not have been totally erased. This step is followed by a slow cooling ramp at a rate that favours the formation of a high β-fraction [15]. The melting cycle observed during the second heating ramp illustrates the crystal formation process which occurred during a controlled slow cooling with limited thermal history.

2.2.10. Mechanical Testing

Tensile and Modulus Tests

Tensile properties were measured on an Instron 5566 (Instron, Norwood, MA, USA) universal testing machine fitted with a 10 kN load cell and an external extensometer, according to ISO 527. The crosshead speed was 5 mm min^{-1}. The gauge length was 115 mm and the extensometer length were set to 25 mm. Ten specimens were tested to failure to obtain the average Young's modulus and maximum tensile stress.

Impact Test

Samples for Izod notched impact strength were prepared and tested in accordance with ASTM D 256. Samples were of dimensions 12.6 mm × 63.5 mm × thickness and a 45° notch was machined into each sample using a Ceast Notchvis (Ceast, Italy). Seven Izod samples for each variable were tested at a velocity of 3.46 m s^{-1}, a 150° angle and a 0.5 J hammer using a Ceast Resil impact testing machine (Ceast, Torino, Italy). Pull-out surfaces were compared between different formulations with SEM pictures taken from the central part of the transverse fracture area.

2.2.11. Water Uptake

ISO test specimens (dogbone) were submerged in water at 20 °C for 70 days. During this time water uptake was monitored by briefly removing the samples from the water bath, wiping the surface dry with a tissue and weighing them. This approach is based on ASTM D1037. Samples were measured in triplicate.

3. Results

3.1. Crystal Structure

The PLOM analysis showed no obvious transcrystallinity around the fibres without NA (Figure 1A,D). Although transcrystallisation is generally expected in cellulose fibres [47–50], it is also known to be impeded by the presence of lignin and hemicelluloses that decrease the frequency of nucleation sites and degree of crystalinity [51,52]. The absence of transcrystallinity with the untreated Tencel™ fibres is likely due to the cellulose II configuration of these artificially produced fibres which reduces nucleation activity [53–56].

Treatment with NA resulted in crystal growth perpendicular to the two types of fibres (Figure 1B,C and Figure 1E,F). The fibres treated with the β-NA (Figure 1C,F) showed better definition of the crystalline structure than those treated with the α-NA (Figure 1B,E).

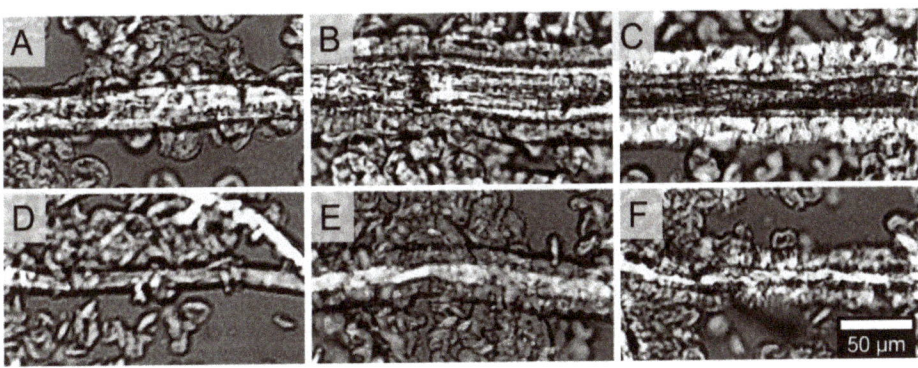

Figure 1. Optical micrographs of iPP trans-crystallisation around High Temperature Thermo-Mechanical Pulp fibres (HT TMP) and Tencel™ fibres with different sizing. The images are taken after a cooling period (10 °C min-1 ramp) and 5 min of isothermal crystallisation at 133 °C. (**A**) HT TMP only, (**B**) HT TMP + α-NA, (**C**) HT TMP + β-NA, (**D**) Tencel™ only, (**E**) Tencel™ + α-NA, (**F**) Tencel™ + β-NA. Scale bar = 50 µm.

SEM of the composites after compounding and injection moulding confirmed that the iPP matrix containing fibres treated with the β-NA had iPP crystals that were fan-shaped, typical of β-crystals, whereas the samples without β-NA only showed α-crystal morphology (Figure 2). The etching process not only removes the amorphous phase of PP but also the fibres from the composite. The images show that the crystals were not specifically present around the space previously occupied by the fibre but seemed to be distributed throughout the matrix.

Compared to the controlled PLOM analysis, the compounded and injection moulded composites investigated by SEM were subjected to shear stresses and faster cooling. Large α-crystals were evident in the pure iPP (Figure 2a). The introduction of HT-TMP (Figure 2d) and Tencel™ (Figure 2g) fibres did not change the crystal type and morphology but their size was decreased by half (Figure 2j). This was the result of steric hindrance as the presence of the fibres increased the number of nucleation sites thereby increasing the crystal density and thus restricting the growth of the α-crystals. The α-NA resulted in a 30-fold size reduction of the α-crystals. This is likely due to the high number of nucleation sites leading to a growth competition. The addition of both types of fibres (Figure 2e,h) did not lead to further reduction of the crystal size (Figure 2j). The fan shaped β-crystals induced by the β-NA (Figure 2c) had an average diameter around 3.3 µm (Figure 2j). The addition of HT TMP and Tencel™

fibres (Figure 2c,f) also did not reduce the size of β-crystals further (Figure 2j). The hollows in the centre of β-spherulites are attributable to the etched DCNDCA particles [57].

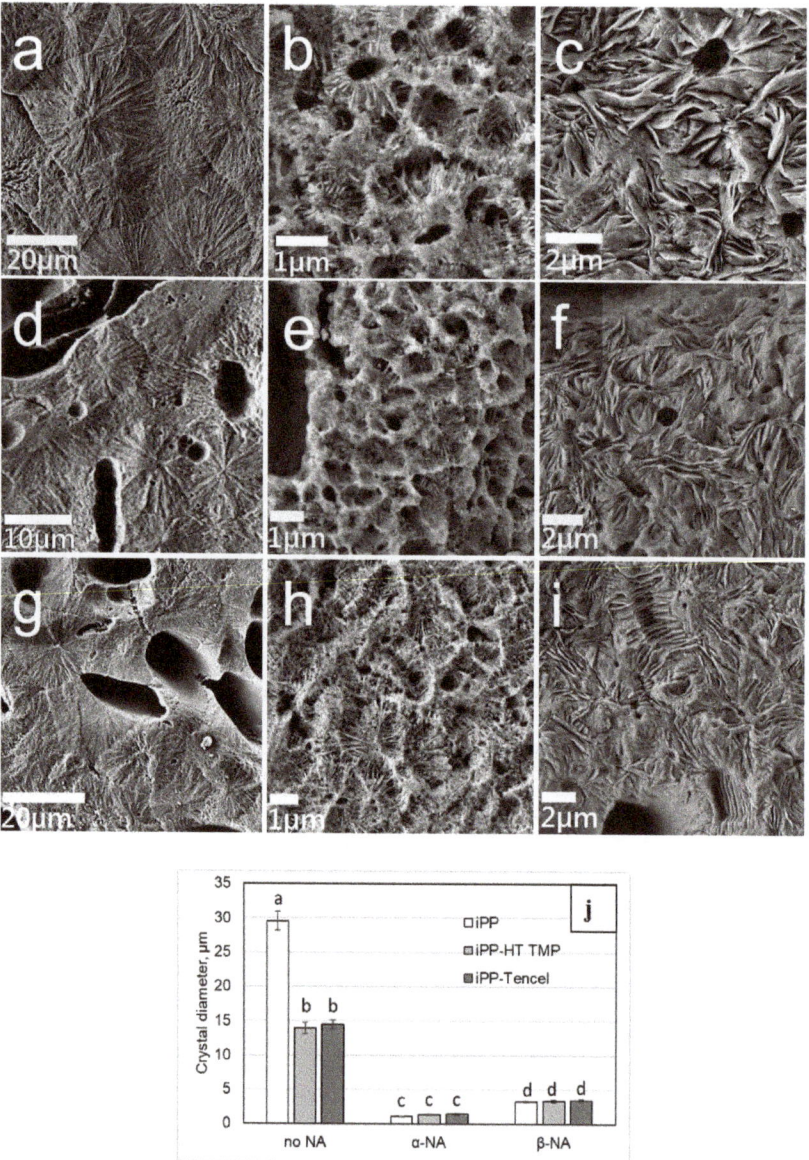

Figure 2. Scanning electron microscopy (SEM) images of iPP crystallisation. Transverse face of injection moulded samples after permanganate etching. (**a**) iPP, (**b**) iPP + α NA, (**c**) iPP + β-NA, (**d**) iPP + HT TMP, (**e**) iPP + HT TMP/α-NA, (**f**) iPP + HT TMP/β-NA, (**g**) iPP + Tencel™, (**h**) iPP+ Tencel™/α-NA, (**i**) iPP + Tencel™ + β-NA. (**j**) Crystal diameters of iPP and iPP composites with various NAs (n ~ 60, different letters indicate significant difference at α = 0.05).

3.2. Differential Scanning Calorimetry (DSC)

Total crystallinity for all samples ranged between 36% and 45% (Table 1). The addition of fibre to the iPP had little effect on the overall crystallinity of the composites. The addition of NA also had little effect on the overall crystallinity of the iPP, however, the type of NA had a significant effect of the type of crystal structure present. As expected, no β-crystals were observed in the composites without the addition of β-NA.

The addition of β-NA, in the absence of fibre, led to an approximately 13% reduction of α-crystal enthalpy. The decrease in the proportion of α-crystals was compensated by the appearance of β-crystals. Therefore, the overall crystallinity was maintained. The same trend was observed in the presence of both types of fibre. Removal of the thermal history led to a larger decrease in α-crystals, down to ~10.5% and the highest level of β-crystals observed 34% and 30.4% for HT TMP and Tencel™, respectively. The presence of fibres coupled with a slow cooling ramp and without shear stress was the most favourable condition observed to generate a high β-crystal proportion. Our results are in accordance with Kang et al., who found that iPP in the presence of the dual selective β-NA (TMB-5) coupled and a slow cooling rate favoured the formation of high β-crystal fraction [15]. However, those results contradict Dong et al., [57] who found that fast cooling is favourable for creating high β-content in iPP/TMB-5 system. In our case, the slow cooling was favourable to the growth of β-crystals, possibly because the β-crystals formed at a higher temperature compared to α-crystals and, therefore, β-crystals had time to grow without competition from the α-crystals.

Evaluating how a combination of a slow cooling ramp and high shear stress (injection moulding) affects β-crystals formation was not technically possible. Indeed, fast cooling of the polymer is central to the injection moulding process to reduce cycle time without compromising on part quality [58].

The potential factors explaining why the composites containing HT TMP fibres had a higher β-crystals content compared to one with Tencel™ fibres are probably related to the fibre surface. Tencel™ fibres have a smooth surface [59] and HT TMP fibres are rough with a coating of lignin. A rough surface increases the interfacial shear strength [60] which is a factor favouring β-interfacial structures [61]. Moreover, lignin particles can increase β-crystallinity [62].

The addition of fibre into iPP without NA did not have a major influence on the melting temperature of the polymer, neither did the addition of the α-NA (Table 1). The melting temperature was around 163 °C for the α-crystals for all composites. The addition of the β-NA added a second melting peak at 152 °C (after removal of thermal history). For the samples with β-NA, the cooling process slightly shifts the β-crystals melting peaks.

Table 1. Comparison of α and β-crystal proportions and effect of α- and β-nucleating agent (NA) on iPP melting properties, after injection moulding or after melting and a slow cooling ramp (10 °C/min).

Fibre Type	Treatment	No NA		α-NA		β-NA			
		Crystal Proportion (%)							
		α	β	α	β	α	β	avg (α+β)	α %/β %
No fibre	After IM	36.4 ± 0.3	-	39.1 ± 2.1	-	23.0 ± 1.0	16.5 ± 0.5	39.5	58.2/41.8
	Slow ramp	39.8 ± 1.9	-	45.1 ± 1.8	-	27.0 ± 1.5	15.5 ± 1.2	42.5	63.5/36.5
HT TMP	After IM	41.6 ± 0.4	-	35.8 ± 3.7	-	23.8 ± 1.0	15.5 ±0.1	39.3	60.5/39.5
	Slow ramp	42.1 ± 1.5	-	41.8 ± 0.5	-	11.1 ± 0.9	34.0 ± 1.1	45.1	24.6/75.4
Tencel™	After IM	42.0 ± 2.7	-	36.0 ± 3.4	-	28.5 ± 2.8	12.7 ± 1.0	41.2	69.2/30.8
	Slow ramp	38.1 ± 0.5	-	37.8 ± 1.8	-	10.1 ± 0.6	30.4 ± 0.9	40.5	24.9/75.1
	Melting Peak Temperature (°C)								
		α	β	α	β	α	β		
No fibre	After IM	166.4 ± 0.3	-	163.0 ± 0.2	-	163.9 ± 0.3	148.9 ± 0.3		
	Slow ramp	163.1 ± 0.2	-	164.1 ± 0.1	-	163.5 ± 0.2	151.5 ± 0.2		
HT TMP	After IM	164.8 ± 0.7	-	163.3 ± 0.2	-	163.7 ± 0.1	146.6 ± 0.0		
	Slow ramp	161.9 ± 0.2	-	165.0 ± 0.1	-	163.1 ± 0.1	151.7 ± 0.1		
Tencel™	After IM	163.9 ± 0.5	-	161.8 ± 0.2	-	164.3 ± 0.3	147.4 ± 0.2		
	Slow ramp	162.2 ± 0.3	-	163.7 ± 0.2	-	164.4 ± 0.9	152.3 ± 0.0		

Compared to IM, after a 10 °C min^{-1} cooling ramp, the melting peak temperature of the β-crystals increases by 2.6 °C, 5.1 °C and 4.9 °C for the iPP with β-NA without fibres, with HT TMP and with

Tencel™ respectively. This indicates that the β-crystals which were grown at high temperature during slow cooling are more stable and therefore melt at higher temperature [63].

Samples were cut from the centre of the transverse surface of a tensile test specimen (dogbone). Proportions are extracted from integration of the melting peaks obtained from the DSC experiment. (n = 3, ± = standard deviation)

For industrial processing, the crystallisation temperature is an important parameter as a higher crystallisation temperature allows faster processing [64]. The presence of fibre in the iPP increases nucleation sites, and thus the crystallisation temperature by 6 and 8 °C for the HT TMP and the Tencel™, respectively (Figure 3).

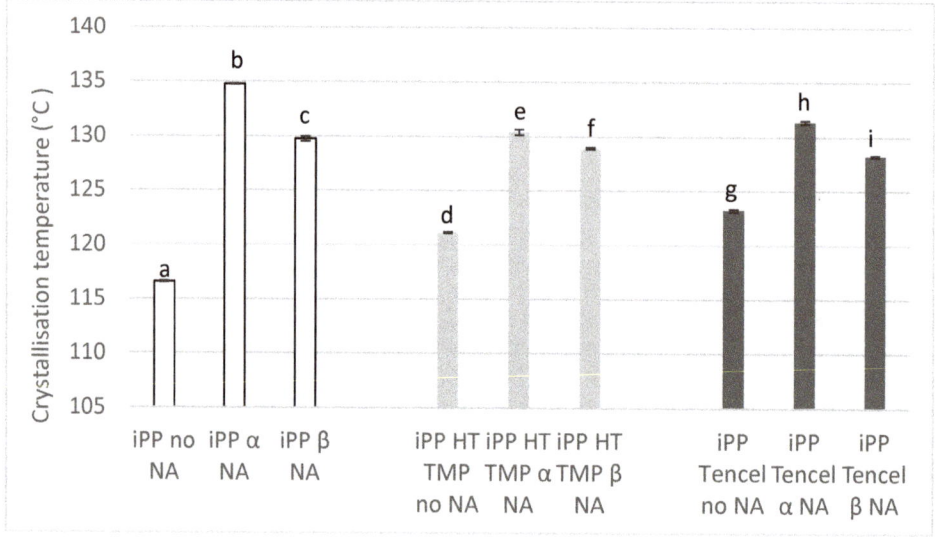

Figure 3. Effect of NA and fibres on iPP crystallisation temperature during a 10 °C min^{-1} cooling ramp (n = 3, different letters indicate significant difference at α = 0.05).

Addition of α-NA does not affect the proportion of α-crystals in the iPP, but its crystallisation temperature is 20 °C higher. This is lowered with the addition of fibre. In presence of α-NA, the crystallisation temperature for HT TMP and Tencel™ composites increased by 12 and 8 °C, respectively. The addition of β-NA leads to a 15 °C rise in crystallisation temperature for the pure iPP. Again, this was lowered with the addition of fibre. The presence of β-NA lead to 8 and 5 °C increase in crystallisation temperature for the HT TMP and Tencel™ iPP composites, respectively. Overall, the pattern of crystallisation temperature increase remains the same for composites filled with either fibre, with a minor antagonist effect when the NA and fibres are present together.

3.3. Mechanical Properties

For commercial application, obtaining a composite with good mechanical properties especially yield strength and impact strength is extremely important [58].

3.3.1. Young's Modulus and Stress

The addition of fibre to iPP led to at least a doubling of the Young's Modulus (Figure 4a). The effect of the NA alone is minimal in comparison, with an increase of 29% and 17% for α-NA and β-NA added to iPP. Additionally, the addition of both NAs in presence of fibres has no significant effect on Young's Modulus of the composites with the exception of α-NA with Tencel™ (+18%).

The situation is relatively similar for tensile maximum stress properties (Figure 4b). The addition of β-NA has no effect on iPP, while the α-NA leads to a moderate increase (+13%). The addition of HT TMP and Tencel™ generates an increase in tensile maximum stress of 46% and 75%, respectively. In all cases, the addition of α-NA lead to a minor increase, while the β-NA lead to an expected small decrease in tensile maximum stress.

3.3.2. Impact Strength

Reinforcing iPP with HT TMP fibre gave an improvement in impact strength of the composite (Figure 4c). Adding an impact strength modifier, such as a β-NA, to the fibre was considered an opportunity to further enhance impact performance. Compared to pure iPP, the sole addition of HT TMP fibre led to a 46% increase in impact strength. When added in conjunction with HT TMP fibre, the α-NA addition reduces the impact strength of the fibre composite compared to the addition of fibre alone. β-NA has no significant influence. The major gain is due to the presence of HT TMP. A major improvement in impact strength comes from the addition of Tencel™ fibres with an increase of 181%. Addition of α-NA led to no significant change, while the addition of the β-NA led to moderate further improvement (+17%).

In summary, the main improvement for mechanical properties comes from the presence of the fibres themselves. In some cases, the NAs are providing a further minor improvement but in other cases, they can also have detrimental effects.

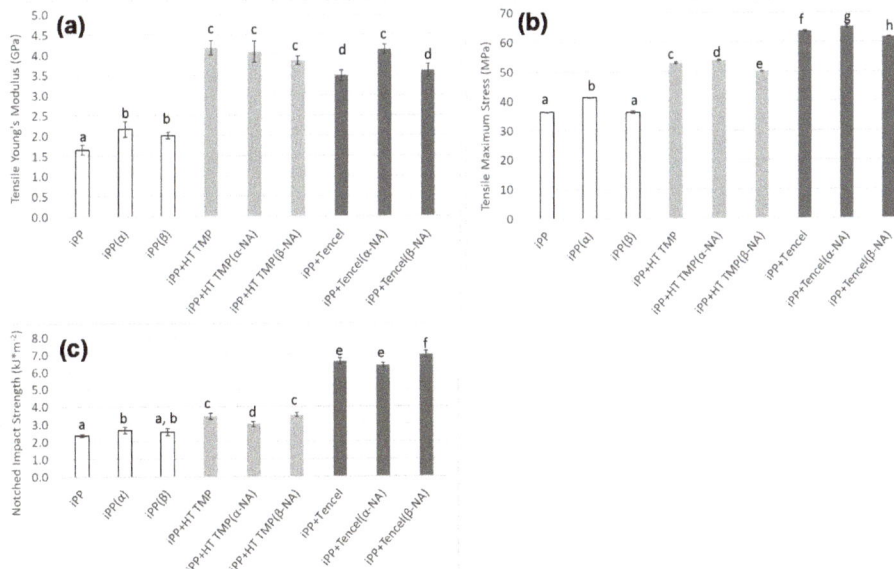

Figure 4. Influence of the α- and β-NA and HT TMP and Tencel™ fibres on (a) Young's modulus, (b) tensile maximum stress, and (c) notched impact strength. (Anova: different letter indicates significant difference at α = 0.05).

3.4. Pull Out

SEM micrographs of transverse fracture surfaces from tensile testing samples show only minor pull-out of fibres for HT TMP filled composites (Figure 5). The few visible fibres appear to be delaminated or covered in matrix. The influence of lignin is complex, although lignin removal was found to improve composite properties [65], the presence of lignin can also have the effect of improving properties due to its interaction with coupling agents [66–69]. There are no visually observable differences in pull-out between the different HT TMP treatments. The fracture surfaces of Tencel™ filled-composites show a large number of fibres being pulled-out across all treatments. No matrix material adheres to the pulled-out fibres. It indicates that the interfacial adhesion between Tencel™ and iPP is either lower than that of HT TMP and iPP or it can be explained by their higher shear strength compared to the HT TMP fibre [70,71]. Those observations agree with the higher impact strength observed for the Tencel-based composites, as a composite containing high strength fibres weakly compatibilized with the matrix will exhibit good impact performance. The TC layer observed by PLOM in the sandwich composite is clearly not covalently attached to the Tencel™ fibres after injection moulding. The NA do not seem to have a strong effect on the interface.

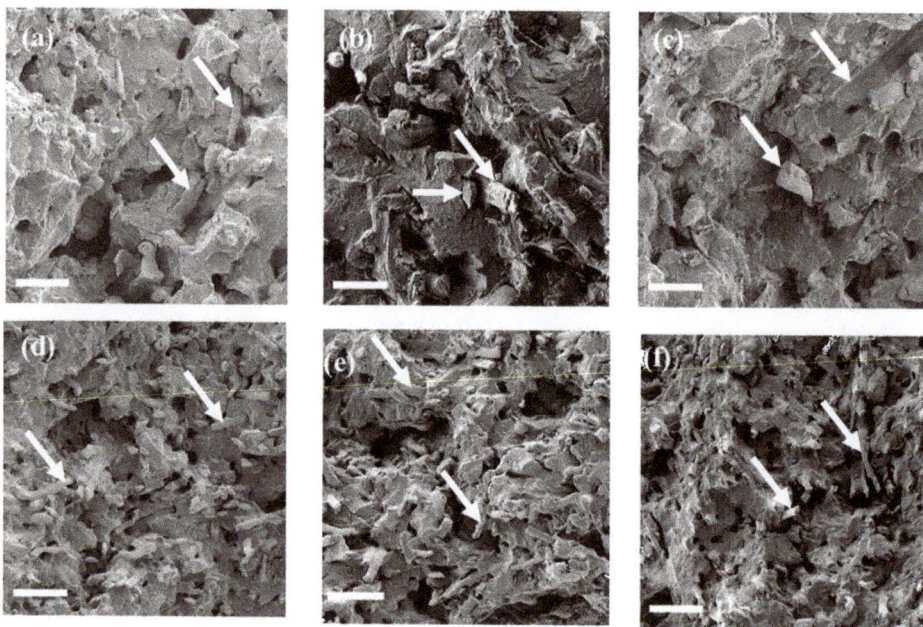

Figure 5. Comparison of pull-out surface between HT TMP and Tencel, from SEM pictures of transverse tensile test specimen fracture area. (**a**) HT TMP only, (**b**) HT TMP + α-NA, (**c**) HT TMP + β-NA, (**d**) Tencel™ only, (**e**) Tencel™ + α-NA, (**f**) Tencel™ + β-NA. Arrows indicate fibres. Scale bar = 100 μm.

3.5. Water Uptake

A slow water absorption rate allows for better dimensional stability of the composite when exposed to humidity. The nine formulations were immersed in water for 70 days. The results are expressed in percentage of water absorbed (Figure 6 and Table 2). The water uptake of the pure hydrophobic iPP is relatively low (0.3%). This value is comparable to previous studies [72]. The addition of β-NA leads to a slight but significant decrease in water absorption.

The hydrophilic character of natural fibres, ascribed to the bonding of water molecules to the free hydroxyl group is responsible for a water uptake increase in the plastic-fibre composites [73]. The addition of HT TMP and Tencel™ to iPP lead to an increase of water absorption to around 2%. This indicates that the two types of fibres have the same type of hydrophilic behaviour. Surface treatment of fibre has often been used to decrease natural fibre composite water uptake [74,75]. Wu et al., [76] compared the water absorption of flax fibres functionalised with MAPP within either α or β crystalline iPP matrix. They did not observe any difference between the two types of crystal matrix. However, when the fibres were not functionalised, the α matrix containing the flax fibres had a slightly higher water absorption than its β crystals counterpart. In our case, the addition of β-NA similarly shows a small but significant reduction in water uptake that is consistent across all treatments.

Unexpectedly, the addition of α-NA shows a significant increase in water uptake for samples containing fibres. The increase is only 0.5% in the case of the slightly hydrophobic HT TMP fibres, but is more pronounced (3.4%) in the case of the hydrophilic Tencel™ fibres.

Additionally, DSC analysis (shown earlier) indicated the crystallinity of the α-NA and fibre composites was lower than both the control and β-NA composites which is correlated to this increase in water uptake. This increase in amorphous structure will create pathways for accelerating the water penetration into the composite. The larger the decrease in crystallinity, the larger the water uptake

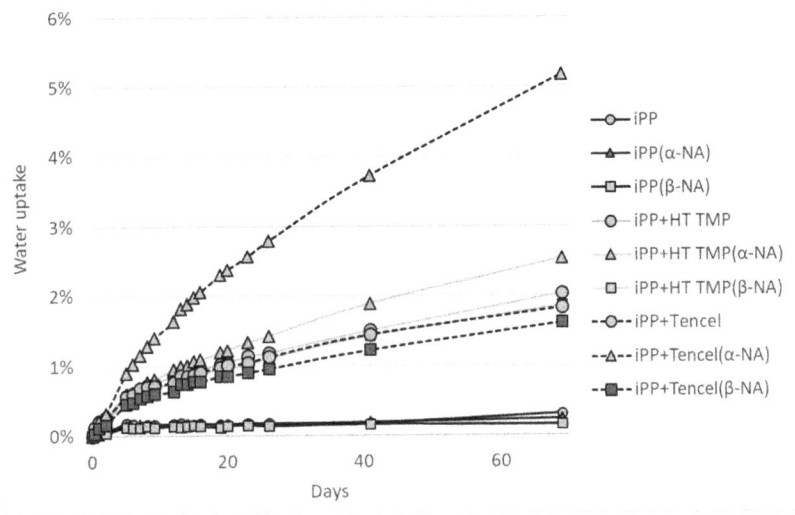

Figure 6. Water uptake of the various iPP composites.

Table 2. Percentage of water uptake of iPP composites after 70 days (n = 3, different letters indicate significant difference at α = 0.05).

Treatment	Avg	Stdev	Anova
iPP no NA	0.3%	0.1%	A
iPP (α-NA)	0.2%	0.0%	A
iPP (β-NA)	0.2%	0.0%	B
iPP/ HT TMP (no NA)	2.0%	0.0%	C
iPP/ HT TMP (α-NA)	2.5%	0.0%	D
iPP/ HT TMP (β-NA)	1.9%	0.1%	E
iPP/Tencel™ (no NA)	1.8%	0.0%	E
iPP/Tencel™ (α-NA)	5.2%	0.1%	F
iPP/Tencel™ (β-NA)	1.6%	0.0%	G

4. Discussion

The main aim of this study was to improve the mechanical properties of HT TMP composites. We developed an original approach by directly adding the NA on the fibres which could be adopted to tailor the lignocellulosic fibre composite characteristics without an additive incorporation step. The sizing of the fibres was successful; however, the hoped-for tensile strength (α-NA) and impact resistance (β-NA) improvements were not observed. Our strategy failed to improve the mechanical properties of the composite. It may be that the interface between lignocellulosic material and polyolefin matrix in presence of MAPP is already sufficiently strong and is not improved by the TC and nucleation of the matrix [77]. The absence of covalent linkage between the NA and the fibre might be one of the reasons for the lack of mechanical properties improvement. Another reason could be that the concentration of β-NA is too high. To allow for losses in the blending process, a relatively high concentration of NA was applied. However, losses were modest and nitrogen analysis revealed the concentration of β-NA in the final composite to be 0.4%. Other studies have reported an optimal concentration of β-NA is 0.1%. This allows a maximal growth of β-spherulite in needle like structure and a concentration as low as 0.2% would lead to β-crystals growth competition leading to incomplete spherulite and a decrease on impact strength [78]. When a composite has a high concentration of β-NA, the ordered structure effect cannot be reached [79]. Another aspect could be that the crystal structure at proximity of natural fibre might impair the MAPP efficiently grafting onto the fibre.

5. Conclusions

This study investigated the effect of coating lignocellulosic fibres with NA on the mechanical properties of the resulting iPP-fibre composite. The influence on the TC layer, crystal morphology and proportion, crystallisation temperature, mechanical properties, surface fracture and water uptake were investigated. The addition of NA led to slight increases in composite crystallinity as well as an increase in temperature of crystallisation. The presence of the NA also leads to a significant decrease in crystal size. Sizing the fibres with NA was shown to be an effective method to manipulate the crystallinity of the natural fibre composite. The presence of NA is deeply modifying the nature of matrix crystallinity and its thermal characteristics. However, under the experimental conditions of this study, the mechanical properties of the iPP-fibre composites were mainly influenced by the characteristics of the fibres while the NA had little influence. This suggests that the NAs probably have little influence on interfacial adhesion. In summary:

- The α- and β-NA as fibre sizings have the expected effect on iPP crystals.
- The α- and β-NA cause transcrystallinity at the interphase fibre-matrix in a sandwich composite.
- The α- and β-NA affects the whole matrix crystallinity after compounding.
- The tensile strength and impact strength of the fibre-iPP composite is not significantly improved by the NA.

Author Contributions: Conceptualization, A.T., A.D., R.R. and M.S.; methodology, A.T., A.D. and M.S.; formal analysis, A.T., and R.R.; resources, M.S.; data curation, A.T., and M.S.; writing—original draft preparation, M.S.; writing—review and editing, A.T. and A.D.; visualization, R.R. and A.T.; supervision, M.S.; project administration, M.S. All authors have read and agreed to the published version of the manuscript.

Funding: The New Zealand Ministry of Business, Innovation and Employment (MBIE) funding of Scion supported this work. Scion core funding C04X1703 Scion Platforms Plan (Strategic Science Investment Fund) has been used for this article.

Acknowledgments: We wish to thank Damien Even, Marc Gaugler, Kate parker, Jeremy Warnes, Doug Gaunt and Elspeth MacRae for their helpful suggestions.

Conflicts of Interest: The authors declare no conflict of interest. The funders had no role in the design of the study; in the collection, analyses, or interpretation of data; in the writing of the manuscript, or in the decision to publish the results.

References

1. Clements, A.; Dunn, M.; Firth, V.; Hubbard, L.; Laonby, J.; Waddington, D. The essential chemical industry. *Chem. Ind. Educ. Cent. Univ. York. UK* **2010**, *511*, 34.
2. Zhang, L.; Qin, Y.; Zheng, G.; Dai, K.; Liu, C.; Yan, X.; Guo, J.; Shen, C.; Guo, Z. Interfacial crystallization and mechanical property of isotactic polypropylene based single-polymer composites. *Polymer* **2016**, *90*, 18–25. [CrossRef]
3. Li, M.; Li, G.; Zhang, Z.; Dai, X.; Mai, K. Enhanced β-crystallization in polypropylene random copolymer with a supported β-nucleating agent. *Thermochim. Acta* **2014**, *598*, 36–44. [CrossRef]

4. Kersch, M.; Schmidt, H.-W.; Altstädt, V. Influence of different beta-nucleating agents on the morphology of isotactic polypropylene and their toughening effectiveness. *Polymer* **2016**, *98*, 320–326. [CrossRef]
5. Zhang, Y.; Zhang, L.; Liu, H.; Du, H.; Zhang, J.; Wang, T.; Zhang, X. Novel approach to tune mechanics of β-nucleation agent nucleated polypropylene: Role of oriented β spherulite. *Polymer* **2013**, *54*, 6026–6035. [CrossRef]
6. Silvestre, C.; Cimmino, S.; Duraccio, D.; Schick, C. Isothermal crystallization of isotactic poly (propylene) studied by superfast calorimetry. *Macromol. Rapid Commun.* **2007**, *28*, 875–881. [CrossRef]
7. De Santis, F.; Adamovsky, S.; Titomanlio, G.; Schick, C. Isothermal nanocalorimetry of isotactic polypropylene. *Macromolecules* **2007**, *40*, 9026–9031. [CrossRef]
8. Grein, C. Toughness of neat, rubber modified and filled β-nucleated polypropylene: From fundamentals to applications. In *Intrinsic Molecular Mobility and Toughness of Polymers ii*; Springer: Berlin/Heidelberg, Germany, 2005; pp. 43–104.
9. Cai, Z.; Zhang, Y.; Li, J.; Xue, F.; Shang, Y.; He, X.; Feng, J.; Wu, Z.; Jiang, S. Real time synchrotron saxs and waxs investigations on temperature related deformation and transitions of β-iPP with uniaxial stretching. *Polymer* **2012**, *53*, 1593–1601. [CrossRef]
10. Chen, Y.; Wu, Z.; Fan, Q.; Yang, S.; Song, E.; Zhang, Q. Great toughness reinforcement of isotactic polypropylene/elastomer blends with quasi-cocontinuous phase morphology by traces of β-nucleating agents and carbon nanotubes. *Compos. Sci. Technol.* **2018**, *167*, 277–284. [CrossRef]
11. Shangguan, Y.; Song, Y.; Peng, M.; Li, B.; Zheng, Q. Formation of β-crystal from nonisothermal crystallization of compression-molded isotactic polypropylene melt. *Eur. Polym. J.* **2005**, *41*, 1766–1771. [CrossRef]
12. Zhang, B.; Chen, J.; Ji, F.; Zhang, X.; Zheng, G.; Shen, C. Effects of melt structure on shear-induced β-cylindrites of isotactic polypropylene. *Polymer* **2012**, *53*, 1791–1800. [CrossRef]
13. Sun, X.; Li, H.; Zhang, X.; Wang, J.; Wang, D.; Yan, S. Effect of fiber molecular weight on the interfacial morphology of iPP fiber/matrix single polymer composites. *Macromolecules* **2006**, *39*, 1087–1092. [CrossRef]
14. Wang, L.; Yang, M.-B. Unusual hierarchical distribution of β-crystals and improved mechanical properties of injection-molded bars of isotactic polypropylene. *Rsc Adv.* **2014**, *4*, 25135–25147. [CrossRef]
15. Kang, J.; Wang, B.; Peng, H.; Li, J.; Chen, J.; Gai, J.; Cao, Y.; Li, H.; Yang, F.; Xiang, M. Investigation on the dynamic crystallization and melting behavior of β-nucleated isotactic polypropylene with different stereo-defect distribution—The role of dual-selective β-nucleation agent. *Polym. Adv. Technol.* **2014**, *25*, 97–107. [CrossRef]
16. Assouline, E.; Pohl, S.; Fulchiron, R.; Gerard, J.-F.; Lustiger, A.; Wagner, H.; Marom, G. The kinetics of α and β transcrystallization in fibre-reinforced polypropylene. *Polymer* **2000**, *41*, 7843–7854. [CrossRef]
17. Sanadi, A.; Young, R.; Clemons, C.; Rowell, R. Recycled newspaper fibers as reinforcing fillers in thermoplastics: Part i-analysis of tensile and impact properties in polypropylene. *J. Reinf. Plast. Compos.* **1994**, *13*, 54–67. [CrossRef]
18. Thomason, J. The influence of fibre length and concentration on the properties of glass fibre reinforced polypropylene: 5. Injection moulded long and short fibre PP. *Compos. Part A Appl. Sci. Manuf.* **2002**, *33*, 1641–1652. [CrossRef]
19. Hyer, M.W. *Stress Analysis of Fiber-Reinforced Composite Materials*; DEStech Publications, Inc.: Lancaster, PA, USA, 2009.
20. Bourmaud, A.; Ausias, G.; Lebrun, G.; Tachon, M.L.; Baley, C. Observation of the structure of a composite polypropylene/flax and damage mechanisms under stress. *Ind. Crop. Prod.* **2013**, *43*, 225–236. [CrossRef]
21. Thomason, J.; Van Rooyen, A. Transcrystallized interphase in thermoplastic composites. *J. Mater. Sci.* **1992**, *27*, 897–907. [CrossRef]
22. Quan, H.; Li, Z.-M.; Yang, M.-B.; Huang, R. On transcrystallinity in semi-crystalline polymer composites. *Compos. Sci. Technol.* **2005**, *65*, 999–1021. [CrossRef]
23. Yan, B.; Wu, H.; Jiang, G.; Guo, S.; Huang, J. Interfacial crystalline structures in injection over-molded polypropylene and bond strength. *Acs Appl. Mater. Interfaces* **2010**, *2*, 3023–3036. [CrossRef] [PubMed]
24. Xu, H.; Xie, L.; Jiang, X.; Li, X.-J.; Li, Y.; Zhang, Z.-J.; Zhong, G.-J.; Li, Z.-M. Toward stronger transcrystalline layers in poly(l-lactic acid)/natural fiber biocomposites with the aid of an accelerator of chain mobility. *J. Phys. Chem. B* **2014**, *118*, 812–823. [CrossRef] [PubMed]
25. Garkhail, S.; Wieland, B.; George, J.; Soykeabkaew, N.; Peijs, T. Transcrystallisation in PP/flax composites and its effect on interfacial and mechanical properties. *J. Mater. Sci.* **2009**, *44*, 510–519. [CrossRef]

26. Borysiak, S. Fundamental studies on lignocellulose/polypropylene composites: Effects of wood treatment on the transcrystalline morphology and mechanical properties. *J. Appl. Polym. Sci.* **2013**, *127*, 1309–1322. [CrossRef]
27. Wang, C.; Liu, F.-H.; Huang, W.-H. Electrospun-fiber induced transcrystallization of isotactic polypropylene matrix. *Polymer* **2011**, *52*, 1326–1336. [CrossRef]
28. Ning, N.; Zhang, W.; Yan, J.; Xu, F.; Wang, T.; Su, H.; Tang, C.; Fu, Q. Largely enhanced crystallization of semi-crystalline polymer on the surface of glass fiber by using graphene oxide as a modifier. *Polymer* **2013**, *54*, 303–309. [CrossRef]
29. Wang, Y.; Tong, B.; Hou, S.; Li, M.; Shen, C. Transcrystallization behavior at the poly (lactic acid)/sisal fibre biocomposite interface. *Compos. Part A: Appl. Sci. Manuf.* **2011**, *42*, 66–74. [CrossRef]
30. Sorieul, M.; Dickson, A.; Hill, S.J.; Pearson, H. Plant fibre: Molecular structure and biomechanical properties, of a complex living material, influencing its deconstruction towards a biobased composite. *Materials* **2016**, *9*, 618. [CrossRef]
31. Markiewicz, E.; Borysiak, S.; Paukszata, D. Polypropylene-lignocellulosic material composites as promising sound absorbing materials. *Polimery* **2009**, *54*, 430–435. [CrossRef]
32. Saheb, D.N.; Jog, J.P. Natural fiber polymer composites: A review. *Adv. Polym. Technol. J. Polym. Process. Inst.* **1999**, *18*, 351–363. [CrossRef]
33. Karimi, K.; Taherzadeh, M.J. A critical review of analytical methods in pretreatment of lignocelluloses: Composition, imaging, and crystallinity. *Bioresour. Technol.* **2016**, *200*, 1008–1018. [CrossRef] [PubMed]
34. Donaldson, L.; Lomax, T.D. Adhesive/fibre interaction in medium density fibreboard. *Wood Sci. Technol.* **1989**, *23*, 371–380. [CrossRef]
35. Shao, Z.; Li, K. The effect of fiber surface lignin on interfiber bonding. *J. Wood Chem. Technol.* **2006**, *26*, 231–244. [CrossRef]
36. Warnes, J.M.; Fernyhough, A.; Anderson, C.R.; Lee, B.J.; Witt, M.R.J. Method for Producing Wood Fibre Pellets. U.S. Patent 9,511,508, 6 December 2016.
37. Thumm, A.; Even, D.; Gini, P.-Y.; Sorieul, M. Processing and properties of mdf fibre-reinforced biopolyesters with chain extender additives. *Int. J. Polym. Sci.* **2018**, *2018*, 9601753. [CrossRef]
38. Bourban, C.; Karamuk, E.; De Fondaumiere, M.; Ruffieux, K.; Mayer, J.; Wintermantel, E. Processing and characterization of a new biodegradable composite made of a phb/v matrix and regenerated cellulosic fibers. *J. Environ. Polym. Degrad.* **1997**, *5*, 159–166.
39. Graupner, N.; Müssig, J. A comparison of the mechanical characteristics of kenaf and lyocell fibre reinforced poly (lactic acid)(pla) and poly (3-hydroxybutyrate)(phb) composites. *Compos. Part A Appl. Sci. Manuf.* **2011**, *42*, 2010–2019. [CrossRef]
40. McDonald, A.; Clare, A.; Dawson, B. Surface characterisation of radiata pine high-temperature TMP fibres by x-ray photo-electron spectroscopy. In Proceedings of the 53rd General APPITA Conference, Rotorua, New Zealand, 19–23 April 1999; pp. 51–57.
41. Dickson, A.; Teuber, L.; Gaugler, M.; Sandquist, D. Effect of processing conditions on wood and glass fiber length attrition during twin screw composite compounding. *J. Appl. Polym. Sci.* **2020**, *137*, 48551. [CrossRef]
42. Wolters, W.S.; Hanssen, R.; Palanisami, T.K.; Dotson, D.L. Nucleating Agent Additive Compositions and Methods. U.S. Patent 7,659,336, 9 February 2010.
43. Ye, J.; Fang, J.; Zhang, L.; Li, C. Transcrystalline induced by mwcnts and organic nucleating agents at the interface of glass fiber/polypropylene. *Polym. Compos.* **2018**, *39*, 3424–3433. [CrossRef]
44. Bassett, D.C.; Olley, R.H. On the lamellar morphology of isotactic polypropylene spherulites. *Polymer* **1984**, *25*, 935–943. [CrossRef]
45. Schindelin, J.; Arganda-Carreras, I.; Frise, E.; Kaynig, V.; Longair, M.; Pietzsch, T.; Preibisch, S.; Rueden, C.; Saalfeld, S.; Schmid, B.; et al. Fiji: An open-source platform for biological-image analysis. *Nat. Methods* **2012**, *9*, 676. [CrossRef]
46. Blaine, R.L. Thermal applications note. *Polym. Heats Fusion*, 2002.
47. Ljungberg, N.; Cavaillé, J.-Y.; Heux, L. Nanocomposites of isotactic polypropylene reinforced with rod-like cellulose whiskers. *Polymer* **2006**, *47*, 6285–6292. [CrossRef]
48. Habibi, Y.; Dufresne, A. Highly filled bionanocomposites from functionalized polysaccharide nanocrystals. *Biomacromolecules* **2008**, *9*, 1974–1980. [CrossRef] [PubMed]

49. Gray, D.G. Transcrystallization of polypropylene at cellulose nanocrystal surfaces. *Cellulose* **2008**, *15*, 297–301. [CrossRef]
50. Nwabunma, D.; Kyu, T. *Polyolefin Composites*; John Wiley & Sons: Hoboken, NJ, USA, 2008.
51. Gray, D.G. Polypropylene transcrystallization at the surface of cellulose fibers. *J. Polym. Sci. Polym. Lett. Ed.* **1974**, *12*, 509–515. [CrossRef]
52. Klapiszewski, Ł.; Grząbka-Zasadzińska, A.; Borysiak, S.; Jesionowski, T. Preparation and characterization of polypropylene composites reinforced by functional zno/lignin hybrid materials. *Polym. Test.* **2019**, *79*, 106058. [CrossRef]
53. Wunderlich, B.; Grebowicz, J. Thermotropic mesophases and mesophase transitions of linear, flexible macromolecules. In *Liquid Crystal Polymers ii/iii*; Springer: Berlin/Heidelberg, Germany, 1984; pp. 1–59.
54. Paukszta, D.; Borysiak, S. The influence of processing and the polymorphism of lignocellulosic fillers on the structure and properties of composite materials—A review. *Materials* **2013**, *6*, 2747–2767. [CrossRef]
55. Borysiak, S. Influence of wood mercerization on the crystallization of polypropylene in wood/PP composites. *J. Therm. Anal. Calorim.* **2012**, *109*, 595–603. [CrossRef]
56. Borysiak, S.; Doczekalska, B. The influence of chemical modification of wood on its nucleation ability in polypropylene composites. *Polimery* **2009**, *54*, 820–827. [CrossRef]
57. Dong, M.; Guo, Z.; Su, Z.; Yu, J. The effects of crystallization condition on the microstructure and thermal stability of istactic polypropylene nucleated by β-form nucleating agent. *J. Appl. Polym. Sci.* **2011**, *119*, 1374–1382. [CrossRef]
58. Khan, M.; Afaq, S.K.; Khan, N.U.; Ahmad, S. Cycle time reduction in injection molding process by selection of robust cooling channel design. *Isrn Mech. Eng.* **2014**, *2014*, 1–8. [CrossRef]
59. Firgo, H.; Schuster, K.; Suchomel, F.; Männer, J.; Burrow, T.; Abu Rous, M. The functional properties of tencel®-a current update. *Lenzing. Ber.* **2006**, *85*, 22–30.
60. Bera, M.; Alagirusamy, R.; Das, A. A study on interfacial properties of jute-PP composites. *J. Reinf. Plast. Compos.* **2010**, *29*, 3155–3161. [CrossRef]
61. Sun, X.; Li, H.; Wang, J.; Yan, S. Shear-induced interfacial structure of isotactic polypropylene (iPP) in iPP/fiber composites. *Macromolecules* **2006**, *39*, 8720–8726. [CrossRef]
62. Canetti, M.; De Chirico, A.; Audisio, G. Morphology, crystallization and melting properties of isotactic polypropylene blended with lignin. *J. Appl. Polym. Sci.* **2004**, *91*, 1435–1442. [CrossRef]
63. Zhang, Z.; Wang, C.; Yang, Z.; Chen, C.; Mai, K. Crystallization behavior and melting characteristics of PP nucleated by a novel supported β-nucleating agent. *Polymer* **2008**, *49*, 5137–5145. [CrossRef]
64. Raka, L.; Bogoeva-Gaceva, G. Crystallization of polypropylene: Application of differential scanning calorimetry part ii. Crystal forms and nucleation. *Contrib. Sect. Nat. Math. Biotech. Sci.* **2017**, *8;29*, 1–2. [CrossRef]
65. Ou, R.; Xie, Y.; Wolcott, M.P.; Sui, S.; Wang, Q. Morphology, mechanical properties, and dimensional stability of wood particle/high density polyethylene composites: Effect of removal of wood cell wall composition. *Mater. Des.* **2014**, *58*, 339–345. [CrossRef]
66. Arbelaiz, A.; Cantero, G.; Fernandez, B.; Mondragon, I.; Ganan, P.; Kenny, J. Flax fiber surface modifications: Effects on fiber physico mechanical and flax/polypropylene interface properties. *Polym. Compos.* **2005**, *26*, 324–332. [CrossRef]
67. Graupner, N.; Fischer, H.; Ziegmann, G.; Müssig, J. Improvement and analysis of fibre/matrix adhesion of regenerated cellulose fibre reinforced PP-, MAPP-and PLA-composites by the use of eucalyptus globulus lignin. *Compos. Part B Eng.* **2014**, *66*, 117–125. [CrossRef]
68. Peltola, H.; Pääkkönen, E.; Jetsu, P.; Heinemann, S. Wood based pla and PP composites: Effect of fibre type and matrix polymer on fibre morphology, dispersion and composite properties. *Compos. Part A Appl. Sci. Manuf.* **2014**, *61*, 13–22. [CrossRef]
69. Peltola, H.; Immonen, K.; Johansson, L.S.; Virkajärvi, J.; Sandquist, D. Influence of pulp bleaching and compatibilizer selection on performance of pulp fiber reinforced pla biocomposites. *J. Appl. Polym. Sci.* **2019**, *136*, 47955. [CrossRef]
70. Bahia, H.S. Process of Making Lyocell Fibre or Film. U.S. Patent 6,258,304, 10 July 2001.
71. Solala, I.; Antikainen, T.; Reza, M.; Johansson, L.-S.; Hughes, M.; Vuorinen, T. Spruce fiber properties after high-temperature thermomechanical pulping (HT-TMP). *Holzforschung* **2014**, *68*, 195–201. [CrossRef]

72. Panthapulakkal, S.; Sain, M. Injection-molded short hemp fiber/glass fiber-reinforced polypropylene hybrid composites—Mechanical, water absorption and thermal properties. *J. Appl. Polym. Sci.* **2007**, *103*, 2432–2441. [CrossRef]
73. Ayrilmis, N.; Kaymakci, A.; Ozdemir, F. Physical, mechanical, and thermal properties of polypropylene composites filled with walnut shell flour. *J. Ind. Eng. Chem.* **2013**, *19*, 908–914. [CrossRef]
74. Arbelaiz, A.; Fernandez, B.; Ramos, J.; Retegi, A.; Llano-Ponte, R.; Mondragon, I. Mechanical properties of short flax fibre bundle/polypropylene composites: Influence of matrix/fibre modification, fibre content, water uptake and recycling. *Compos. Sci. Technol.* **2005**, *65*, 1582–1592. [CrossRef]
75. Sreekala, M.; Thomas, S. Effect of fibre surface modification on water-sorption characteristics of oil palm fibres. *Compos. Sci. Technol.* **2003**, *63*, 861–869. [CrossRef]
76. Wu, C.-M.; Lai, W.-Y.; Wang, C.-Y. Effects of surface modification on the mechanical properties of flax/β-polypropylene composites. *Materials* **2016**, *9*, 314. [CrossRef]
77. Dickson, A.R.; Even, D.; Warnes, J.M.; Fernyhough, A. The effect of reprocessing on the mechanical properties of polypropylene reinforced with wood pulp, flax or glass fibre. *Compos. Part A: Appl. Sci. Manuf.* **2014**, *61*, 258–267. [CrossRef]
78. Sheng, Q.; Zhang, Y.; Xia, C.; Mi, D.; Xu, X.; Wang, T.; Zhang, J. A new insight into the effect of β modification on the mechanical properties of iPP: The role of crystalline morphology. *Mater. Des.* **2016**, *95*, 247–255. [CrossRef]
79. Kang, J.; Weng, G.; Chen, Z.; Chen, J.; Cao, Y.; Yang, F.; Xiang, M. New understanding in the influence of melt structure and β-nucleating agents on the polymorphic behavior of isotactic polypropylene. *Rsc Adv.* **2014**, *4*, 29514–29526. [CrossRef]

© 2020 by the authors. Licensee MDPI, Basel, Switzerland. This article is an open access article distributed under the terms and conditions of the Creative Commons Attribution (CC BY) license (http://creativecommons.org/licenses/by/4.0/).

Article

Ultralight Industrial Bamboo Residue-Derived Holocellulose Thermal Insulation Aerogels with Hydrophobic and Fire Resistant Properties

Hanxiao Huang [1,2,3], Yunshui Yu [1], Yan Qing [1], Xiaofeng Zhang [1,2,3], Jia Cui [1] and Hankun Wang [2,3,*]

1. College of Material Science and Engineering, Central South University of Forestry and Technology, Changsha 410004, China; huanghx@icbr.ac.cn (H.H.); yuyunshui@csuft.edu.cn (Y.Y.); qingyan0429@163.com (Y.Q.); zhangxf@icbr.ac.cn (X.Z.); cuijia9417@163.com (J.C.)
2. Institute of New Bamboo and Rattan Based Biomaterials, International Center for Bamboo and Rattan, Beijing 100102, China
3. SFA and Beijing Co-built Key Laboratory of Bamboo and Rattan Science & Technology, State Forestry and Grassland Administration, Beijing 100102, China
* Correspondence: wanghankun@icbr.ac.cn; Tel.: +86-10-8478-9909; Fax: +86-10-8423-8052

Received: 8 December 2019; Accepted: 15 January 2020; Published: 19 January 2020

Abstract: In this study, water-soluble ammonium polyphosphate- (APP) and methyl trimethoxysilane (MTMS)-modified industrial bamboo residue (IBR)-derived holocellulose nanofibrils (HCNF/APP/MTMS) were used as the raw materials to prepare aerogels in a freeze-drying process. Synthetically modified aerogels were confirmed by Fourier transform infrared spectroscopy, X-ray diffraction, and thermal stability measurements. As-prepared HCNF/APP/MTMS aerogels showed themselves to be soft and flexible. The scanning electron microscopy (SEM) analysis showed that the foam-like structure translates into a 3D network structure from HCNF aerogels to HCNF/APP/MTMS aerogels. The compressive modules of the HCNF/APP/MTMS aerogels were decreased from 38 kPa to 8.9 kPa with a density in the range of 12.04–28.54 kg/m^3, which was due to the structural change caused by the addition of APP and MTMS. Compared with HCNF aerogels, HCNF/APP/MTMS aerogels showed a high hydrophobicity, in which the water contact angle was 130°, and great flame retardant properties. The peak of heat release rate (pHRR) and total smoke production (TSP) decreased from 466.6 to 219.1 kW/m^2 and 0.18 to 0.04 m^2, respectively, meanwhile, the fire growth rate (FIGRA) decreased to 8.76 kW/s·m^2. The thermal conductivity of the HCNF/APP/MTMS aerogels was 0.039 W/m·K. All results indicated the prepared aerogels should be expected to show great potential for thermally insulative materials.

Keywords: industrial bamboo residue; holocellulose aerogel; hydrophobicity; fire resistance; thermal insulation material

1. Introduction

In recent years, significant industrial waste from cellulose has been produced annually [1]. Typically, the industrial waste is landfilled or burned, which has caused a large amount of cellulose resources to be wasted as well as environmental pollution [2,3]. As the people began to pay attention to sustainable development of the environment, some industrial waste was then recycled and used for low-cost preparation of bio-materials. Bamboo is a biomass resource with fast growth and vast availability [4,5]. It is also the second most important lignocellulosic material behind wood [6,7]. With the development of the bamboo industry, the parenchyma cells of bamboo are often discarded because of their loose structure, leading to a large amount of bamboo waste [8]. However, the loose structure of parenchyma tissue is beneficial to chemical treatment and mechanical fibrillation. 80%

of industrial bamboo residue (IBR) is parenchyma tissue, which can be used to prepare cellulose nanofibers inexpensively [9]. Therefore, the application of IBR will effectively increase the added value of bamboo.

One potential method for increasing usage of IBR is to prepare cellulose aerogels, which is a solid biomass material that replaces the liquid in gels with gas without changing the 3D network structure or volume of the gel [10]. As a next generation material, cellulose aerogels overcome the fragility of silicon aerogels and can be self-assembled without a crosslinking agent. Over the past few decades, cellulose aerogels have garnered significant attention in the field of thermal insulation, because of their specific surface areas, low density, high porosity, favorable biodegradability, and biocompatibility [11]. However, the abundant hydroxyl groups and flammable properties of cellulose aerogels result in poor water and fire resistance. In thermal insulation applications, this inherent defect hinders the utilization of bio-based aerogels. But then again, abundant hydroxyl groups in cellulose aerogels provide powerful conditions for modification [12]. To improve water and fire resistance, physical or chemical modifications via reinforcing components are effective modification methods. Many methods have been reported regarding flame retardant and hydrophobic modification, including: (1) cellulose nanofibril (CNF) aerogels were modified with methylene diphenyl diisocyanate (MDI) by solvent exchange [13]; (2) CNF aerogels were crosslinked with ionic liquid 1-allyl-3-methylimidazolium chloride for hydrophobic modification [14]; (3) CNF aerogels were coated on a silane modifier by chemical vapor deposition (CVD) [15]; (4) and CNF aerogels were modified with cationic chitosan (Ch), anionic poly(vinylphosphonic acid) (PVPA), and anionic montmorillonite clay (MMT) by a layer-by-layer technique [16]. However, these methods have complicated modification processes, and often do not pay attention to the durability of modifiers. These modifiers easily lose their functionality when they are affected by environment changes. Therefore, current research goals should focus on a simple process that is controllable and enables environmental protection.

Water-soluble ammonium polyphosphate (APP) is an efficient commercial flame retardant with excellent performance that is non-toxic and emits no gases or drips during a flame test and has been shown to improve the flame retardancy of poly(vinyl alcohol) (PVA) aerogels [17]. However, water-soluble APP can be dissolved in water and be removed when it encounters water. This decreases the durability of APP in aerogels for practical applications. If aerogels are given hydrophobicity, this problem can be solved. Currently, some hydrophobic aerogels were proposed in a process where acid-hydrolyzed methyltrimethoxysilane (MTMS) sol modified CNF suspensions were generated via freeze-drying [18]. This modification was carried out in the aqueous phase without chemical post-treatment, which improved the hydrophobicity and mechanical properties of the CNF aerogel. MTMS has strong hydrolysis activity and good chemical stability that can be used in the reaction under aqueous conditions. It provides the conditions necessary so that the modification of MTMS and APP can be carried out simultaneously under aqueous conditions, which avoids the complicated modification process. On the other hand, the MTMS modification can effectively improve the durability of APP, providing a possibility for the application of aerogels in the field of thermal insulation.

Although numerous studies have reported on cellulose aerogels for thermal insulation materials with various water and fire resistance modifications, few have focused on using holocellulose nanofibrils (HCNF) as raw materials to prepare aerogels. Cellulose aerogels most commonly have been prepared from CNF, which uses high-pressure homogenizers. However, the energy consumption of this kind of preparation is very high and causes easy clogging of the homogenizers. Thus, the use of various pre-treatment methods, such as oxidation by 2,2,6,6-tetramethylpiperidiny loxyl or cellulose modification by the introduction of charged groups [19], were necessary for commercial exploitation of CNF production. However, this also created a complicated preparation process. Currently, it has been reported that a high content of hemicellulose might lead to an easy nanofiber fibrillation tendency [20]. Compared to TEMPO-CNF (2.5 nm), the diameter of CNF which has been prepared by holocellulose is 4.2 nm [21]. In addition, the morphology and structure of TEMPO-CNF and holocellulose-derived CNF showed no significant difference. Using holocellulose to prepare aerogels can allow one to skip the steps of TEMPO-oxidation and alkali treatment. Also, it can greatly improve the utilization of raw materials. Thus, in this study, instead of the traditional cellulose aerogel preparation methods, the IBR-derived holocellulose was directly used as the raw material to prepare aerogels via freeze-drying. To improve the properties of the holocellulose aerogel composites, the HCNF solutions were freeze-dried in the presence of APP and acid-hydrolyzed MTMS, resulting in ultralight aerogels with good hydrophobicity and fire resistance. The mechanical, thermal, hydrophobic, and flame properties of the modified aerogels were characterized, in addition to their thermal insulative properties. This eco-friendly method for the preparation of cellulose-based thermal insulation aerogels from IBR not only simplified the preparation of cellulose aerogels, but also shows its practical application prospects.

2. Materials and Methods

2.1. Materials

Industrial bamboo residues (IBR) were collected from Youzhu Technology Co. Ltd.(Yongan, China) without further treatment. Water-soluble APP (76%) was purchased from Shandong Usolf Chemical Technology Co. Ltd. (Qingdao, China). Glacial acetic acid was obtained from Beihua Fine Chemicals Co., Ltd. (Beijing, China). Sodium chlorite (80%) and MTMS (98%) was obtained from Aladdin Chemistry Co. Ltd. (Shanghai, China). Potassium bromide (KBr) and Congo red were obtained from Guangfu Technology Development Co. Ltd (Tianjin, China). All chemical reagents were used as received without further purification. Deionized water was used in all experiments.

2.2. Synthesis of HCNF/APP/MTMS Aerogels

The IBR was processed with chemical pretreatments of 3 wt% sodium chlorite (75 °C, 6 h) to remove the lignin, resulting in the holocellulose samples, and then, the holocellulose samples were dispersed in deionized water with a concentration of 1 wt% and nanofibrillated using industrial high-power ultrasonication (Scientz-08, Ningbo Scientz Biotechnology Co., Ltd., Ningbo, China) for 20 min at 30% power. The resulting HCNF solution was placed in a 4 °C environment before further utilization.

Dissolving APP (0.1 g) in HCNF suspension (100 g, 1 wt%) formed APP/HCNF solutions, which were then stirred for 10 min. The obtained HCNF/APP solutions were adjusted to a pH of 4 with a 0.5 M hydrochloric acid (HCl) solution. Then, the HCNF/APP/MTMS solutions were prepared dropwise by adding MTMS (2.78 g, 20 mmol/$g_{(CNF)}$) to HCNF/APP solutions (pH = 4) and stirred at room temperature for 2 h.

As-prepared HCNF/APP/MTMS solutions were frozen in the refrigerator at −80 °C for 6 h. After freezing, the samples were immediately transferred to a freeze-drier under 0.6 Pa and −80 °C for 36 h. The obtained aerogels were sealed in plastic bags for further characterization. With respect to nomenclature, the aerogels prepared from neat HCNF solutions, HCNF/APP solutions, and HCNF/APP/MTMS solutions were called HCNF, HCNF/APP and HCNF/APP/MTMS aerogels, respectively. The mechanism scheme for HCNF/APP/MTMS aerogels is shown in Figure 1.

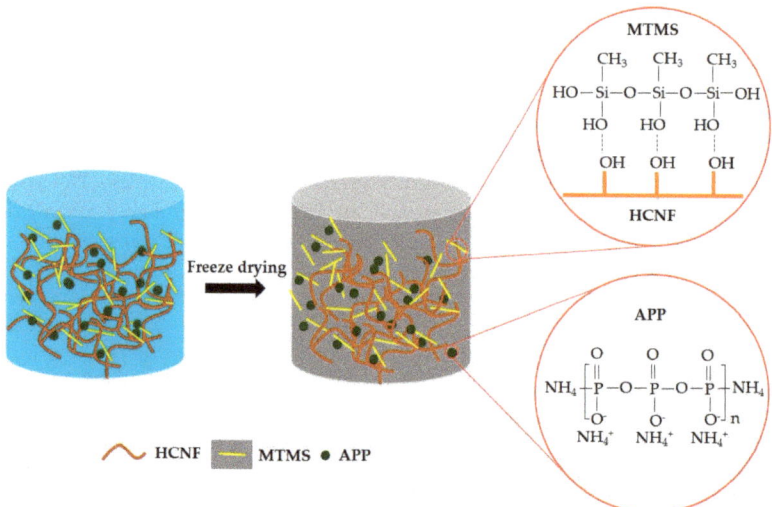

Figure 1. The mechanism scheme for HCNF/APP/MTMS aerogels.

2.3. Characterization

2.3.1. Chemical Composition

Before the tests, the samples were dried in an oven until the mass change was less than 0.02 g. The test methods of chemical composition on lignin, holocellulose, and α-cellulose were according to Chinese Standards of GB/T 2677-8 [22], GB/T 2677-10 [23], and GB/T 744 [24], respectively.

2.3.2. Density

The densities of various aerogels were calculated by measuring their masses and dimensions. The detailed calculation uses the following equation:

$$\rho = \frac{m}{V} \tag{1}$$

where ρ, m, and V are the density, quality, and bulk of aerogels, respectively.

2.3.3. Scanning Electron Microscopy (SEM)

The morphologies of the aerogels were observed by scanning electron microscopy (SEM, XL30, FEI Ltd., Hillsboro, OR, USA). Aerogels were cut with a blade (Leica 819, Leica Microsystems Ltd., Wetzlar, Hessen, Germany) in liquid nitrogen, and then fixed with conductive carbon tape and coated with a platinum layer for 90 s. The morphologies were observed with an accelerating voltage of 7 kV.

2.3.4. Fourier Transform Infrared (FT-IR)

Before the test, all samples were dried in an oven to eliminate the effects of moisture. The Fourier transform-infrared (FT-IR) spectra were recorded by a Nicolet IS10 FT-IR spectrometer (Thermo Fisher Scientific, Waltham, MA, USA). Aerogels were mixed in KBr with proportion of 1:100, and then ground by ball milling (ST-M200, Xuxin Instrument Co. Ltd. Beijing, China) at 1500 r/min, for 5 min. The FT-IR spectra were recorded in the range of 400–4000 cm^{-1} with a resolution of 4 cm^{-1}.

2.3.5. X-ray Diffraction (XRD)

A wide angle X-ray diffractometer (X PERTPRO-30X, PHILIPS Ltd., Almelo, the Netherlands) was used to determine the crystal characteristics of HCNF, HCNF/APP, and HCNF/APP/MTMS aerogels powders. The aerogel powders were smashed and sieved with more than 40 mesh. The X-rays were operated at 40 kV and 40 mA. The X-ray diffractograms were recorded at 0.02°/s over a 2θ scan in the range of 5–45°.

2.3.6. Compressive Properties

The compressive properties of the aerogels (20 mm × 20 mm × 25 mm) were measured on an Instron 5848 testing machine (Instron Co. Ltd., Canton, MA, USA) with a load cell of 500 N. The stress-strain curves were measured with a compression speed of 5 mm/min to 80% strain of aerogel under a controlled atmosphere of 25 °C and 50% humidity.

2.3.7. Hydrophobicity and Contact Angle

The hydrophobicity of aerogels was measured by deionized water, which was dyed with Congo red. The surface wettability of the aerogels was measured by static contact angle analysis using a contact angle goniometer (OCA20, Dataphysics Instrument, Filderstadt, Germany). The volume of the water droplet was 3 µL, and five positions were tested.

2.3.8. Thermal Stability

Thermal stability measurements were obtained using a thermogravimetric analyzer (TGA, Q 50 TA Instruments, New Castle, DE, USA) from room temperature to 700 °C at a 10 °C/min heating rate under N_2 protection. The quality of the tested aerogels was between 7–10 mg.

2.3.9. Flammability and Cone Calorimetry

The flame-retardant properties were evaluated by measuring the combustion with a butane blowtorch (~1000 °C) under a fuming cupboard. The combustion of aerogels was investigated under a cone calorimeter device (FTT, Fire Testing Technology Ltd., West Sussex, RH19 2HL, UK) with heat flux of 50 kW/m^2 in accordance with the ISO 5660-2 [25]. The aerogel (100 mm × 100 mm × 10 mm) was placed in a horizontal configuration.

2.3.10. Thermal Conductivity

Thermal conductivities were tested with a hot disk thermal constant analyzer (TPS2500S, Kegonas Co. Ltd., Uppsala, Sweden), which used a transient plane source method at 24 °C. The probe (R = 3.189 mm) was sandwiched between two aerogels (100 mm × 100 mm × 10 mm) to measure the changes in temperature. The output power and time of tests were 100 mW and 10 s, respectively.

3. Results and Discussion

3.1. Characterization of HCNF, HCNF/APP, and APP/MTMS/HCNF Aerogels

The morphology of the industrial bamboo residues (IBR) are shown in Figure 2a. The chemical composition of IBR is summarized in Table 1. The contents of lignin, holocellulose, and α-cellulose were 22.70, 69.08, and 39.09%, respectively. It is noteworthy that the IBR contained amounts of holocellulose, which means the utilization of IBR will be greatly improved if holocellulose can be used reasonably. Based on the results, the IBR directly went through a bleach treatment and high intensity ultrasonication without any shredding, resulting in HCNF, followed by APP and MTMS modification. The morphology of the as-obtained HCNF was characterized by SEM image (Figure 2b), indicating the HCNF was successfully prepared.

 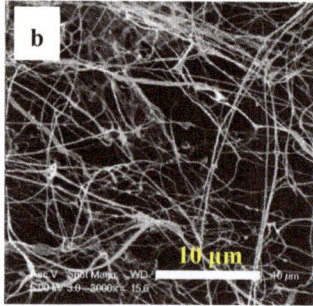

Figure 2. (**a**) Optical images of industrial bamboo residues (IBR) and (**b**) SEM images of as-prepared HCNF suspension with 0.01 wt%.

Table 1. The chemical composition of industrial bamboo residues.

Sample	Lignin (%)	Holocellulose (%)	α-Cellulose (%)
Industrial bamboo residues	22.70 ± 0.66	69.08 ± 0.22	39.09 ± 0.10

To investigate the modified aerogels, FTIR, XRD, and TGA were used to characterize the changes in composition and thermal stability. Figure 3 shows the FT-IR spectra of HCNF, HCNF/APP, and HCNF/APP/MTMS aerogels. The peaks at 1235 cm^{-1} and 1730 cm^{-1} in all aerogels are respectively attributed to the C=O stretching vibration in the acetyl groups and C–O stretching vibration in the glucuronic acid unit of the hemicellulose, indicating that hemicellulose exists in the aerogels [26]. In the HCNF/APP aerogels, the region at 3400~3030 cm^{-1} was broadened, which was attributed to the N–H asymmetric stretching vibration of the NH$_4^+$ in the APP [27]. This indicated that APP was homogeneously distributed in the aerogels. A new band, which indicated stretching vibrations of Si-C/Si-O-Si at ca. 770 cm^{-1}, appeared in the HCNF/APP/MTMS aerogels, and the C–H deformation vibrations of –CH$_3$ at ca. 1272 cm^{-1} increased significantly. Meanwhile, the amount of –OH in the HCNF/APP/MTMS aerogel was shown with a dramatic decline in the stretching vibration peak of –OH observed at ca. ~ 3314 cm^{-1} [28]. This behavior was similar to that of MTMS-modified oil-water separation materials [18].

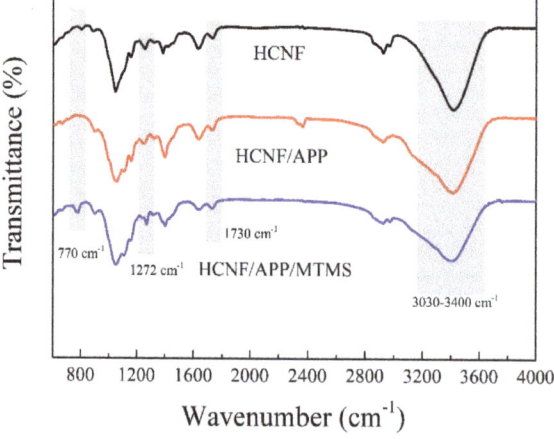

Figure 3. The FT-IR spectra of HCNF, HCNF/APP, and HCNF/APP/MTMS aerogels.

The XRD patterns of HCNF, HCNF/APP, and HCNF/APP/MTMS aerogels are shown in Figure 4. The two main diffraction peaks appeared at 2θ = 16.5°, 22.5°, and 34.6° in HCNF and HCNF/APP aerogels, which represent the crystalline area (110, 200 and 004) in the cellulose I pattern [29]. Thus, the cellulose I crystal integrity was maintained with the addition of APP. Note that with the HCNF/APP/MTMS aerogels, an additional strong diffraction peak near 2θ ≈ 10° appeared. The new peak covered the (110) crystalline area in the cellulose I crystal. However, the positions of the diffraction peaks that belonged to the cellulose I crystal were not changed. This indicates that the cellulose I crystal was also not affected by the modification of MTMS. Furthermore, the new diffraction peaks strongly resemble that of organic-inorganic phyllosilicates (001) [30], which shows the existence of MTMS.

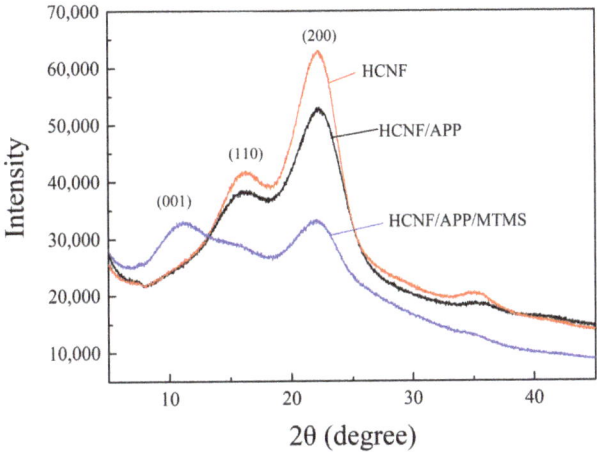

Figure 4. The XRD pattern of HCNF, HCNF/APP, and HCNF/APP/MTMS aerogels.

The TGA and DTA curves of HCNF, HCNF/APP, and HCNF/APP/MTMS aerogels are shown in Figure 5, and the degradation data is summarized in Table 2. It was observed that the temperature at which 30% weight loss of the HCNF/APP aerogels occurred was decreased compared to HCNF aerogels. The lower values in both $T_{30\%}$ and T_{MAX} of HCNF/APP aerogels was main due to water-soluble APP decomposing when heated to 180 °C, leading to dehydration and carbonization of the HCNF substrate in the aerogels. However, the pyrolysis rate of the HCNF/APP aerogels was significantly decreased, and the value of residue at 700 °C increased from 13.03% to 34.57%. The heat decomposition of APP could dehydrate and carbonize the aerogel matrix to produce C=C and N–P–C structures. This could result in a charred layer, which could hinder heat transfer and prevent further decomposition [31]. It is noteworthy that, compared to HCNF/APP aerogel, the thermal stability at 30% weight loss was further improved in HCNF/APP/MTMS aerogels. Tjos could be related to the high heat stability of MTMS. From a previous report, it is known that the weight loss could eliminate the low-molecular-weight species which adsorbed at the surface of cellulosic substrate if the polysiloxane is not bonded strongly to the CNF substrate [32]. Furthermore, the effective coating of the cellulosic substrate by a modifier can form an excellent physical barrier to prevent CNF from combustion [33]. Thus, the behaviors seen in this study, in which the temperature at 30% weight loss of HCNF/APP/MTMS aerogel was significantly delayed, indicate that the improvement of thermal stability was assigned to the inherent heat resistance of the polysiloxane bonded at the HCNF surface. Moreover, the Si–O–C solid residues formed by the modified aerogel at high temperatures further enhance the effects of carbon sequestration, making the mass residues increase to 36.25%.

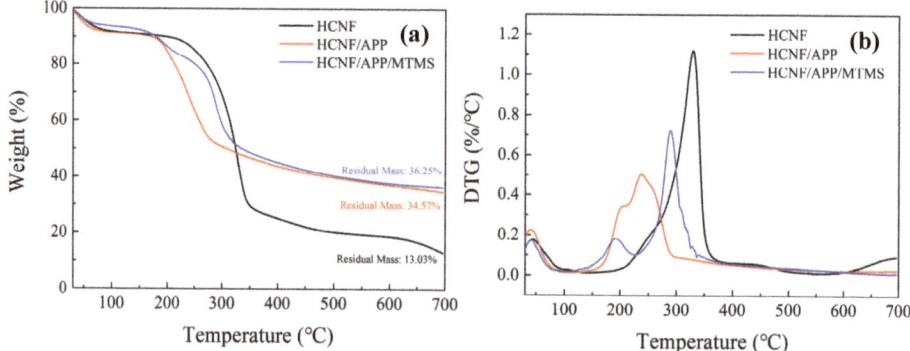

Figure 5. The TGA (**a**) and DTG (**b**) curves of HCNF, HCNF/APP, and HCNF/APP/MTMS aerogels.

Table 2. TGA data of HCNF, HCNF/APP, and HCNF/APP/MTMS aerogels.

Samples	$T_{30\%}$ [1] (°C)	T_{MAX} [2] (°C)	Residual Mass [3] (%)
HCNF	298.26	329.87	13.03
HCNF/APP	236.62	237.49	34.57
HCNF/APP/MTMS	282.47	288.46	36.25

[1] $T_{30\%}$ is temperature at 30% weight loss of aerogels. [2] T_{MAX} is temperature at the maximum rate of aerogel degradation. [3] Residual mass after heating up to 700 °C.

To further study the properties of the modified aerogels, the morphology of HCNF and HCNF/APP/MTMS aerogels are shown in Figure 6. The HCNF/APP/MTMS aerogels were more soft and flexible than HCNF aerogels. It was indicated that the structure of HCNF/APP/MTMS aerogels clearly changes. The aerogel structures were analyzed by SEM images and are shown in Figure 7. The macropores were clearly observed in both aerogels. However, from HCNF to HCNF/APP/MTMS aerogels, the foam-like structure translates into a 3D network structure and an amount of inorganic particles coated on the surface of the aerogel matrix. Interestingly, HCNFs and bamboo fibers were intertwined, forming a porous structure in both aerogels (Figure 7c,d). This is primarily due to the structural differences of the fibers and parenchymal cells of bamboo [8]. Under the same ultrasonic treatment conditions, the parenchymal cells are more easily microfibrillated than bamboo fibers.

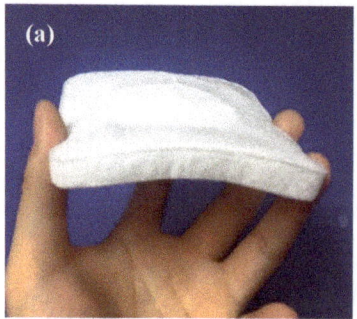
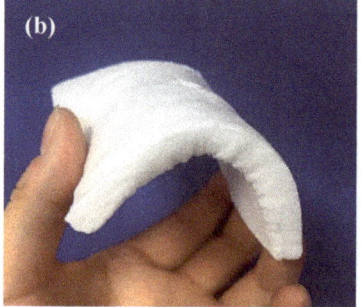

Figure 6. Optical images of the (**a**) HCNF aerogel, and (**b**) HCNF/APP/MTMS aerogel with different features.

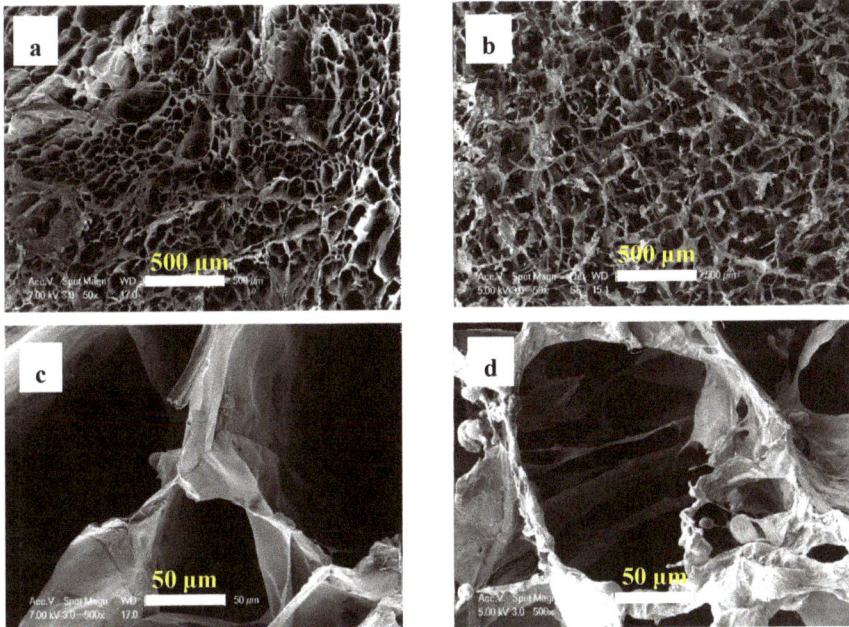

Figure 7. SEM images of the HCNF (**a**,**c**) and HCNF/APP/MTMS (**b**,**d**) aerogels.

3.2. Compressive Properties

Mechanical properties are important for thermal insulation materials. The densities of the HCNF, HCNF/APP, and HCNF/APP/MTMS aerogels were 12.04 (0.25), 16.27 (0.29), and 28.54 (0.50) kg/m^3, respectively. The compressive properties of HCNF, HCNF/APP, and HCNF/APP/MTMS aerogels are shown in Figure 8. The typical compression curves can be divided into three stages, namely the elastic, yield, and densification stages. From the stress-strain curves, a distinct linear elastic region can be observed in the HCNF aerogels, and those of the HCNF/APP and HCNF/APP/MTMS aerogels were not boundaries between the elastic and yield regions (Figure 8a). These kinds of behaviors are similar to the compression curves of other porous materials [34].

The compressive modulus is the slope of the linear region of the stress-strain curve in the elastic stage. In this study, the compressive modulus can be determined at strains below 5%, because the aerogels are elastic in this stage. As shown in Figure 8b, a compressive modulus of 38 kPa was determined for the HCNF aerogels at a density of 12.04 kg/m^3. However, the compressive modulus of the HCNF/APP aerogels was only 1.4 kPa at density a 16.27 kg/m^3, and that of the HCNF/APP/MTMS aerogels was 8.9 kPa at a density of 28.54 kg/m^3. There was a substantial reduction compared to the HCNF aerogels. Based on the structural changes of the aerogels, with the addition of the APP and MTMS, the foam-like structure translates into a 3D network structure, which indicated that the bonding points of substrate showed a substantial reduction in the HCNF/APP/MTMS aerogels, causing a weak crosslinking between HCNFs. The tightly-bound foam-like structure results in a relatively high resistance to compression, and the weak bonding in the 3D network structure induces lower stress values at the same deformation [35]. Thus, the mechanical properties of HCNF/APP/MTMS aerogels decreased clearly. To further understand the influence of APP and MTMS on modified fibers against density increase, the specific moduli of HCNF, HCNF/APP and HCNF/APP/MTMS aerogels were measured by using densities for normalization. The specific modulus of the HCNF aerogel was 3.16 kPa/(kg/m^3), and that of HCNF/APP and HCNF/APP/MTMS aerogels decreased to 0.086 and 0.312 kPa/(kg/m^3), respectively, which shows a substantial reduction. It was indicated that the

APP and MTMS seriously affected the crosslinking of HCNFs, which proved the above conclusion. However, the specific moduli of HCNF/APP/MTMS aerogels were better than that of HCNF/APP aerogels, which could be due to the silane layer bonding enhancing the skeleton structure of aerogels, making HCNF/APP/MTMS aerogels have better mechanical properties than HCNF/APP aerogels.

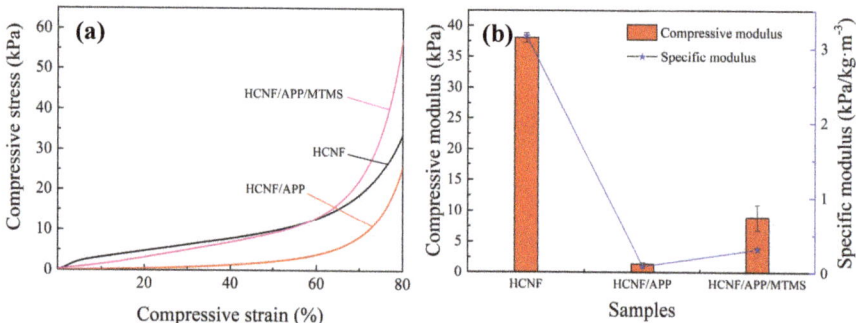

Figure 8. The (a) stress-strain curves and (b) compressive moduli and specific moduli of the HCNF, HCNF/APP, and HCNF/APP/MTMS aerogels.

3.3. Hydrolytic and Flame Resistance

For thermal insulation materials, water and fire resistance are important. Water seepage can enable thermal insulation materials to lose insulative properties. Additionally, combustible thermally-insulative materials can easily to form a "chimney effect" so that the fire cannot be controlled. To research the application feasibility of thermal insulation materials, hydrophobic and fire resistance properties were utilized. As shown in Figure 9a, the HCNF aerogels, which contain abundant hydroxyl groups, quickly absorbed water and sunk in 3 s. However, from Figure 9b, the HCNF/APP/MTMS aerogels became hydrophobic as water formed droplets on the surface, and the aerogels could float on water. Furthermore, a HCNF/APP/MTMS aerogel could refloat on water when it was pressed into the water completely. Thus, the HCNF/APP/MTMS aerogels exhibited good hydrophobic properties. Furthermore, by the contact angle test, the 3 μL water droplet on the surface of the modified aerogel maintained a round shape with high contact angles of 130° for more than 10 min (Figure 9c). This was primarily due to the MTMS covering the HCNF surface, causing the aerogel to form an air shield at the interface between the water and the aerogel. The reduced availability of free hydroxyl groups in the matrix can effectively reduce water absorption [36]. Thus, the HCNF/APP/MTMS aerogel exhibited good water resistance, and meanwhile proved that the aerogel was well bound to MTMS. This behavior not only protects the aerogel from water damage, but also solves the problem of the dissolution of APP in water.

Moreover, the high-temperature fire resistance properties of the aerogel were assessed using a butane blowtorch (~1000 °C). It was observed that the HCNF aerogels immediately burned and the structure collapsed when contacted by fire until complete carbonization (Video S1). However, the HCNF/APP/MTMS aerogels did not burn when they came in contact with fire by butane blowtorch. Furthermore, during 60 s of fire treatment at 1000 °C, the HCNF/APP/MTMS aerogels slowly carbonized from the side near the fire source. After the flame treatment, the fire did not spread to the remainder of the aerogel (Video S2). This is due to the decomposition of APP, enabling the aerogel to form a char layer quickly. Meanwhile, the released NH_3 and H_2O diluted the flammable gas and oxygen to prevent further flame spreading [37].

Cone calorimetry has been widely used for flame resistance testing and can provide significant amounts of data, including time to ignition (TTI), release rate (HRR), total heat release (THR), smoke production rate (SPR), and total smoke production (TSP). To gain insight into fire resistance, a cone calorimeter device, according to the standard of ISO 5660-2, was used in this study. As seen in Table 3, the TTI was increased from 0 (burns as soon as it is ignited) to 3 s after APP and MTMS modification. Although all samples were flammable under a heat flux of 50 kW/m^2, the addition of APP and MTMS extended the TTI of the samples. As shown in Figure 10a, the HRR of HCNF/APP/MTMS aerogels was observably reduced compared with HCNF aerogels. The peak heat release rates (pHRR) of the HCNF, HCNF/APP aerogels were measured to be 466.6k W/m^2 and 341 kW/m^2, respectively, and that of the HCNF/APP/MTMS aerogels was 219.1 kW/m^2, which showed a 47% decrease. Moreover, the THR of the HCNF/APP/MTMS sample was reduced from 6.9 to 6.1 MJ/m^2 (Figure 10b). This is due to the decomposition of APP caused the high efficiency carbonization effect leading to incomplete combustion of the aerogels during testing.

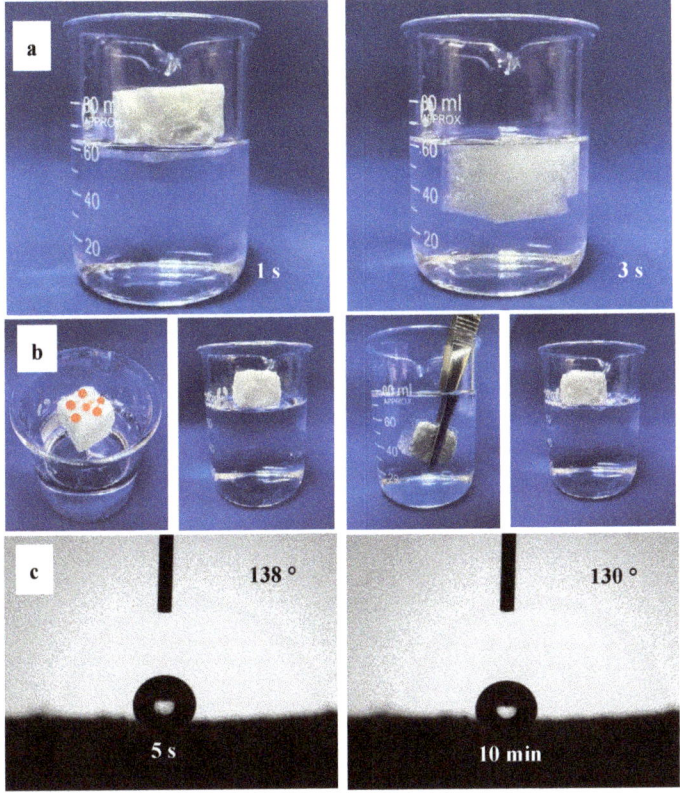

Figure 9. Hydrophobicity of (**a**) HCNF and (**b**) HCNF/APP/MTMS aerogels, and (**c**) water contact angles of the HCNF/APP/MTMS aeroge measured at 5 s and 10 min.

Table 3. Cone calorimeter data of the HCNF, HCNF/APP, HCNF/APP/MTMS aerogels.

Samples	TTI (s)	pHRR (kW/m^2)	T$_{pHRR}$ (s)	FIGRA (kW/s·m^2)	THR (MJ/m^2)	pSPR (m^2/s)	TSP (m^2)
HCNF	0 [1]	466.6	20	23.30	6.9	0.024	0.18
HCNF/APP	1	341.0	20	17.05	6.6	0.013	0.09
HCNF/AP-P/MTMS	3	219.1	25	8.76	6.1	0.006	0.04

[1] the HCNF aerogel was immediately burned, so the TTI cannot be measured.

SPR and TSP are important factors for insulation materials. Figure 10c,d show SPR and TSP curves during sample burning. It can be clearly observed that the SPR and TSP decreased with the addition of APP and MTMS. The peak of smoke production (pSPR) decreased from 0.024 to 0.006 m^2/s, and TSP decreased from 0.18 to 0.04 m^2. Together with Figure 11, it can be seen that there was no residue after cone calorimeter testing from the HCNF aerogels, while a little residue and a char layer were obtained after cone calorimeter testing from the HCNF/APP aerogels. However, more continuous char layers and the stable porous structure was preserved in the HCNF/APP/MTMS aerogels after cone calorimeter testing. These behaviors showed that the APP and MTMS played a key role in enhancing the fire resistance of HCNF/APP/MTMS aerogels. The continuous char layer can form a protective layer to prevent the spread of flame, ensuring that the cellulose skeleton is not destroyed and produces less smoke [38]. And the stable porous structure of the HCNF/APP/MTMS aerogels can effectively absorb some of the smoke [39].

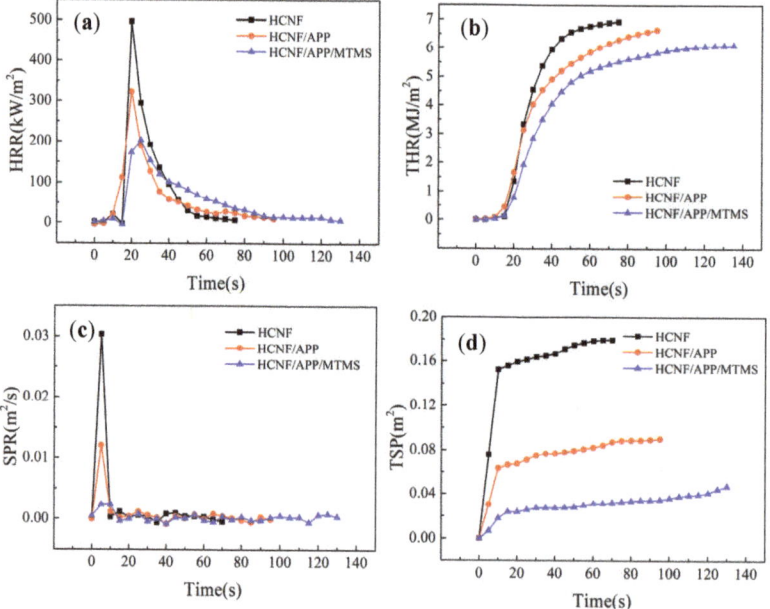

Figure 10. The (**a**) HRR, (**b**) THR, (**c**) SPR, and (**d**) TSP curves of the HCNF, HCNF/APP, and HCNF/APP/MTMS aerogels.

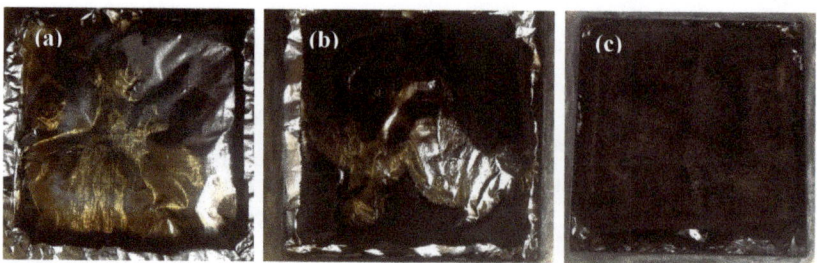

Figure 11. Digital photographs of residues of (**a**) HCNF, (**b**) HCNF/APP, and (**c**) HCNF/APP/MTMS aerogels after the cone calorimeter test.

The aerogel fire growth rate (FIGRA) is given in Table 3. FIGRA is an important parameter used to assess fire risks. The lower the FIGRA index, the higher the possibility that humans are able to survive in a fire. For the HCNF/APP/MTMS aerogel, the FIGRA index decreased from 23.3 to 8.76 kW/s·m^2. Compared to the HCNF aerogel, the HCNF/APP/MTMS aerogel provides better chances for human survival in a fire incident.

3.4. Thermal Insulation Properties

Thermal conductivity is an important parameter for thermal insulation materials. The total thermal conductivities of HCNF and HCNF/APP/MTMS aerogels were measured using a hot disk thermal constant analyzer at room temperature. The results are given in Table 4.

Table 4. The thermal conductivity of HCNF, and HCNF/APP/MTMS aerogels.

Samples	Temperature (°C)	Thermal Conductivity (W/m·K)	Standard Deviation	Specific Thermal Conductivity (W/m·K)/(kg/m^3)
HCNF	25	0.0285	0.0055	0.0024
HCNF/APP/MTMS	25	0.0398		0.0014

The aerogels made from HCNF aerogels with a bulk density of 12.04 kg/m^3 had a thermal conductivity of 0.0285 W/m·K, slightly higher than the thermal conductivity of air (0.023 W/m·K). The thermal insulative properties of HCNF aerogels are believed to be due to morphology and molecular structure. Due to the fact that HCNF aerogels exhibit low density and high porosity, the thermal conductivities of the solids in the HCNF aerogels were decreased. Furthermore, with the addition of APP and MTMS, the thermal conductivities of HCNF/APP/MTMS aerogels were increased to 0.039 W/m·K. The thermal conductivities were normalized by density to measure the specific thermal conductivity of HCNF and HCNF/APP/MTMS aerogels. The specific thermal conductivity of the HCNF aerogel was 0.0024 (W/m·K)/(kg/m^3), and that of HCNF/APP/MTMS aerogels was 0.0014 (W/m·K)/(kg/m^3). These behaviors indicated that the density of HCNF/APP/MTMS aerogels of up to 28.54 kg/m^3 is the main reason for the increase of thermal conductivity. In addition, the APP and MTMS changed the structure of the HCNF/APP/MTMS aerogels, making it form an opening 3D network structure. This disordered structure with a macropore (>70 nm) can reduce the mean free paths of air molecules [40]. On the whole, the thermal conductivities of the HCNF and HCNF/APP/MTMS aerogels were in the range in 0.0285 to 0.039 W/m·K. For comparison, the expanded polystyrene of 0.030–0.045 W/m·K, and conventional biomass-derived materials such as wood or hemp fiber insulation boards, have thermal conductivities of 0.040 W/m·K or higher [41]. Moreover, conventional biomass-derived materials have poor water and fire resistance, while expanded polystyrene, although it has less water absorption, still has poor fire resistance. Thus, HCNF/APP/MTMS aerogel with ultralight, hydrophobic, and fire resistant properties makes it possible to prepare ultralight and functional thermal insulation materials in a manner which can easily be exploited in the field of building energy efficiency.

4. Conclusions

In this study, holocellulose aerogels were successfully prepared by freeze-drying, using IBR-derived holocellulose which allows one to skip steps of alkali treatment, and greatly improves the utilization of raw materials compared to regular CNF aerogels. Followed by APP and MTMS modification, the aerogels were given properties of hydrophobicity and fire resistance. All aerogels had low densities, ranging from 12 to 28 kg/m^3, including the thermal conductivity of 0.028–0.039 W/m·K. The mechanical properties, microstructures, chemical compositions, hydrophobicities, thermal stabilities, and flame retardancies of the aerogels were determined. The FTIR, XRD, and TGA analyses have shown that APP and MTMS were effectively attached by the aerogels' substrates. The HCNF/APP/MTMS aerogels showed decreased compressive strength, which was due to the structural changes after APP and MTMS additions. However, the HCNF/APP/MTMS aerogels showed higher hydrophobicity, with a contact angle up to 130° and the ability to float on water. Moreover, the fire tests, as measured by butane blowtorch (~1000 °C), showed tjat the HCNF/APP/MTMS aerogels exhibit good flame retardancy. The pHRR of 219.1 kW/m^2, pSPR of 0.006 m^2/s, and TSP of 0.04 m^2 were characterized by cone calorimetry. Furthermore, the FIGRA index decreased from 23.3 to 8.76 kW/s·m^2, which indicates that the HCNF/APP/MTMS aerogels should exhibit better performance during fire incidents. As described above, the HCNF/APP/MTMS aerogels solved problems such as poor water and fire resistance, and meanwhile solved the problem that flame retardants can be easily lost in water, showing improvement in service durability. The as-prepared aerogels could be useful for energy efficient building.

Supplementary Materials: The following are available online at http://www.mdpi.com/1996-1944/13/2/477/s1, Video S1: The fire test of the HCNF aerogel by a butane blowtorch, Video S2: The fire test of the HCNF/APP/MTMS aerogel by a butane blowtorch.

Author Contributions: Data curation, H.H.; Funding acquisition, H.W.; Investigation, H.H. and H.W.; Methodology, Y.Y., X.Z. and J.C.; Supervision, Y.Y. and Y.Q.; Writing-original draft, H.H.; Writing-review & editing, H.H. and H.W. All authors have read and agreed to the published version of the manuscript.

Funding: This research was funded by National Key R&D Program of China (2017YFD0600804); The Open-end Fund for Forestry Engineering of Central South University of Forestry and Technology.

Acknowledgments: The authors gratefully acknowledge the Science and Technology Team of ICBR for Institute of New Bamboo and Rattan Based Biomaterials, and particularly grateful to Jing Yuan for her great help in the experiment.

Conflicts of Interest: The authors declare no conflict of interest.

References

1. Rajinipriya, M.; Nagalakshmaiah, M.; Robert, M.; Elkoun, S. Importance of Agricultural and Industrial Waste in the Field of Nanocellulose and Recent Industrial Developments of Wood Based Nanocellulose: A Review. *ACS Sustain. Chem. Eng.* **2018**, *6*, 2807–2828. [CrossRef]
2. Jonoobi, M.; Mathew, A.P.; Oksman, K. Producing Low-Cost Cellulose Nanofiber from Sludge as New Source of Raw Materials. *Ind. Crop Prod.* **2012**, *40*, 232–238. [CrossRef]
3. Freitas, N.; Aurélio Pinheiro, J.; Brígida, A.; Saraiva Morais, J.P.; Filho, M.; Rosa, M. Fibrous Residues of Palm Oil as a Source of Green Chemical Building Blocks. *Ind. Crop Prod.* **2016**, *94*, 480–489. [CrossRef]
4. Huang, C.; Ma, J.; Liang, C.; Li, X.; Yong, Q. Influence of Sulfur Dioxide-Ethanol-Water Pretreatment on the Physicochemical Properties and Enzymatic Digestibility of Bamboo Residues. *Bioresour. Technol.* **2018**, *263*, 17–24. [CrossRef] [PubMed]
5. Xin, D.; Yang, Z.; Liu, F.; Xu, X.; Zhang, J. Comparison of Aqueous Ammonia and Dilute Acid Pretreatment of Bamboo Fractions: Structure Properties and Enzymatic Hydrolysis. *Bioresour. Technol.* **2015**, *175*, 529–536. [CrossRef]
6. Scurlock, J.M.O.; Dayton, D.C.; Hames, B. Bamboo: An Overlooked Biomass Resource? *Biomass Bioenergy* **2000**, *19*, 229–244. [CrossRef]
7. Chang, F.; Lee, S.-H.; Toba, K.; Nagatani, A.; Endo, T. Bamboo Nanofiber Preparation by HCW and Grinding Treatment and Its Application for Nanocomposite. *Wood Sci. Technol.* **2012**, *46*, 393–403. [CrossRef]
8. Wang, H.; Zhang, X.; Jiang, Z.; Li, W.; Yu, Y. A Comparison Study on the Preparation of Nanocellulose Fibrils from Fibers and Parenchymal Cells in Bamboo (Phyllostachys Pubescens). *Ind. Crop Prod.* **2015**, *71*, 80–88. [CrossRef]
9. Wang, H. Preparation, Characterization and Application of Nano Cellulose Fibrils from Bamboo. Ph.D Thesis, Chinese Academy of Forest Sciences, Beijing, China, 2013. [CrossRef]
10. Kistler, S.S. Coherent Expanded Aerogels. *J. Phys. Chem.* **1932**, *36*, 52–64. [CrossRef]
11. Lavoine, N.; Bergström, L. Nanocellulose-Based Foams and Aerogels: Processing, Properties, and Applications. *J. Mater. Chem. A* **2017**, *5*, 16105–16117. [CrossRef]
12. De France, K.J.; Hoare, T.; Cranston, E.D. Review of Hydrogels and Aerogels Containing Nanocellulose. *Chem. Mater.* **2017**, *29*, 4609–4631. [CrossRef]

13. Jiang, F.; Hsieh, Y.L. Cellulose Nanofibril Aerogels: Synergistic Improvement of Hydrophobicity, Strength, and Thermal Stability via Cross-Linking with Diisocyanate. *ACS Appl. Mater. Interfaces* **2017**, *9*, 2825–2834. [CrossRef] [PubMed]
14. Granström, M.; Pääkkö, M.; Jin, H.; Kolehmainen, E.; Kilpeläinen, I.; Ikkala, O. Highly Water Repellent Aerogels Based on Cellulose Stearoyl Esters. *Polym. Chem.* **2011**, *2*, 1789–1796. [CrossRef]
15. Nguyen, D.D.; Vu, C.M.; Vu, H.T.; Choi, H.J. Micron-Size White Bamboo Fibril-Based Silane Cellulose Aerogel: Fabrication and Oil Absorbent Characteristics. *Materials* **2019**, *12*, 1407. [CrossRef] [PubMed]
16. Köklükaya, O.; Carosio, F.; Wågberg, L. Superior Flame-Resistant Cellulose Nanofibril Aerogels Modified with Hybrid Layer-by-Layer Coatings. *ACS Appl. Mater. Interfaces* **2017**, *9*, 29082–29092. [CrossRef] [PubMed]
17. Wang, L.; Sánchez-Soto, M.; Maspoch, M.L. Polymer/Clay Aerogel Composites with Flame Retardant Agents: Mechanical, Thermal and Fire Behavior. *Mater. Des.* **2013**, *52*, 609–614. [CrossRef]
18. Zhang, X.; Wang, H.; Cai, Z.; Yan, N.; Liu, M.; Yu, Y. Highly Compressible and Hydrophobic Anisotropic Aerogels for Selective Oil/Organic Solvent Absorption. *ACS Sustain. Chem. Eng.* **2019**, *7*, 332–340. [CrossRef]
19. Klemm, D.; Cranston, E.D.; Fischer, D.; Gama, M.; Kedzior, S.A.; Kralisch, D.; Kramer, F.; Kondo, T.; Lindström, T.; Nietzsche, S.; et al. Nanocellulose as a Natural Source for Groundbreaking Applications in Materials Science: Today's State. *Mater. Today* **2018**, *21*, 720–748. [CrossRef]
20. Sharma, P.R.; Chattopadhyay, A.; Sharma, S.K.; Geng, L.; Amiralian, N.; Martin, D.; Hsiao, B.S. Nanocellulose from Spinifex as an Effective Adsorbent to Remove Cadmium(II) from Water. *ACS Sustain. Chem. Eng.* **2018**, *6*, 3279–3290. [CrossRef]
21. Tanaka, R.; Saito, T.; Hänninen, T.; Ono, Y.; Hakalahti, M.; Tammelin, T.; Isogai, A. Viscoelastic Properties of Core-Shell-Structured, Hemicellulose-Rich Nanofibrillated Cellulose in Dispersion and Wet-Film States. *Biomacromolecules* **2016**, *17*, 2104–2111. [CrossRef]
22. *Fibrous Raw Material—Determination of Acid-Insoluble Lignin*; GB/T 2677-8; State Bureau of Technical Supervision: Beijing, China, 1994.
23. *Fibrous Raw Material—Determination of Holocellulose*; GB/T 2677-10; State Bureau of Technical Supervision: Beijing, China, 1995.
24. *Pulps—Determination of Alkali Resistance*; GB/T 744; SAS/TC141: Tianjin, China, 2004.
25. *Reaction-to-Fire Tests—Heat Release, Smoke Production and Mass Loss Rate*; ISO 5660; ISO copyright office: Geneva, Switzerland, 2015.
26. Chen, W.; Yu, H.; Liu, Y.; Chen, P.; Zhang, M.; Hai, Y. Individualization of Cellulose Nanofibers from Wood Using High-Intensity Ultrasonication Combined with Chemical Pretreatments. *Carbohydr. Polym.* **2011**, *83*, 1804–1811. [CrossRef]
27. Behniafar, H.; Haghighat, S. Thermally Stable and Organosoluble Binaphthylene-Based Poly(Urea-ether-imide)s: One-Pot Preparation and Characterization. *Polym. Adv. Technol.* **2008**, *19*, 1040–1047. [CrossRef]
28. Zhao, S.; Zhang, Z.; Sèbe, G.; Wu, R.; Rivera Virtudazo, R.V.; Tingaut, P.; Koebel, M. Multiscale Assembly of Superinsulating Silica Aerogels Within Silylated Nanocellulosic Scaffolds: Improved Mechanical Properties Promoted by Nanoscale Chemical Compatibilization. *Adv. Funct. Mater.* **2015**, *25*, 2326–2334. [CrossRef]
29. Yadav, M.; Liu, Y.; Chiu, F. Fabrication of Cellulose Nanocrystal/Silver/Alginate Bionanocomposite Films with Enhanced Mechanical and Barrier Properties for Food Packaging Application. *Nanomaterials* **2019**, *9*, 1523. [CrossRef]
30. Malfait, W.J.; Jurányi, F.; Zhao, S.; Arreguin, S.A.; Koebel, M.M. Dynamics of Silica Aerogel's Hydrophobic Groups: A Quasielastic Neutron Scattering Study. *J. Phys. Chem. C* **2017**, *121*, 20335–20344. [CrossRef]
31. Wang, Y.T.; Liao, S.F.; Shang, K.; Chen, M.J.; Huang, J.Q.; Wang, Y.Z.; Schiraldi, D.A. Efficient Approach to Improving the Flame Retardancy of Poly(Vinyl Alcohol)/Clay Aerogels: Incorporating Piperazine-Modified Ammonium Polyphosphate. *ACS Appl. Mater. Interfaces* **2015**, *7*, 1780–1786. [CrossRef]
32. Zhang, Z.; Tingaut, P.; Rentsch, D.; Zimmermann, T.; Sèbe, G. Controlled Silylation of Nanofibrillated Cellulose in Water: Reinforcement of a Model Polydimethylsiloxane Network. *ChemSusChem* **2015**, *8*, 2681–2690. [CrossRef]
33. Yang, L.; Mukhopadhyay, A.; Jiao, Y.; Yong, Q.; Chen, L.; Xing, Y.; Hamel, J.; Zhu, H. Ultralight, Highly Thermally Insulating and Fire Resistant Aerogel by Encapsulating Cellulose Nanofibers with Two-Dimensional MoS_2. *Nanoscale* **2017**, *9*, 11452–11462. [CrossRef]
34. Wong, J.C.H.; Kaymak, H.; Brunner, S.; Koebel, M.M. Mechanical Properties of Monolithic Silica Aerogels Made from Polyethoxydisiloxanes. *Microporous Mesoporous Mater.* **2014**, *183*, 23–29. [CrossRef]

35. Jiménez-Saelices, C.; Seantier, B.; Cathala, B.; Grohens, Y. Spray Freeze-Dried Nanofibrillated Cellulose Aerogels with Thermal Superinsulating Properties. *Carbohydr. Polym.* **2017**, *157*, 105–113. [CrossRef]
36. Yadav, M.; Chiu, F.C. Cellulose Nanocrystals Reinforced κ-Carrageenan Based UV Resistant Transparent Bionanocomposite Films for Sustainable Packaging Applications. *Carbohydr. Polym.* **2019**, *211*, 181–194. [CrossRef]
37. Shao, Z.-B.; Deng, C.; Tan, Y.; Chen, M.-J.; Chen, L.; Wang, Y.-Z. An Efficient Mono-Component Polymeric Intumescent Flame Retardant for Polypropylene: Preparation and Application. *ACS Appl. Mater. Interfaces* **2014**, *6*, 7363–7370. [CrossRef]
38. Wang, L.; Wu, S.; Dong, X.; Wang, R.; Zhang, L.; Wang, J.; Zhong, J.; Wu, L.; Wang, X. A Pre-Constructed Graphene-Ammonium Polyphosphate Aerogel (GAPPA) for Efficiently Enhancing the Mechanical and Fire-Safety Performances of Polymers. *J. Mater. Chem. A* **2018**, *6*, 4449–4457. [CrossRef]
39. Lazar, P.; Karlický, F.; Jurecka, P.; Kocman, M.; Otyepkova, E.; Safarova, K.; Otyepka, M. Adsorption of Small Organic Molecules on Graphene. *J. Am. Chem. Soc.* **2013**, *135*, 6372–6377. [CrossRef] [PubMed]
40. Sakai, K.; Kobayashi, Y.; Saito, T.; Isogai, A. Partitioned Airs at Microscale and Nanoscale: Thermal Diffusivity in Ultrahigh Porosity Solids of Nanocellulose. *Sci. Rep.* **2016**, *6*, 20434. [CrossRef] [PubMed]
41. Zhao, S.; Malfait, W.J.; Guerrero-Alburquerque, N.; Koebel, M.M.; Nyström, G. Biopolymer Aerogels and Foams: Chemistry, Properties, and Applications. *Angew. Chem. Int. Ed.* **2018**, *57*, 7580–7608. [CrossRef]

© 2020 by the authors. Licensee MDPI, Basel, Switzerland. This article is an open access article distributed under the terms and conditions of the Creative Commons Attribution (CC BY) license (http://creativecommons.org/licenses/by/4.0/).

Article

Synthesis and Characterization of PLA-Micro-structured Hydroxyapatite Composite Films

Andreea Madalina Pandele [1,2], Andreea Constantinescu [3], Ionut Cristian Radu [1], Florin Miculescu [3], Stefan Ioan Voicu [1,2,*] and Lucian Toma Ciocan [4]

1. Advanced Polymer Materials Group, Faculty of Applied Chemistry and Material Science, University Polytehnica of Bucharest, str. Gheorghe Polizu 1-7, 011061 Bucharest, Romania; pandele.m.a@gmail.com (A.M.P.); radu.ionut57@yahoo.com (I.C.R.)
2. Faculty of Applied Chemistry and Materials Science, University Politehnica of Bucharest, Gheorghe Polizu 1-7, 011061 Bucharest, Romania
3. Faculty of Materials Science, University Politehnica of Bucharest, Splaiul Independentei 313, 011061 Bucharest, Romania; andreeaelena01c@gmail.com (A.C.); f_miculescu@yahoo.com (F.M.)
4. "Carol Davila" University of Medicine and Pharmacy, Prosthetics Technology and Dental Materials Department, 37, Dionisie Lupu Street, District 1, 020022 Bucharest, Romania; tciocan@yahoo.com
* Correspondence: stefan.voicu@upb.ro; Tel.: +40-721165757

Received: 27 November 2019; Accepted: 23 December 2019; Published: 8 January 2020

Abstract: This article presents a facile synthesis method used to obtain new composite films based on polylactic acid and micro-structured hydroxyapatite particles. The composite films were synthesized starting from a polymeric solution in chloroform (12 wt.%) in which various concentrations of hydroxyapatite (1, 2, and 4 wt.% related to polymer) were homogenously dispersed using ultrasonication followed by solvent evaporation. The synthesized composite films were morphologically (through SEM and atomic force microscopy (AFM)) and structurally (through FT-IR and Raman spectroscopy) characterized. The thermal behavior of the composite films was also determined. The SEM and AFM analyses showed the presence of micro-structured hydroxyapatite particles in the film's structure, as well as changes in the surface morphology. There was a significant decrease in the crystallinity of the composite films compared to the pure polymer, this being explained by a decrease in the arrangement of the polymer chains and a concurrent increase in the degree of their clutter. The presence of hydroxyapatite crystals did not have a significant influence on the degradation temperature of the composite film.

Keywords: polylactic acid; hydroxyapatite; composite films

---

1. Introduction

Polymeric films represent a particular domain of materials due to their controlled and directed selectivity capacity [1,2]. Recently, besides the common applications which imply the separation of constituents from a complex composition [3,4], less conventional applications such as adjuvant films for osteointegration have also been studied. In this case, the membrane can be placed at the interface between a metal implant and bone. These materials aim to improve and to accelerate the integration of the metal implant into the bone [5–7]. Composite membranes and polymeric films based on hydroxyapatite (HA) have lately seen great development, especially due to their potential applications in orthopedics [8]. Among the most commonly used polymers employed to obtain the aforementioned composites, biocompatible and bioresorbable polymers are usually preferred due to their ability to be desorbed over time into the human body and to promote osteoblast proliferation during bone implant welding. Hence, composites films based on hydroxyapatite and various polymers such as cellulose derivatives [9], starch [10], and polylactic acid [11] have been synthesized. Polylactic acid (PLA) is a synthetic polymer that can have both a crystalline [12] and an amorphous [13] structure. There are several studies which have indicated the use of PLA/HA composites for different applications [14]. Composites of PLA/HA with a percentage of 80 wt.% HA loading have been synthesized by adding filler into the polymer melt [15]. The obtained composites have been investigated in terms of their mechanical properties, and a significant improvement of Young's modulus has been observed. Moreover, the reported values are close to those of natural bone. In order to synthesize composite films with improved mechanical and thermal properties, new composite films based on polylactic acid/hydroxyapatite/graphene oxide (PLA/HA/GO) have also been prepared. In this case the HA and GO have been homogenously dispersed into a polymer solution using dimethylformamide and methylene chloride as solvents. The final composite films were obtained after the evaporation of the solvents at 40 °C [16]. The as-synthesized composite films were tested for potential use in tissue regeneration [17]. It seems that the presence of GO promotes the dispersion of HA into the polymer matrix and also has a positive impact on the biocompatibility of the final material.

Due to PLA's good biocompatibility [18] and bioresorbability [19], polymer composites with HA nanoparticles are focused on the synthesis of materials that improve the osteointegration of the implantable scaffold. Rakmae et al. have reported the modification of the surface of HA particles with 3-aminopropyltriethoxysilane (APES) or 3-methacryloxypropyltrimethoxysilane (MPTS) to increase the compatibility between the inorganic filler and the polymer matrix [20]. The researchers observed that the surface modification of HA nanoparticles significantly improved the mechanical and thermal properties of the synthesized composites, preventing the cleavage of PLA chains and increasing the degree of biocompatibility of the final material. At the same time, the mechanical and thermal resistances were further improved by adding a third polarizer to the synthesized composites, namely, poly-caprolactone [21]. The presence of poly-caprolactone allows an easier adjustment of some properties of the membrane by controlling the pore diameter and distribution.

Furthermore, PLA polymeric composites with very high mechanical strength have been synthesized by adding epoxy resin to the composite matrix [22]. Superior mechanical properties were observed in the case of the addition of HA and silver nanoparticles within a melted PLA matrix [23]. The use of this synthesis procedure allows for the obtaining of composites with a high percentage of hydroxyapatite (18 wt.%), and, at the same time, the existence of silver particles provides excellent antibacterial properties. High antibacterial activity for *Escherichia coli* and moderate antibacterial activity for *Staphylococcus aureus* have been observed. Generally, all synthesized composites have exhibited a non-cytotoxic character and can be used both in vitro and in vivo [24–26]. The optimal mechanical properties for implantable materials based on PLA/HA have been achieved for composite materials with a very high content of HA (70–80 wt.%) [15]. These percentages are conditioned by a few synthetic methods that can lead to uniform materials which imply such high inorganic filler content.

The current article presents an easy method with which to synthesize compact PLA/HA composite films by dispersing various concentrations of micro-structured HA particles in PLA solution using chloroform as a solvent, followed by evaporation. The obtained composite films were characterized by SEM microscopy, atomic force microscopy (AFM) microscopy, FT-IR spectroscopy, Raman spectroscopy, and thermal analysis. The novelty degree of the present research is given by the synthesis method used, which involves solvent evaporation for obtaining the films, the use of chloroform as a solvent (with a low boiling point and rapid formation of films by evaporation of the solvent) and also the use of the biogenic source HA.

2. Materials and Methods

2.1. Synthesis of Hydroxyapatite Particles

The hydroxyapatite particles were synthesized from spongy bone samples which were obtained after removal of the cortical components in accordance with a previously described procedure reported in the literature [27–29]. Firstly, bovine bone samples were mechanically cleaned, followed by a heat treatment at 500 °C for one hour in order to remove organic components. Proper thermal treatment was performed at 1200 °C for 6 h with a heating rate of 10 °C/min. After cooling in the air, the samples were ground in an agate ball mill and then sieved. For the present study, hydroxyapatite particles with sizes smaller than 40 µm were used.

2.2. Synthesis of PLA/HA Composite Films

Polylactic acid was dissolved in chloroform at a concentration of 12 wt.%. The hydroxyapatite particles were dispersed by ultrasonication for 30 min into the polymer solution at three different concentrations (1, 2, and 4 wt.%, respectively) in order to obtain a homogenous solution. The composite films were prepared after casting the obtained mixture into a Petri glass 60 mm in diameter (LabBox, Barcelona, Spain) and evaporating the solvent at 40 °C for 72 h in an oven. The long evaporation time was necessary due to the low porosity of the obtained films and the slower solvent evaporation during polymer precipitation. After synthesis, the composite films were washed with distilled water and ethanol to remove any chloroform and polymer residues and kept dry until use.

2.3. Characterization of Obtained Materials

FT-IR spectra were recorded on a Bruker VERTEX 70 spectrometer (Bruker, Massachusetts, United States) using 32 scans with a resolution of 4 cm^{-1} in the 4000–600 cm^{-1} region. The samples were analyzed using Attenuated Total Reflectance annex (ATR) (Bruker, Massachusetts, United States).

Raman spectra were registered on a DXR Raman Microscope (Thermo Fischer, Waltham, Massachusetts, USA) from Thermo Scientific using a 532 nm laser line and a number of 10 scans. The laser beam was focused using the 10× objective of the Raman microscope.

Scanning electron microscopy was performed on a FEI XL 30 ESEM TMP microscope equipped with an EDAX Sapphire device (FEI/Philips, Hillsboro, OR, USA).

Thermogravimetric analysis (TGA) curves were registered on Q500 TA Instruments (TA Instruments, New Castle, DA, USA) equipment using a nitrogen atmosphere from room temperature to 800 °C and a heating rate of 10 °C/min.

Differential scanning calorimetry (DSC) curves were recorded using Netzsch DSC 204 F1 Phoenix equipment (Netzsch, Selb, Germany). The sample was heated from room temperature (RT) to 300 °C using a heating rate of 5 °C/min under nitrogen (20 mL/min flow rate).

AFM analyses were performed using a multimode apparatus Agilent 5500 (Santa Clara, California, United States) equipped with an AC mode III controller. The contact mode images were obtained using a triangular silicon nitride cantilever with a typical spring constant of 0.06 N/m and a resonant frequency of 10 kHz. This type of cantilever is a very soft one and is specific for contact mode and used to obtain images without damaging the sample. The analyses were performed under ambient conditions with a calibrated piezo-scanner with a maximum xy range of 90 µm × 90 µm and a z scan range of 7 µm.

Mechanical tests were performed according to the European Standard EN ISO 527-3 part 3 (for films and tests) using a universal mechanical tester (Instron, Model 3382, Massachusetts, USA) at a relative humidity of ~50 % and a speed of 3 mm/min. The dimensions of the samples were 10 cm × 1 cm × 0.038 cm. For each composite film a minimum of five specimens were tested and the average values and standard deviation (±SD) were reported.

3. Results and Discussion

Figure 1 presents the FT-IR spectra of the HA, PLA, and PLA/HA composite films. In the HA spectrum, the bands corresponding to the stretching vibration of PO_4^{3-} are observed at 603 cm^{-1}, 957 cm^{-1}, and 1041 cm^{-1} respectively. In the PLA spectrum, the bands at 2998 cm^{-1} and 2947 cm^{-1} can be attributed to the asymmetrical and symmetrical stretching vibrations of the C–H bond, the band at 1755 cm^{-1} can be attributed to the stretching vibrations of the C=O bond, the band at 1457 cm^{-1} can be attributed to the deformation vibration of the CH_3 group, the bands at 1184 cm^{-1} and 1089 cm^{-1}, respectively, can be assigned to the C–O–C binding vibrations, and the band at 871 cm^{-1} corresponds to the vibrations of the C–COO bond [30]. The spectra of the composite membranes are similar to the PLA spectrum, exhibiting only the vibration bands corresponding to the polymer structure. This is due, on the one hand, to the overlapping of the vibration bands of HA in the range of 600–1100 cm^{-1} with the bands corresponding to the polymer, and on the other hand because of the small amount of phosphate groups present in the composite film composition compared to the C–H bonds from the polymer structure [9]. However, slight changes in the peak intensity can be identified. Moreover, when considering as the reference band the band at 1088 cm^{-1} (ν_{C-O-C}), whose intensity is unchanged across all the spectra, and calculating the ratio between this band and the band at 1045 cm^{-1} (δ_{C-CH3}), it can be observed that there is an increase in the intensity ratio in the case of the composite films compared with the pure polymeric film. This could suggest the presence of some interferences between HA and the polymer chain [31].

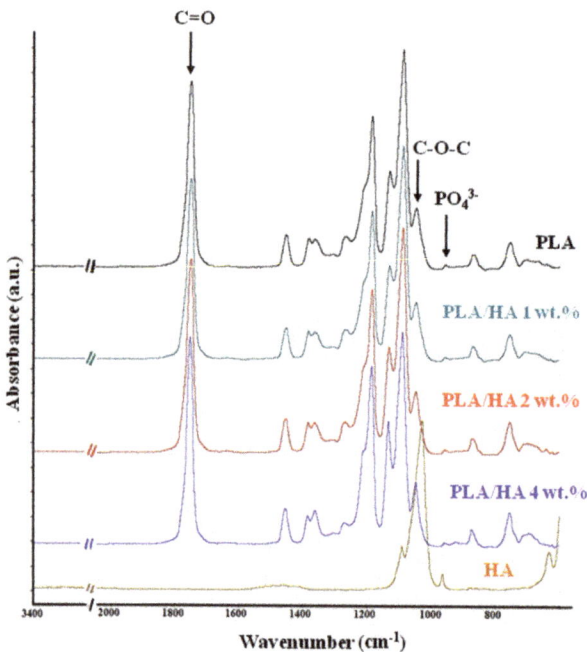

Figure 1. FT-IR spectra of hydroxyapatite (HA) and polylactic acid (PLA)/HA composite films.

The presence of HA in the composite films was also observed using Raman spectroscopy. Figure 2 shows the Raman spectra for PLA and the PLA/HA composite films. According to the figure, the PLA spectrum presents characteristic bands observed at 2995 cm^{-1}, 2990 cm^{-1}, and 2878 cm^{-1}, which can be attributed to asymmetric and symmetrical stretching vibrations of the C–H ($\nu_{as/sCH3}$) bond of the PLA chain; 1762 cm^{-1}, which can be attributed to the stretching vibration of the C=O ($\nu_{C=O}$) bond; 1447 cm^{-1}, which can be attributed to the asymmetric deformation vibration of the CH_3 bound (δ_{asCH3}); 887 cm^{-1}, which can be attributed to the stretching vibration of the C–COO (ν_{C-COO}) bond [16]. Compared to the PLA spectrum, the spectra of the composite films with 2 and 4 wt.% HA loaded onto them contain all the polymer-specific bands as well as the presence of an additional band around 960 cm^{-1} which corresponds to the PO_4^{3-} stretching vibration. This band comes from the HA structure and represents a piece of quite solid evidence for the presence of the inorganic compound in the polymer matrix. The absence of this band in the case of the composite membrane's spectrum with 1 wt.% HA loaded on is due to a small amount of HA being introduced into the polymer matrix which could be homogenously covered by the polymer. Moreover, for the membrane with 4 wt.% HA loaded onto it, due to the high amount of HA introduced into the polymer matrix, the presence of two additional bands at 585 cm^{-1} and 428 cm^{-1} which can be attributed to the asymmetrical deformation vibrations of PO_4^{3-} in the HA structure can be observed [32].

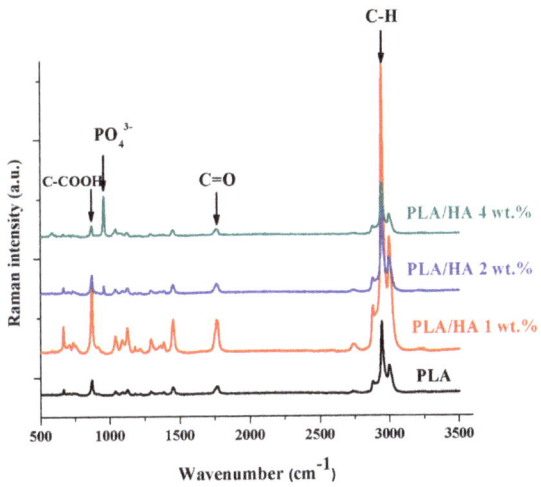

Figure 2. Raman spectra of PLA and PLA/HA composite films.

Thermogravimetric analysis was used to study the effect of HA nanoparticles on the thermostability of the polymer. Figure 3 illustrates the TGA and Differential Thermal Gravimetry (DTG) curves of PLA and the composite films with different concentrations of HA added. According to Figure 3, all membranes exhibit a similar profile with two degradation steps. The thermostability of the materials was tested by measuring the decomposition temperature at 10% mass lost (Td$_{10\%}$). An increase in the thermostability of the composite films with 1 and 2 wt.% HA loaded on was observed at about 53 °C compared to pure PLA. By further increasing the HA contents to 4 wt.%, the Td$_{10\%}$ was shifted to lower values; this was due to the formation of some HA agglomerates in the polymer matrix. However, the thermostability of the HA composite film with 4 wt.% HA can be seen to be superior to the pure polymer films, and this increase in thermal stability is due to the formation of strong hydrogen interactions and Van der Walls forces between inorganic particles and the polymeric chain during the homogenization process. More recent studies have suggested that when using higher concentrations of HA in the polymeric matrix it is very difficult to achieve a homogenous dispersion of inorganic filler into the matrix, leading to the formation of aggregates and diminishing the shielding effect of the particles [33]. On the other hand, the maximum degradation temperature of the first degradation step is almost invariable and the decomposition temperatures report the same value for all the samples (approximately 350 °C).

The thermal properties of the materials were also studied by DSC. Figure 4 shows the DSC curves for PLA and the composite films with 1, 2, and 4 wt.% HA added. The results obtained from the DSC curves for both the pure polymer and the composite films are summarized in Table 1. According to the figure, all samples presented two endothermic peaks around 142 °C and 150 °C due to the gradual melting of different sized polymer blades [31]. Moreover, on the DSC curves the presence of an exothermic peak around 106 °C can be observed which corresponds to the crystallinity of the polymer (a cold crystallization—temperature of crystallization—Tc). After intercalation of HA particles to the polymeric matrix, the Tc values tends to decrease due to the fact that HA particles act as nucleation centers for PLA crystals. Similar results have been reported in the literature by Maria Persson et al. [34].

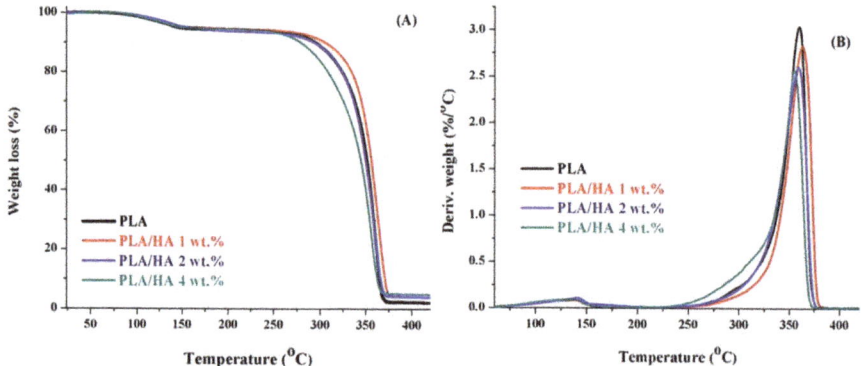

Figure 3. (**A**) Thermogravimetric analysis (TGA) and (**B**) DTG curves of PLA and PLA/HA composite films.

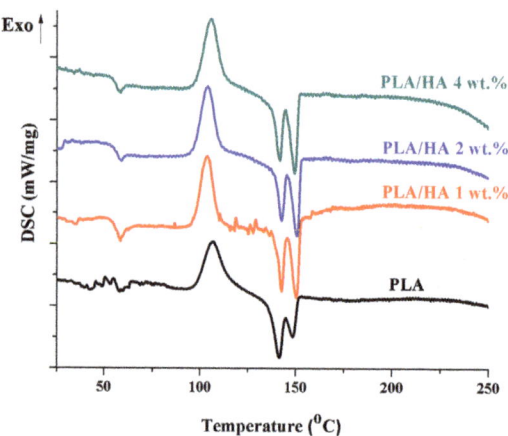

Figure 4. Differential scanning calorimetry (DSC) curves of PLA and PLA/HA composite films.

Table 1. Thermal characteristics of obtained materials (where Tm1 and Tm2 are melting temperatures for low respectively high-temperature endotherms, ΔHc is crystallization enthalpy of the sample and ΔHm is melting enthalpy of the sample).

Sample Name	wt.%	Td10% (°C)	Tmax (°C)	Tc	Tm1	Tm2	ΔHc (J/g)	ΔHm (J/g)	Xc (%)
PLA	100	221	352	106.9	141.4	148.4	18.99	21.82	3.02
PLA/HA 1 wt.%	99	226	350	103.9	142.6	150.3	22.63	23.3	0.72
PLA/HA 2 wt.%	98	274	351	104.2	142.6	150.6	22.63	23.78	1.23
PLA/HA 4 wt.%	96	254	351	106	141.8	149.5	24.75	26.37	1.73

The degree of crystallinity was further calculated according to the equation below, assuming an ideal melt heat of 93.7 J (Equation (1)).

$$X_c = 100 \times (\Delta H_m - \Delta H_c)/93.7 \tag{1}$$

After the calculations, values of crystallinity percentage ranging between 3 and 0.7% were obtained. It can be observed that at a low HA concentration (1 wt.%) the crystallinity of the composite films decreased significantly compared to the pure polymer. By increasing the amount of HA added, the value began to increase, reaching 1.7% for the composite by 4 wt.%. However, this value was below the crystallinity obtained in the case of the pure polymer. This could be explained by the fact that the presence of a small amount of HA (less than 4 wt.%) leads to a decrease in the crystallinity of the polymer as the presence of HA decreases the orientation of the polymer chains by increasing their degree of disorder. Furthermore, it appears that the presence of HA had no significant effect on the melting temperature, but a slight decrease indicates that the crystal size was less stable.

From the DSC assays we can conclude that the addition of HA particles in the polymeric matrix has an influence on the polymer chain arrangement, which further leads to a change in the polymer behavior when heated. This is due to the fact that HA particles act as nucleation centers in order to obtain a rigid phase that has a significant effect on the final polymer properties.

The presence of HA in the polymer matrix has also been shown to have an effect on the morphology of composite films. Scanning electron microscopy (Figure 5) revealed some differences between the analyzed samples. The pure PLA film displayed a smooth surface with no polymeric formation.

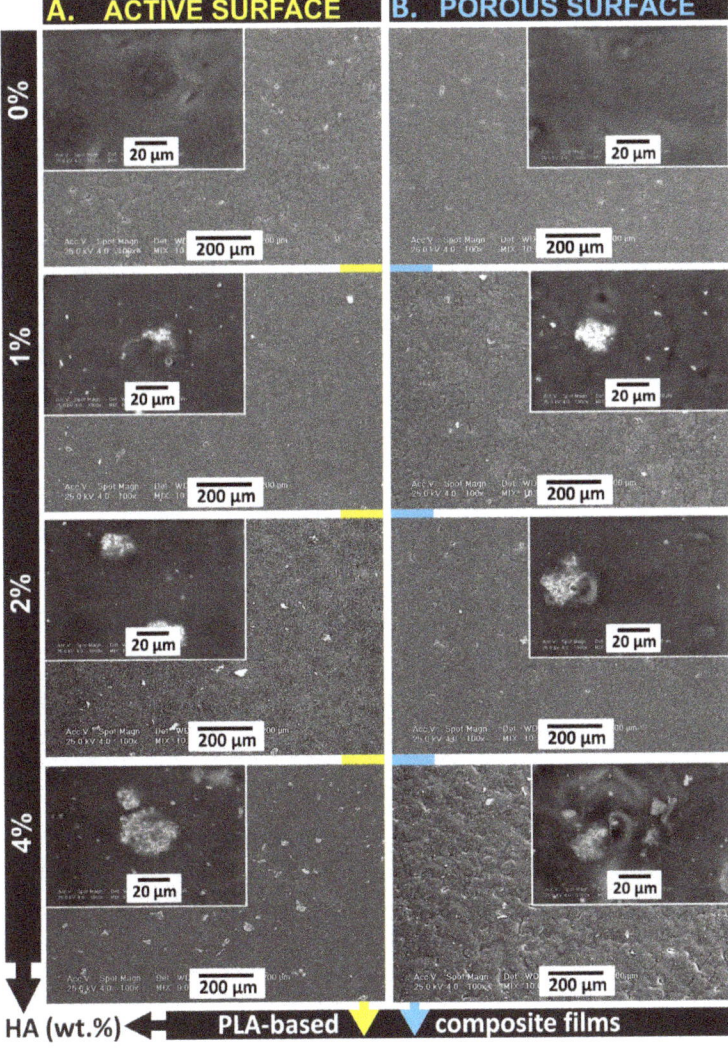

Figure 5. SEM images of PLA and PLA/HA composite films. (**A**) Active surface of the membranes, (**B**) porous surface of the films.

The pores were shown to have very small diameters but a difference was observed between one side and the other, indicating the asymmetry of the membrane. Contrary to polymeric membranes obtained by phase inversion (precipitation with a non-solvent), the films obtained by solvent evaporation can be seen to have much smaller pores due to the slow disappearance of the solvent from the polymer solution. In this case, the porosity is given by the solvent molecules in the film structure that diffuse outside of the film during the evaporation process [2,7]. At the same time, the polymer chains are entrained with the solvent from the base of the solution film to its surface, generating pores [34–36]. In the case of HA composite membranes, hydroxyapatite crystals appear both on the active and porous surfaces. These are observed both in a dispersed form and in the form of large crystals (agglomerates). As the amount of the HA particles in the film's structure increases, large crystals have a higher volume [29,37,38]. Increasing the size of the crystals is a consequence of the poor dispersion of the inorganic filler in the polymer solution, the ultrasonication time being the same for all the samples. The presence of the HA on the porous surface can be explained by the weight of the particles, which are gravitationally deposited to the base of the polymer solution film. This behavior is also observed in the AFM.

The samples were morphologically characterized by atomic force microscopy in order to reveal the surface structure modification appearing with the variation of the composition. Figure 6 reveals the 3D morphology of the PLA and PLA/HA 4 wt.% samples. The PLA sample shows a surface topography with very high roughness areas and smooth areas. The morphological differences can be attributed to the polymer's spatial arrangement in crystalline and amorphous zones. An ordered crystalline arrangement led to high roughness areas with an average roughness of about 41.1 nm, and a disordered amorphous arrangement led to a smooth area. The PLA/HA 4 wt.% composite film shows a decrease in the average roughness, with the average roughness being about 28.6 nm. This decrease appears because of the formation of HA aggregates on the membrane active surface, showing similar results to those observed in the SEM images. The above surface of the composite membrane reveals a more ordered structure due to a better HA dispersion with the average roughness decreasing to 10.8 nm.

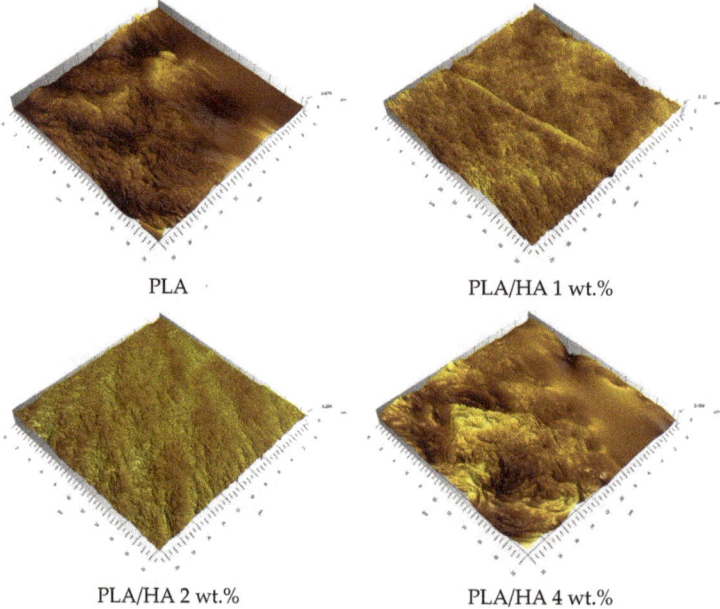

Figure 6. Atomic force microscopy (AFM) images of PLA and PLA/HA composite films.

The composite samples with 1 and 2 wt.% (Figure 6) highlight lower average roughness values of about 7.44 nm and 9.46 nm, respectively. The roughness decreasing assumes a more ordered surface structure due to a high HA dispersion. The higher HA dispersion for the composite samples with 1 and 2 wt.% HA loaded on can lead to the conclusion that these concentrations represent the right amount of inorganic phase within the polymeric matrix. Increasing the HA amount can overcome the system balance and generate HA aggregates.

Figure 7 displays tensile stress versus tensile strain curves for PLA and PLA/HA composite films. According to the figure and Table 2, a slight decrease in Young's modulus in the case of the composite films with 2 and 4 wt.% HA loaded within the polymer matrices can be observed. This behavior may be explained on the one hand by the presence of HA, which acts as a local strain concentrator within the final material, and, on the other hand by the formation of the aggregates at 4 wt.% HA. The agglomeration of HA nanoparticles at 4 wt.% was also observed in SEM and AFM images and led to a low interfacial interaction between HA and the polymer. A similar trend has been observed by H.Y. Mi et al. In their study, Young's modulus tended to decrease in the case of thermoplastic polyurethane/hydroxyapatite electrospun scaffolds with a concentration of >2 wt.% HA, and this was able to be explained by the nonuniform dispersion of HA within the polymer matrix, which led to an inhomogeneous stress distribution [39].

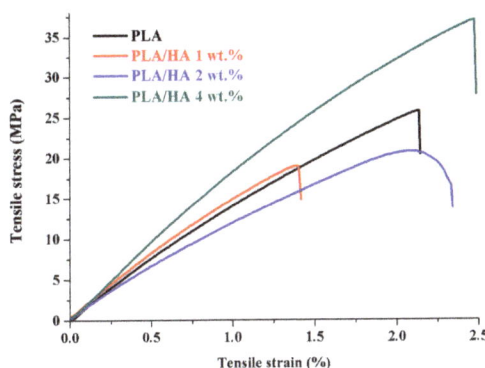

Figure 7. Mechanical tests of PLA and PLA/HA composite films.

Table 2. Mechanical tests of obtained materials.

Sample Name	Young's Modulus (MPa)
PLA	17 ± 0.59
PLA/HA 1 wt.%	17 ± 0.55
PLA/HA 2 wt.%	15 ± 0.14
PLA/HA 4 wt.%	14 ± 1.44

Previous reported research has shown an increased thermal resistance in the case of metal-doped HA with different cations like cerium [40], iron [41], zinc [42], and silver [43]. The higher thermal resistance is given in this case by the presence of cations inside the structure of HA, and also by the increased percentage of HA content (in all cases >10% wt.). Our observed small differences in the thermal behavior of the obtained films can be explained by the lower percent of HA (1–4% wt.), which is more suitable for potential application in osseointegration and is enough to influence the proliferation of pre-osteoblasts through the pores of polymeric films [5,7]. With potential application in osseointegration as films at the interface between a metallic implant and bone, porous films are preferred due to their more bioresorbable behavior under physiological conditions, with in this case the mechanical and thermal properties not being very important [8–10]. Higher mechanical properties can be obtained in the case of scaffolds obtained by 3D printing due to the large amount of polymer in the volume of the obtained material [44]. Also, in terms of mechanical properties, these can be significantly improved by the use of an additional polymer during the preparation of composite films, but with a much higher percent of HA. In the case of the composite film chitosan-PLA-HA at a 50% content HA amount at a compressive strength of 25,682 MPa, the strain to failure has been observed to be 70% with an elastic modulus of 857 MPa. With an increase in HA content (to 80%), the elastic modulus decreased to 660 MPa [45]. In these composites, PLA plays an important role, greatly influencing the nucleation and the growth of HA crystalline. An important factor in defining the mechanical and thermal properties is the synthesis method. When using cryomilling [46], high mechanical resistance and also improved thermal properties can be achieved via a Young's modulus of the composite of 6 GPa and a compressive strength of 110 MPa, which are quite similar to the values for natural bone. In the case of electrospun fibers, the elastic modulus increase has been found to be approximately 40% for a system filled with micrometric HA (µHA) at 10%, 70% for systems of randomly oriented (R) PLA/µHA of 20%, 100% for a system with a nanometric randomly oriented HA (R PLA/nHA) of 10%, and up to 140% for an R PLA/nHA 20% composite [47]. All these methods assure a higher quantity of polymer respective to HA in the structure of composites, which can explain the higher mechanical properties in comparison with our films. Furthermore, even by solvent evaporation, asymmetric polymeric films are obtained with membrane structures which are characterized by a large free volume filled with air inside the material [1]. Osseointegration evaluation is more dependent on the synthesis method than other properties [48], being more suitable in the case of plasma discharge with sputter deposition at the surface of implants [49], Janus membranes [50], or membranes obtained by phase inversion/precipitation from PLA dissolved in acetone [51]. In comparison with membranes obtained from acetone by phase inversion, our films present the advantage of a very low diameter of pores, which is more suitable for further biomedical applications [25]. Also, the use of chloroform as a solvent can assure a better manipulation of pore diameter through the temperature of evaporation (a higher temperature will assure an increased speed of evaporation with a lower diameter of pores and a lower temperature of evaporation will decrease the speed of film formation and will also imply a lower pore diameter).

4. Conclusions

In this work, PLA/HA composite films were synthesized starting from a polymer solution in chloroform (12 wt.%) in which various concentrations of hydroxyapatite (1, 2, and 4 wt.% reportable to the polymer) were dispersed by ultrasonic methods, followed by the synthesis of membrane materials by solvent evaporation, resulting in polymeric composite films. The synthesized membranes were morphologically (by SEM and AFM) and structurally (by FT-IR and Raman spectroscopy) characterized, and the thermal behavior of the synthesized composite films was studied. SEM and AFM images showed the presence of micro-structured hydroxyapatite particles in the composite film structure, as well as changes in the surface morphology. There was a significant decrease in the crystallinity of the composite films compared to the pure polymer, this being explained by the decrease in the arrangement of the polymer chains and a concurrent increase in their degree of disorder. Also, the composite films were characterized in terms of their thermal behavior and it was observed that the presence of hydroxyapatite crystals did not have a significant influence on the degradation temperature of the composite films. Future research will study the influence of HA particle dimensions on composite polymer films using the same conditions (solvent and evaporation method) and in vitro tests related to the synthesized materials.

Author Contributions: Conceptualization, S.I.V., A.M.P., F.M. and L.T.C.; methodology, A.M.P. and S.I.V.; formal analysis, A.M.P., F.M. and I.C.R.; investigation, A.C.; resources, S.I.V.; data curation, A.M.P.; writing—original draft preparation, A.M.P.; writing—review and editing, S.I.V.; supervision, S.I.V.; project administration, S.I.V.; funding acquisition, S.I.V. All authors have read and agreed to the published version of the manuscript.

Funding: This research was funded by UEFISCDI Romania, grant number PN-III-P1-1.1-TE-2016-0542 New generation of membrane systems with visual control of separation process efficiency based on modification of membrane colour surface.

Conflicts of Interest: The authors declare no conflict of interest. The funders had no role in the design of the study; in the collection, analyses, or interpretation of data; in the writing of the manuscript, or in the decision to publish the results.

References

1. Thakur, V.K.; Voicu, S.I. Recent advances in cellulose and chitosan based membranes for water purification: A concise review. *Carbohydr. Polym.* **2016**, *146*, 148–165. [CrossRef] [PubMed]
2. Miculescu, M.; Thakur, V.K.; Miculescu, F.; Voicu, S.I. Graphene-based polymer nanocomposite membranes: A review. *Polym. Adv. Technol.* **2016**, *27*, 844–859. [CrossRef]
3. Ionita, M.; Crica, L.E.; Voicu, S.I.; Pandele, A.M.; Iovu, H. Fabrication of Cellulose Triacetate/Graphene Oxide Porous Membrane. *Polym. Adv. Technol.* **2016**, *27*, 350–357. [CrossRef]
4. Ionita, M.; Vasile, E.; Crica, L.E.; Voicu, S.I.; Pandele, A.M.; Dinescu, S.; Predoiu, L.; Galateanu, B.; Hermenean, A.; Costache, M. Synthesis, characterization and in vitro studies of polysulfone/graphene oxide composite membranes. *Compos. Part B Eng.* **2015**, *72*, 108–115. [CrossRef]
5. Neacsu, P.; Staras, A.I.; Voicu, S.I.; Ionascu, I.; Soare, T.; Uzun, S.; Cojocaru, V.D.; Pandele, A.M.; Croitoru, S.M.; Miculescu, F.; et al. Characterization and In Vitro and In Vivo Assessment of a Novel Cellulose Acetate-Coated Mg-Based Alloy for Orthopedic Applications. *Materials* **2017**, *10*, 686. [CrossRef] [PubMed]
6. Pandele, A.M.; Neacsu, P.; Cimpean, A.; Staras, A.I.; Miculescu, F.; Iordache, A.; Voicu, S.I.; Thakur, V.K.; Toader, O.D. Cellulose acetate membranes functionalized with resveratrol by covalent immobilization for improved Osseointegration. *Appl. Surf. Sci.* **2018**, *438*, 2–13. [CrossRef]
7. Voicu, S.I.; Condruz, R.M.; Mitran, V.; Cimpean, A.; Miculescu, F.; Andronescu, C.; Miculescu, M.; Thakur, V.K. Sericin Covalent Immobilization onto Cellulose Acetate Membranes. *ACS Sustain. Chem. Eng.* **2016**, *4*, 1765–1774. [CrossRef]
8. Corobea, M.S.; Albu, M.G.; Ion, R.; Cimpean, A.; Miculescu, F.; Antoniac, I.V.; Raditoiu, V.; Sirbu, I.; Stoenescu, M.; Voicu, S.I.; et al. Advanced modification of titanium surface with collagen and doxycycline, a new approach in dental implants. *J. Adhes. Sci. Technol.* **2015**, *29*, 2537–2550. [CrossRef]
9. Pandele, A.M.; Comanici, F.E.; Carp, C.A.; Miculescu, F.; Voicu, S.I.; Thakur, V.K.; Serban, B.C. Synthesis and characterization of cellulose acetate-hydroxyapatite micro and nano composites membranes for water purification and biomedical applications. *Vacuum* **2017**, *146*, 599–605. [CrossRef]
10. Miculescu, F.; Maidaniuc, A.; Voicu, S.I.; Thakur, V.K.; Stan, G.; Ciocan, L.T. Progress in Hydroxyapatite-Starch Based Sustainable Biomaterials for Biomedical Bone Substitution Applications. *ACS Sustain. Chem. Eng.* **2017**, *5*, 8491–8512. [CrossRef]
11. Voicu, S.I.; Pandele, M.A.; Vasile, E.; Rughinis, R.; Crica, L.; Pilan, L.; Ionita, M. The impact of sonication time through polysulfone graphene oxide composite films properties. *Dig. J. Nanomater. Biostructures* **2013**, *8*, 1389–1394.
12. Tábi, T.; Sajó, I.E.; Szabó, F.; Luyt, A.S.; Kovács, J.G. Crystalline structure of annealed polylactic acid and its relation to processing. *Express Polym. Lett.* **2010**, *4*, 659–668. [CrossRef]
13. Voicu, S.I.; Ninciuleanu, C.M.; Muhulet, O.; Miculescu, M. Cellulose acetate membranes with controlled porosity and their use for the separation of amino acids and proteins. *J. Optoelectron. Adv. Mater.* **2014**, *16*, 903–908.
14. Miculescu, M.; Muhulet, A.; Nedelcu, A.; Voicu, S.I. Synthesis and characterization of polysulfone-carbon nanotubes -polyethylene imine composite membranes. *Optoelectron. Adv. Mater. Rapid Commun.* **2014**, *8*, 1072–1076.
15. Russias, J.; Saiz, E.; Nalla, R.K.; Gryn, K.; Ritchie, R.O.; Tomsia, A.P. Fabrication and mechanical properties of PLA/HA composites: A study of in vitro degradation. *Mater. Sci. Eng. C* **2006**, *26*, 1289–1295. [CrossRef]
16. Gong, M.; Zhao, Q.; Dai, L.; Li, Y.; Jiang, T. Fabrication of polylactic acid/hydroxyapatite/graphene oxide composite and their thermal stability, hydrophobic and mechanical properties. *J. Asian Ceram. Soc.* **2017**, *5*, 160–168. [CrossRef]
17. Ma, H.; Su, W.; Tai, Z.; Sun, D.; Yan, X.; Liu, B.; Xue, Q. Preparation and cytocompatibility of polylactic acid/ hydroxyapatite/graphene oxide nanocomposite fibrous membrane. *Chin. Sci. Bull.* **2012**, *57*, 3051–3058. [CrossRef]
18. Dumitriu, C.; Voicu, S.I.; Muhulet, A.; Nechifor, G.; Popescu, S.; Ungureanu, C.; Carja, A.; Miculescu, F.; Trusca, R.; Pirvu, C. Cellulose acetate-titanium dioxide nanotubes membrane fraxiparinized through polydopamine. *Carbohydr. Polym.* **2018**, *181*, 215–223. [CrossRef]

19. Muhulet, A.; Miculescu, F.; Voicu, S.I.; Schütt, F.; Thakur, V.K.; Mishra, Y.K. Fundamentals and Scopes of Doped Carbon Nanotubes towards Energy and Biosensing Applications. *Mater. Today Energy* **2018**, *9*, 154–186. [CrossRef]
20. Rakmae, S.; Ruksakulpiwat, Y.; Sutapun, W.; Suppakarn, N. Physical properties and cytotoxicity of surface-modified bovine bone-based hydroxyapatite/poly (lactic acid) composites. *J. Compos. Mater.* **2011**, *45*, 1259–1269. [CrossRef]
21. Azzaoui, K.; Mejdoubi, E.; Lamhamdi, A.; Hammouti, B.; Akartasse, N.; Berrabah, M.; Elidrissi, A.; Jodeh, S.; Hamed, O.; Abidi, N. Novel Tricomponenets composites Films from Polylactic Acid/ Hydroxyapatite/ Poly-Caprolactone Suitable For Biomedical Applications. *J. Mater. Environ. Sci.* **2016**, *7*, 761–769.
22. Monmaturapoj, N.; Srion, A.; Chalermkarnon, P.; Buchatip, S.; Petchsuk, A.; Noppakunmongkolchai, W.; Mai-Ngam, K. Properties of poly(lactic acid)/hydroxyapatite composite through the use of epoxy functional compatibilizers for biomedical application. *J. Biomater. Appl.* **2017**, *32*, 175–190. [CrossRef] [PubMed]
23. Liu, C.; Chan, K.W.; Shen, J.; Wong, H.M.; Yeung, K.W.K.; Tjong, S.C. Melt-compounded polylactic acid composite hybrids with hydroxyapatite nanorods and silver nanoparticles: Biodegradation, antibacterial ability, bioactivity and cytotoxicity. *RSC Adv.* **2015**, *5*, 72288–72299. [CrossRef]
24. Sanchez-Arevalo, F.M.; Munoz-Ramırez, L.D.; Alvarez-Camacho, M.; Rivera-Torres, F.; Maciel-Cerda, A.; Montiel-Campos, R.; Vera-Graziano, R. Macro- and micromechanical behaviours of poly(lactic acid)–hydroxyapatite electrospun composite scaffolds. *J. Mater. Sci.* **2017**, *52*, 3353–3367. [CrossRef]
25. Salerno, A.; Fernandez-Gutierrez, M.; San Roman del Barrio, J.; Pascual, C.D. Macroporous and nanometre scale fibrous PLA and PLA–HA composite scaffolds fabricated by a biosafe strategy. *RSC Adv.* **2014**, *4*, 61491. [CrossRef]
26. Jeong, S.I.; Ko, E.K.; Yum, J.; Jung, C.H.; Lee, Y.M.; Shin, H. Nanofibrous Poly(lactic acid)/Hydroxyapatite Composite Scaffolds for Guided Tissue Regeneration. *Macromol. Biosci.* **2008**, *8*, 328–338. [CrossRef]
27. Miculescu, F.; Mocanu, A.C.; Dascalu, C.A.; Maidaniuc, A.; Batalu, D.; Berbecaru, A.; Voicu, S.I.; Miculescu, M.; Thakur, V.K.; Ciocan, L.T. Facile synthesis and characterization of hydroxyapatite particles for high value nanocomposites and biomaterials. *Vacuum* **2017**, *146*, 614–622. [CrossRef]
28. Maidaniuc, A.; Miculescu, F.; Voicu, S.I.; Andronescu, C.; Miculescu, M.; Matei, E.; Mocanu, A.C.; Pencea, I.; Csaki, I.; Machedon-Pisu, T.; et al. Induced wettability and surface-volume correlation of composition for bovine bone derived hydroxyapatite particles. *Appl. Surf. Sci.* **2018**, *438*, 147–157. [CrossRef]
29. Maidaniuc, A.; Miculescu, M.; Voicu, S.I.; Ciocan, L.T.; Niculescu, M.; Corobea, M.C.; Rada, M.E.; Miculescu, F. Effect of micron sized silver particles concentration on the adhesion induced by sintering and antibacterial properties of hydroxyapatite microcomposites. *J. Adhes. Sci. Technol.* **2016**, *30*, 1829–1841. [CrossRef]
30. Voicu, S.I.; Dobrica, A.; Sava, S.; Ivan, A.; Naftanaila, L. Cationic surfactants-controlled geometry and dimensions of polymeric membrane pores. *J. Optoelectron. Adv. Mater.* **2012**, *14*, 923–928.
31. Persson, M.; Lorite, G.S.; Cho, S.W.; Tuukkanen, J.; O Skrifvars, M. Melt Spinning of Poly(lactic acid) and Hydroxyapatite Composite Fibers: Influence of Filler Content on the Fiber Properties. *ACS Appl. Mater. Interfaces* **2013**, *5*, 6864–6872. [CrossRef] [PubMed]
32. Cukrowski, I.; Popović, L.; Barnard, W.; Paul, S.O.; van Rooyen, P.H.; Liles, D.C. Modeling and spectroscopic studies of bisphosphonate–bone interactions. The Raman, NMR and crystallographic investigations of Ca–HEDP complexes. *Bone* **2007**, *41*, 668–678. [CrossRef]
33. Bikiaris, D. Can nanoparticles really enhance thermalstability of polymers? Part II: An overview on thermal decomposition of polycondensation polymers. *Thermochim. Acta* **2011**, *523*, 25–45. [CrossRef]
34. Senatov, F.S.; Niaza, K.V.; Zadorozhnyy, M.Y.; Maksimkin, A.V.; Kaloshkin, S.D.; Estrin, Y.Z. Mechanical properties and shape memory effect of 3D-printed PLA-based porous scaffolds. *J. Mech. Behav. Biomed. Mater.* **2016**, *57*, 139–148. [CrossRef] [PubMed]
35. Corobea, M.C.; Muhulet, O.; Miculescu, F.; Antoniac, I.V.; Vuluga, Z.; Florea, D.; Vuluga, D.M.; Butnaru, M.; Ivanov, D.; Voicu, S.I.; et al. Novel Nanocomposite Membranes from Cellulose Acetate and Clay-Silica Nanowires. *Polym. Adv. Technol.* **2016**, *27*, 1586–1595. [CrossRef]
36. Satulu, V.; Mitu, B.; Pandele, A.M.; Voicu, S.I.; Kravets, L.; Dinescu, G. Composite polyethylene terephthalate track membranes with thin teflon-likelayers: Preparation and surface properties. *Appl. Surf. Sci.* **2019**, *476*, 452–459. [CrossRef]
37. Miculescu, F.; Ciocan, L.; Miculescu, M.; Ernuteanu, A. Effect of heating process on micro structure level of cortical bone prepared for compositional analysis. *Dig. J. Nanomater. Biostruct.* **2011**, *6*, 225–233.

38. Miculescu, F.; Jepu, I.; Porosnicu, C.; Lungu, C.P.; Miculescu, M.; Burhala, B. A Study on the Influence of the Primary Electron Beam on Nanodimensional Layers Analysis. *Dig. J. Nanomater. Biostruct.* **2011**, *6*, 307–317.
39. Mi, H.Y.; Palumbo, S.; Jiang, S.; Turng, L.S.; Li, W.J.; Peng, X.F. Thermoplastic polyurethane/hydroxyapatite electrospun scaffolds for bone tissue engineering: Effects of polymer properties and particle size. *J. Biomed. Mater. Res. Part B Appl. Biomater.* **2014**, *102*, 1434–1444. [CrossRef]
40. Yuan, Q.; Qin, C.; Wu, J.; Xu, A.; Zhang, Z.; Liao, J.; Lin, S.; Ren, X.; Zhang, P. Synthesis and characterization of Cerium-doped hydroxyapatite/polylactic acid composite coatings on metal substrates. *Mater. Chem. Phys.* **2016**, *182*, 365–371. [CrossRef]
41. Morsi, M.A.; Hezma, A.E.M. Effect of iron doped hydroxyapatite nanoparticles on the structural, morphological, mechanical and magnetic properties of polylactic acid polymer. *J. Mater. Res. Technol.* **2019**, *8*, 2098–2106. [CrossRef]
42. Yuan, Q.; Wu, J.; Qin, C.; Xu, A.; Zhang, Z.; Lin, S.; Ren, X.; Zhang, P. Spin-coating synthesis and characterization of Zn-doped hydroxyapatite/ polylactic acid composite coatings. *Surf. Coat. Technol.* **2016**, *307*, 461–469. [CrossRef]
43. Liu, F.; Wang, X.; Chen, T.; Zhang, N.; Wei, Q.; Tian, J.; Wang, Y.; Ma, C.; Lu, Y. Hydroxyapatite/silver electrospun fibers for anti-infection and osteoinduction. *J. Adv. Res.* **2020**, *21*, 91–102. [CrossRef]
44. Mondal, S.; Nguyen, T.P.; Pham, V.H.; Hoang, G.; Manivasagan, P.; Kim, M.H.; Nam, S.Y.; Oh, J. Hydroxyapatite nano bioceramics optimized 3D printed poly lactic acid scaffold for bone tissue engineering application. *Ceram. Int.* **2019**. [CrossRef]
45. Cai, X.; Tong, H.; Shen, X.; Chen, W.; Yan, J.; Hu, J. Preparation and characterization of homogeneous chitosan–polylactic acid/hydroxyapatite nanocomposite for bone tissue engineering and evaluation of its mechanical properties. *Acta Biomater.* **2009**, *5*, 2693–2703. [CrossRef] [PubMed]
46. Pietrzykowska, E.; Mukhovskyi, R.; Chodara, A.; Wojnarowicz, J.; Koltsov, I.; Chudoba, T.; Łojkowski, W. Composites of polylactide and nano-hydroxyapatite created by cryomilling and warm isostatic pressing for bone implants applications. *Mater. Lett.* **2019**, *236*, 625–628. [CrossRef]
47. Lopresti, F.; Pavia, F.C.; Vitrano, I.; Kersaudy-Kerhoas, M.; Brucato, V.; La Carrubba, V. Effect of hydroxyapatite concentration and size on morpho-mechanical properties of PLA-based randomly oriented and aligned electrospun nanofibrous mats. *J. Mech. Behav. Biomed. Mater.* **2020**, *101*, 103449. [CrossRef]
48. Prasad, A.; Bhasney, S.M.; Sankar, M.R.; Katiyar, V. Fish Scale Derived Hydroxyapatite reinforced Poly (Lactic acid) Polymeric Bio-films: Possibilities for Sealing/locking the Internal Fixation Devices. *Mater. Today Proc.* **2017**, *4*, 1340–1349. [CrossRef]
49. Tverdokhlebov, S.I.; Bolbasov, E.N.; Shesterikov, E.V.; Antonova, L.V.; Golovkin, A.S.; Matveeva, V.G.; Petlin, D.G.; Anissimov, Y.G. Modification of polylactic acid surface using RF plasma discharge with sputter deposition of a hydroxyapatite target for increased biocompatibility. *Appl. Surf. Sci.* **2015**, *329*, 32–39. [CrossRef]
50. Ma, B.; Han, J.; Zhang, S.; Liu, F.; Wang, S.; Duan, J.; Sang, Y.; Jiang, H.; Li, D.; Ge, S.; et al. Hydroxyapatite nanobelt/polylactic acid Janus membrane with osteoinduction/barrier dual functions for precise bone defect repair. *Acta Biomater.* **2018**, *71*, 108–117. [CrossRef]
51. Talal, A.; Waheed, N.; Al-Masri, M.; McKay, I.J.; Tanner, K.E.; Hughes, F.J. Absorption and release of protein from hydroxyapatite-polylactic acid (HA-PLA) membranes. *J. Dent.* **2009**, *37*, 820–826. [CrossRef] [PubMed]

© 2020 by the authors. Licensee MDPI, Basel, Switzerland. This article is an open access article distributed under the terms and conditions of the Creative Commons Attribution (CC BY) license (http://creativecommons.org/licenses/by/4.0/).

Article

Preparation and Performance of Radiata-Pine-Derived Polyvinyl Alcohol/Carbon Quantum Dots Fluorescent Films

Li Xu [1,2,3,*], Yushu Zhang [1,2,3], Haiqing Pan [1], Nan Xu [1,2,3], Changtong Mei [1,2,3], Haiyan Mao [1,2,3,4,5], Wenqing Zhang [6], Jiabin Cai [1,2,3] and Changyan Xu [1,2,3,*]

1. College of Materials Science and Engineering, Nanjing Forestry University, Nanjing 210037, China; zhangyushu605@163.com (Y.Z.); phq625238728@163.com (H.P.); xunan@njfu.edu.cn (N.X.); mei@njfu.edu.cn (C.M.); maohaiyan@berkeley.edu (H.M.); nldfloor@163.com (J.C.)
2. Jiangsu Co-Innovation Center of Efficient Processing and Utilization of Forest Products, Nanjing Forestry University, Nanjing 210037, China
3. Jiangsu Province Key Laboratory of Green Biomass-based Fuels and Chemicals, Nanjing 210037, China
4. Department of Chemical and Biomolecular Engineering, University of California, Berkeley, CA 94720, USA
5. Jiangsu Chenguang Coating Co., Ltd., Changzhou 213164, China
6. Jiangsu Province Taizhou Efficient Processing Engineering Technology Research Center for Radiata Pine, Taizhou 214500, China; 13817330966@163.com
* Correspondence: njxl@njfu.edu.cn (L.X.); changyanxu1999@163.com (C.X.); Tel.: +86-0258-542-7519 (C.X.)

Received: 15 November 2019; Accepted: 19 December 2019; Published: 21 December 2019

Abstract: In this study, the low-cost processing residue of Radiata pine (*Pinus radiata D. Don*) was used as the lone carbon source for synthesis of CQDs (Carbon quantum dots) with a QY (The quantum yield of the CQDs) of 1.60%. The CQDs were obtained by the hydrothermal method, and +a PVA-based biofilm was prepared by the fluidized drying method. The effects of CQDs and CNF (cellulose nanofibers) content on the morphology, optical, mechanical, water-resistance, and wettability properties of the PVA/CQDs and PVA/CNF/CQDs films are discussed. The results revealed that, when the excitation wavelength was increased from 340 to 390 nm, the emission peak became slightly red-shifted, which was induced by the condensation between CQDs and PVA. The PVA composite films showed an increase in fluorescence intensity with the addition of the CNF and CQDs to polymers. The chemical structure of prepared films was determined by the FTIR spectroscopy, and no new chemical bonds were formed. In addition, the UV transmittance was inversely proportional to the change of CQDs content, which indicated that CQDs improved the UV barrier properties of the films. Furthermore, embedding CQDs Nano-materials and CNF into the PVA matrix improved the mechanical behavior of the Nano-composite. Tensile modulus and strength at break increased significantly with increasing the concentration of CQDs Nano-materials inside the Nano-composite, which was due to the increased in the density of crosslinking behavior. With the increase of CQDs content (>1 mL), the water absorption and surface contact angle of the prepared films decreased gradually, and the water-resistance and surface wettability of the films were improved. Therefore, PVA/CNF/CQDs bio-nanocomposite films could be used to prepare anti-counterfeiting, high-transparency, and ultraviolet-resistant composites, which have potential applications in ecological packaging materials.

Keywords: tensile properties; UV barrier; water-resistance

1. Introduction

Polyvinyl alcohol (PVA) is a kind of semi-crystalline polymer with a linear structure and strong hydrogen bonds intermolecular. It has been widely used in papermaking [1], textiles [2], coatings [3], adhesives [4], packaging [5], and biomedicine [6] because of its toughness [7], adhesiveness [8], biocompatibility [9], swelling behavior [10], lack of toxicity [11], and sufficient thermal stability; it is also odorless and tasteless. However, PVA is highly hydrophilic and water-soluble [12], and its light transmittance is not suitable for light-barrier packaging [13]. In order to expand the application of PVA, much attention has been paid to fabricating functional PVA composites. The existence of hydrogen-bonding groups in PVA structure and the ability to form hydrogen bonds makes PVA suitable for mixing with other materials to improve its functional properties [14,15]. In previous studies, the prepared silica in situ enhanced PVA/chitosan biodegradation films [16], PVA/tea polyphenol composite films [17], PVA reinforced with cellulose nanocrystals or cellulose nanofibers (CNF) [18], and CNF/PVA-borax hybrid foams [19] had better functional properties than PVA itself, from the perspective of packaging-industry application.

In recent years, carbon materials have been paid much attention for the preparation of functional composites. As a typical carbon material, nanocellulose (CNF) has been successfully used to improve the properties of PVA films because of its nanometer size in diameter, high strength, excellent stiffness, renewability, and high surface area [20–23]. In addition, the raw materials that can be used to extract CNF are inexhaustible in nature [24]. In particular, PVA contains a lot of hydroxyl, which are accessible for surface modification [23,25]. Carbon quantum dot (CQD) is a new kind of carbon material that emerged in recent years. Since its discovery, CQD has been known as an excellent candidate for diverse applications such as optoelectronics [26], detection of transition metal ions [27], fluorescent inks [28], and cellular imaging [29] due to its strong photoluminescence [30], high photostability [31], good water-solubility [32], low cytotoxicity [33], excellent biocompatibility [34], and environmental friendliness [35]. With persistent efforts, researchers have successfully extracted CQDs from agricultural and forestry waste, including mustard seeds [36], *Actinidia deliciosa* [37], and unripe peach [38]. Recently a few researchers have aimed to make PVA/CQDs composite films. Dong et al. prepared CQDs by hydrothermal reaction. It was added to the PVA solution, and the poly (vinylidene fluoride) film pretreated with an alkaline solution was immersed in the prepared CQD/PVA solution, to coat the surface of the film with a UV-shielding layer, and the composite film could shield the UV light completely [39]. El-Shamy et al. successfully fabricated novel P-type thermoelectric PVA/CQDs nanocomposite films through the solution casting technique. Thermoelectric properties such as electrical conductivity, Seebeck coefficient, and thermal conductivity of the prepared films were studied. These nanocomposites have a promising potential application in thermoelectric devices because they are economical and easy to scale up [40].

Radiata pine is most widely distributed in New Zealand, Australia, and Spain, and it is a major cultivated species in Argentina, Chile, Uruguay, Kenya, and South Africa. Radiata pine wood can be used in construction, wood-based panel, papermaking, furniture, railway sleepers, and other aspects because of its medium density, uniform structure, strong stability, nail holding strength, and good permeability. However, radiata pine is still a new tree species in China. Data from the production line show that about 13% of sawdust is generated when processing 1 m^3 of logs into sawdust, and about 18% of the shavings and sawdust is produced while making wood structures with 1 m^3 of sawn timber. Such a large amount of processing residue not only causes great waste of resources, but also pollutes the factory environment and becomes a potential fire hazard. This study provides a new way for efficient utilization of these wastes in extracting carbon quantum dots (CQDs) from them with a simple hydrothermal method [41,42] and preparing PVA/CNF/CQDs fluorescent films. The influence of CQDs load on photoluminescent properties, UV-vis absorption characteristics, tensile property, surface contact angle to distilled water, and water-resistance of PVA/CQDs films and PVA/CNF/CQDs composites were investigated.

2. Materials and Methods

2.1. Materials

Radiata pine wood (*Pinus radiata* D. Don) processing residue was procured from Jingjiang Guolin Forest Co., Ltd. (Jiangsu, China). After being oven-dried at 105 °C, until its water content was less than 15%, the residue was crushed with a pulverizer (400 Y, Yongkang boao hardware products Co. Ltd., Zhejiang, China) and then passed through a 90-mesh stainless-steel sieve. The resulting wood powder was sealed in a plastic bag, for later use. Coniferous nanocellulose suspension was offered by Zhongshan Nan Fiber New Material Co., Ltd. (Guangdong, China). Its solid content, fiber diameter, and aspect ratio were 2.5 ± 0.5 wt.%, 30 nm, and ≥20, respectively. Both PVA (degree of polymerization, 750 ± 50) and quinine sulfate (99.4%) were purchased from Aladdin Biochemical Co., Ltd. (Shanghai, China). All chemicals were used as received, without further purification.

2.2. Synthesis of CQDs from Radiata Pine Processing Waste

CQDs were synthesized by a hydrothermal method. Firstly, the Radiata pine woody powder (1.3 g) was dispersed into purified water (70 mL) under a strong stirring. Then, the mixture solution was transferred into a high-pressure Teflon-lined stainless-steel autoclave (100 mL, Taizhou, China) and heated at 200 °C for 8 h in an oil bath. After being cooled down to room temperature, the resultant light-yellow product (Figure 1a) was removed out from the autoclave and then filtered through a piece of microporous membrane (pore size, 0.22 μm), to remove large particles. Brown–yellow CQDs (Figure 1b) were obtained by a dialysis treatment for 48 h, with a dialysis membrane (MWCO, 500–1000 Da, Spectrum Labs, Los Angeles, CA, USA). The resulting CQDs (with a concentration of 0.1 wt.%) were stored in a refrigerator (4 °C) for later use.

Figure 1. (a) Filtered CQDs solution; (b) dialysis CQDs solution.

2.3. Fabrication of PVA-Based Films

2.3.1. Experimental Method

Table 1 shows the experimental method for preparation of PVA-based films. No. 1 (PVA film) and No. 6 (PVA/CNF film) are the control to No. 2–5 and No. 7–10, respectively, for investigating the influence of CQDs on the properties of PVA films and PVA/CNF composites.

Table 1. Experimental method for preparation of PVA-based films.

Film No.	CQDs (0.1 wt.%)/mL	PVA (10%)/mL	CNF (1%)/mL
No. 1	0	10	0
No. 2	0.2	10	0
No. 3	1	10	0
No. 4	2	10	0
No. 5	4	10	0
No. 6	0	10	10
No. 7	0.2	10	10
No. 8	1	10	10
No. 9	2	10	10
No. 10	4	10	10

2.3.2. Fabrication of PVA-Based Films

A 10 wt.% PVA solution was obtained by mixing PVA (10 g) and purified water (90 mL) at 95 °C for 2 h, with stirring, and 1 wt.% CNF solution was prepared by diluting the purchased CNF solution (8 g, 2.5 wt.%) with deionized water (12 mL), under strong stirring at room temperature for 1 h. PVA solution (10 wt.%) and CNF solution (1 wt.%) were mixed according to the formula in Table 1, and then ultrasonic-processed (XO-1200, Nanjing Xianqu Biological Technology Co., Ltd., China) in an ice/water bath for 40 min at 20–25 kHz frequency, with an output power of 960 W, resulting in PVA/CNF mixture for later use.

PVA film (No. 1, shown in Figure 2) was obtained by placing 10 mL of PVA solution (10 wt.%) in a petri dish and drying it at 40 °C for 24 h (101-2BS, Beijing Hengnuolixing Technology Co., Ltd., China). PVA/CNF film (No. 6, shown in Figure 2) was prepared in the same manner.

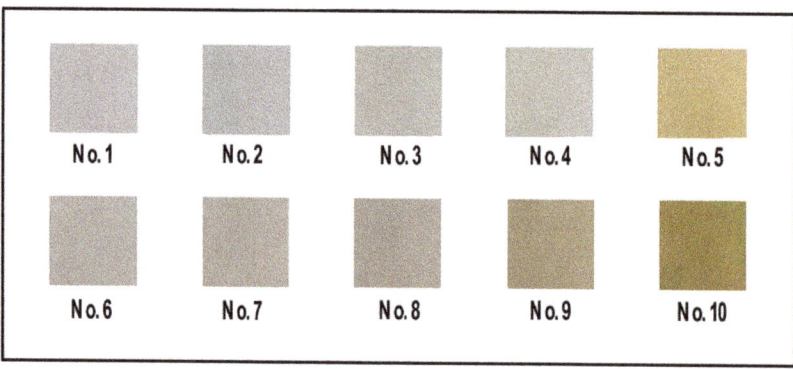

Figure 2. Images of PVA, PVA/CQDs, PVA/CNF, and PVA/CNF/CQDs films under daylight.

According to the formulation in Table 1, a certain amount of CQDs were mixed with PVA solution or PVA/CNF solution. After ultrasonic processing (XO-1200, Nanjing Xianqu Biological Technology Co., Ltd., China) in an ice/water bath for 30 min at 20–25 kHz frequency, with an output power of 960 W, the mixture was cast in a glass petri dish with a diameter of 90 mm and oven-dried with ventilation (101-2BS, Beijing Hengnuolixing Technology Co., Ltd., China) at 60 °C, to a state of smooth demolding, resulting in PVA/CQDs films (No. 2–5, shown in Figure 2) and PVA/CNF/CQDs films (No. 7–10, shown in Figure 2).

2.3.3. Characterizations of the Obtained CQDs

A three-use ultraviolet analyzer (ZF-1, Li Chen, Zhejiang, China) with an emission wavelength of 365 and 254 nm was used to obtain the fluorescence photographs of the prepared CQDs. The morphology of the CQDs was investigated with a cold field-emission scanning electron microscope (FE-SEM, S-4700, Hitachi, Japan) with an accelerating voltage of 200 kV, and the photoluminescence (PL) spectra and photoluminescent intensity were recorded, using an LS 55 fluorescence spectrophotometer (F-7000, Hitachi, Japan) with a band pass for excitation and emission of 10.

The quantum yield (QY) of the CQDs in this study was determined by referring to the method reported in the literature [43]. Quinine sulfate was dissolved in 0.1 mL of H_2SO_4 (Φ = 54%) as a standard. In order to minimize inner filter effect, both absorbance of the CQDs and the quinine sulfate solutions were adjusted to below 0.1. The QY was calculated by Equation (1):

$$QY_X = QY_{ST}\left(\frac{I_X}{I_{ST}}\right)\left(\frac{A_{ST}}{A_X}\right)\left(\frac{\eta_X}{\eta_{ST}}\right)^2 \tag{1}$$

where I and A are the fluorescence integral intensity and absorbance, respectively; η the refractive index of the solvent is 1.33; and the subscripts x and ST correspond to CQDs and quinine sulfate, respectively. According to the PL spectra of the CQDs and the quinine sulfate, the QY of our CQDs is 1.60%.

2.3.4. Characterizations of PVA-Based Films

The photoluminescent spectra and photoluminescent intensity of the PVA-based films were tested by using the same method as in Section 2.3.3. In addition, ZF-1 Ultraviolet Analyzer was used to study whether the prepared films had fluorescence effect and the influence of CQDs on fluorescence intensity of the films. The Fourier Transform infrared (FTIR) spectra of the PVA-based films were recorded on a spectrometer (VERTEX 80V, Bruker, Hamburg, Germany), using the KBr pellet technique, with a resolution of 2 cm^{-1} in the range of 500–4000 cm^{-1}. The ultraviolet-visible absorption characteristics of the films were measured with an UV-vis spectrophotometer (Lambda 950, Perkin Elmer, Waltham, MA, USA). The PVA-based film was cut to pieces with a diameter of 30 mm, after drying in an oven (Electric blast drying box, 101-2BS, Beijing Hengnuolixing Technology Co., Ltd., Beijing, China) at 90 °C for 24 h. The surface contact angles measurement of the prepared films to distilled water was tested with an automatic single fiber contact angle measuring instrument (OCA40, Data Physics Instruments GmbH, Feldstadt, German). In this study, we used the water absorption of the film to quantify its water-resistance. The size of the sample was 20 × 20 (mm, length × width). After being oven-dried at 90 °C for 24 h, the sample was weighed (Wi) and then immersed in distilled water (50 mL) for 24 h, at room temperature. After we dried the water on the sample's surface with napkins, the sample was weighed again (Wf). The water absorption of the sample was calculated by Equation (2):

$$A = \left[\left(W_f - W_i\right)/W_i\right] \times 100\% \tag{2}$$

The tensile test was done by using a universal mechanical tester (CMT 4204, 220V, Shenzhen Sans Testing Machine Co., Ltd., Guangdong, China), at a relative humidity of 50% and a loading speed of 10 mm/min. The size of the sample was 50 × 20 (mm, length × width). Three specimens were tested for each film, and the average values were reported.

3. Results and Discussion

3.1. Optical Performance of the Prepared CQDs

The Radiata-pine-derived CQDs synthesized by hydrothermal method shows good fluorescence photoluminescence, and the fluorescence intensity of the prepared CQDs was affected by excitation waves (340–390 nm), as shown in Figure 3. When the excitation wavelength was increased from 340 to 390 nm, the emission peak became slightly red-shifted, which was induced by the condensation between CQDs and PVA [44]. When the excitation wavelength of ultraviolet ray was 340–390 nm, the wavelength of the strongest emission peak of the CQDs was 400–450 nm, which belongs to blue–violet light. Moreover, when the excitation wavelength was 365 nm, the CQDs presented the strongest fluorescence intensity (563 a.u.), and a corresponding emission wavelength was 423 nm, which belongs to violet light. This indicates that, in the PL spectrum of our CQDs, there appears an excitation wavelength-dependent feature. The previous study also found similar results [43,45].

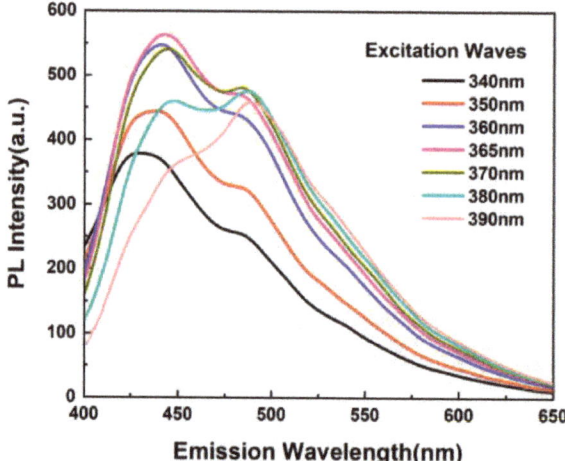

Figure 3. PL intensity of the diluted CQDs (0.1 wt.%) solution at various excitation wavelengths.

Figure 4 shows the optical images of the prepared CQDs under daylight (a) and UV light (c). The CQDs solution is yellowish and transparent under daylight; however, it emits violet fluorescence under UV light, with a wavelength of 365 nm. This is consistent with the finding in Figure 3. As for the mechanism of CQDs with fluorescence photoluminescence, it is not well understood [44]. It may be affected by such factors as electronic conjugate structures [46], emissive traps [47], and free zig-zag sites [48]. Cunjin Wang, et al. believed that S-CQDs reduced the non-radiative transition of electrons due to fewer surface defects, causing more electrons to radiate in the form of photons, therefore resulting in strong blue fluorescence of S-CQDs under 365 nm ultraviolet lamps [49].

Figure 4. Optical images of the prepared CQDs solution under daylight (**a**) and UV light (**b**).

3.2. PL Property of the Prepared PVA-Based Films

Figure 5 presents the images of the prepared PVA, PVA/CQDs, PVA/CNF, and PVA/CNF/CQDs films under ultraviolet rays (UV-rays) with a wavelength of 365 nm. The sample No. 1 (pure PVA film) has no fluorescence under UV-rays. When the load of CQDs in PVA/CQDs composites increases from 0.2 to 2 mL (from No. 2 to No. 4), the fluorescence intensity of the films increases sequentially; this is due to the increasing dosage of CQDs in the composites. However, further increasing the content of CQDs to 4 mL (No. 5) leads to a decrease of fluorescence intensity. This is due to the conjugation effect and the fluorescence self-quenching caused by too many CQDs in the composites [50,51]. Thus, introducing our CQDs into PVA films can endow PVA/CQDs composites with certain photoluminescence, but the content of CQDs in the composite has a threshold, which is 2 mL.

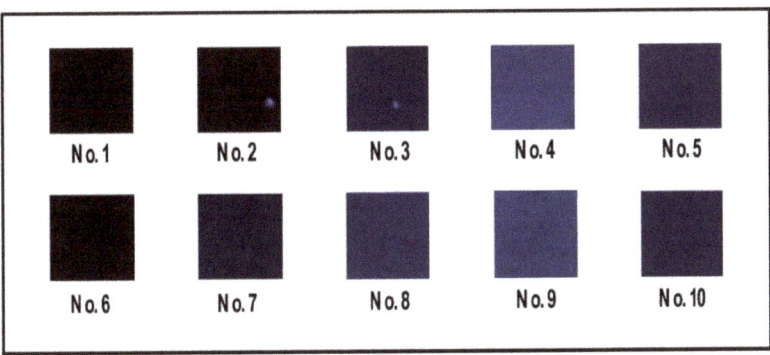

Figure 5. Images of PVA, PVA/CQDs, PVA/CNF, and PVA/CNF/CQDs films under ultraviolet rays (365 nm).

Similarly, the introduction of our CQDs into PVA/CNF composites can endow PVA/CNF/CQDs films with certain fluorescence photogenic effect. Sample No. 6 (PVA/CNF film) has no fluorescence under UV-rays, indicating that the CNF in the PVA/CNF composites has no contribution to fluorescence photogenic effect of PVA/CNF films. The fluorescence intensity of the prepared PVA/CNF/CQDs films gradually enhances with increasing the dosage of the CQDs from 0.2 to 2 mL (from No. 7 to No. 9). When this dosage exceeds 2 mL (No. 10), the fluorescence intensity of the PVA/CNF/CQDs films no longer increases, or even starts to decrease. This trend is consistent with some previous research results; it is the effect of the Mg–N-CQDs concentration was investigated, showing that the quenching efficiency of Hg(II) was decreased with increasing concentration of Mg–N-CQDs, which confirmed by Liu et al. [52]. Thus, just like PVA/CQDs films, there is also a threshold value for the CQDs content in PVA/CNF/CQDs composites, which is 3 mL. It indicates that CNF may delay the occurrence of quenching phenomenon due to too many CQDs in the composites. In the structure of the composites, the CNF and PVA are cross-linked by hydrogen bonds in Figure 6 to generate the tension, and the CQDs surface atoms are subjected to external strain from their relaxation positions, thus generating new energy states in the CQDs band-gap. These states provide a non-radiate recombination path for quenching [53].

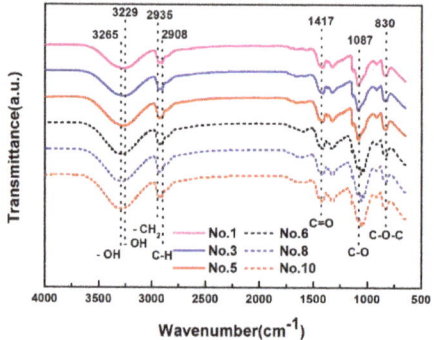

Figure 6. FTIR spectra of PVA, PVA/CQDs, PVA/CNF, and PVA/CNF/CQDs films.

3.3. Optical Performance of the Prepared PVA-Based Films

3.3.1. PL Spectra of the Prepared PVA-Based Films

Figure 7 displays the optical performance of the prepared films with various contents of CQDs (No. 1 to No. 10), at the excitation wavelength of 365 nm. It is observed that the fluorescence spectra of all the films present a similar changing trend. The fluorescence intensity decreases significantly with the increase of emission wavelength from 350 to 375 nm, which belongs to the band of ultraviolet light. However, when the emission wavelength increases from 425 nm to 575 nm, the fluorescence intensity of the films decreases slowly in the visible light band, and tends to be almost the same, even quenching for the emission wavelength greater than 575 nm. Moreover, the fluorescence spectra are clearly visible with convex peaks in the emission wavelength of 375 to 450 nm, which is the lipid oxidation fluorescence peak. The reason is that 2p2 nonbonding electrons of the oxygen atom occur n → π* electron transition in the free hydroxyl group of the molecular conformation of PVA [54]. From this, we conclude that PVA-based films dopes with CNF and CQDs. It shows that the addition of CNF and CQDs does not affect the characteristic emission peak of PVA-based films; however, the fluorescence intensity is affected.

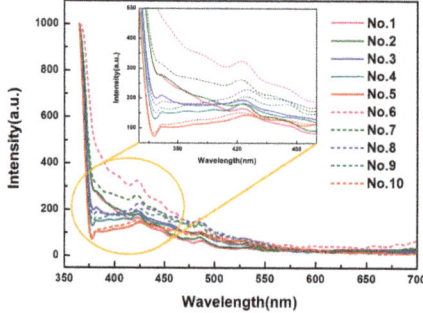

Figure 7. PL spectra of PVA, PVA/CQDs, PVA/CNF, and PVA/CNF/CQDs films (No. 1–10).

With the increase of CQDs content from 0 to 4 mL, the light intensity of emission waves with a wavelength of 375–410 nm emitted by the films gradually decreases, showing that the CQDs content has a significant effect on the attenuation of the fluorescence intensity of composites, which is consistent with the result of Ling [55]. One possible reason is due to the increase of electron-withdrawing group (carbonyl group) in the films, which can weaken the fluorescence [54]. Another possibility may be due to the $\pi \to \pi^*$ and $n \to \pi^*$ electron transitions to improve after doping CQDs in PVA, PVA/CNF films (No. 1 and No. 6), thus weakening the fluorescence intensity of the films [56].

In general, the emission intensity (with same wavelength) of the PVA/CNF/CQDs films is higher than that of the PVA/CQDs composites, indicating that introducing CNF into the PVA/CQDs films leads to an increase of emission intensity. The reason is maybe that the absorption ability of excited photons is enhanced by adding CNF, or due to existing the PVA/CNF hydrogel [57] with excellent fluorescence [58], which leads to the increase of fluorescence intensity of PVA/CNF/CQDs films [59]. In addition, the optimum emission peak of the PVA/CQDs films is in 418–421 nm, while that of PVA/CNF/CQDs films is in 422–425 nm, which creates the red-shift phenomenon. It is due to the introduction of CNF.

3.3.2. FTIR Spectra of the PVA-Based Films

Figure 6 shows FTIR spectra of the PVA, PVA/CQDs, PVA/CNF, and PVA/CNF/CQDs films, and Table 2 lists the typical absorption bands in the spectra and their corresponding functional groups. The broad and intense peaks at 3265 and 3229 cm^{-1} correspond to the –OH stretching vibration [60], while the peak at 2935 cm^{-1} is attributed to the asymmetric bending vibration of –CH$_2$ [61]. The absorption band at 2908 cm^{-1} is related to the C–H stretching vibration, which is similar to the result of Saikia [62]. The peaks at 1417, 1087, and 830 cm^{-1} are associated with the stretching vibration of nonconjugated C=O [63], the stretching vibration of C–O [61], and the asymmetric aromatic ring skeleton vibration of C–O–C [30], respectively. The vibration frequency of the peaks at 2908, 1417, and 1087 cm^{-1} distributes sparsely due to the electron transition, which leads to the weaker absorption bands. However, the stronger absorption bands appear at 2935 and 830 cm^{-1} due to a large variation of dipole moment caused by the asymmetric molecular vibrations. The presence of these surface-functional groups, as shown in Table 2, results in the high hydrophilic solubility of the prepared films (No. 1, No. 3, No. 5, No. 6, No. 8, and No. 10 in Table 1).

Table 2. Functional group absorption bands of the PVA, PVA/CQDs, PVA/CNF, and PVA/CNF/CQDs films.

Wavenumber (cm^{-1})	Functional Groups	Vibrations
3265, 3229	–OH	stretching
2935	–CH$_2$	bending
2908	C–H	stretching
1417	C=O	stretching
1087	C–O	stretching
830	C–O–C	asymmetric aromatic ring skeleton

There are no significant differences among the spectra of the pure PVA, PVA/CQDs, PVA/CNF, and PVA/CNF/CQDs films (No. 1, No. 3, No5, No. 6, No. 8, and No. 10 in Table 1) in Figure 6, indicating that the introduction of CNF (0.8 wt.%) or CQDs (1 and 4 mL, as shown in Table 1) into PVA does not cause significant functional group changes in PVA films. This is due to the fact that the infrared spectra of the prepared films do not deviate greatly, because the main component of the prepared films is PVA.

Compared with the spectra of the PVA, PVA/CQDs, PVA/CNF, and PVA/CNF/CQDs films (No. 1, No. 3, No. 5, No. 6, No. 8, and No. 10 in Table 1), the absorption peaks are shifted from 3229 to 3265 cm^{-1}, which may be due to the hydrogen bonds between crosslinking [64]. Compared with the spectra of the PVA and PVA/CQDs films (No. 1, No. 3, and No. 5 in Table 1), the peak at 1087 cm^{-1} in infrared spectra of the PVA/CNF and PVA/CNF/CQDs films (No. 6, No. 8, and No. 10 in Table 1) shifts lightly to right. There are two possible reasons for this shift: The one reason is the existence of π-π^* conjugate effect due to the sp2 Hybrid when the C–O group in the films links to the larger groups of conjugate system; the other is that it is easily formed through the intramolecular hydrogen bonding between hydroxyl groups and carbonyl groups [65]. It shows that the absorption peaks are strongly related to the oxidation functional groups, which is that the PVA/CNF/CQDs films is basically composed of oxygen atoms with amorphous carbon bonds [60].

The results show that no new chemical bonds are produced in the PVA/CQDs, PVA/CNF, and PVA/CNF/CQDs films (No. 3, No. 5, No. 6, No. 8, and No. 10 in Table 1).

3.4. Barrier Property to Light of the Prepared PVA-Based Films

Transmittance curves of the films with different CQDs contents are shown in Figure 8. All the films (No. 1–10) show a common trend; that is, when the wavelength changes from 290 nm to 800 nm, the transmittance increases gradually until it becomes stable, reaching about 90%. It indicates that the films have worse barrier property to light after the 700 nm wavelength. However, a strong absorption peak appears at the wavelength from 240 to 290 nm, which belongs to C–H bend peak of a vibrational structure. It is due to the dipole–dipole interaction between molecules [66], or n → π * transition, because of unsaturated (C=O) bond break, as reported earlier [67].

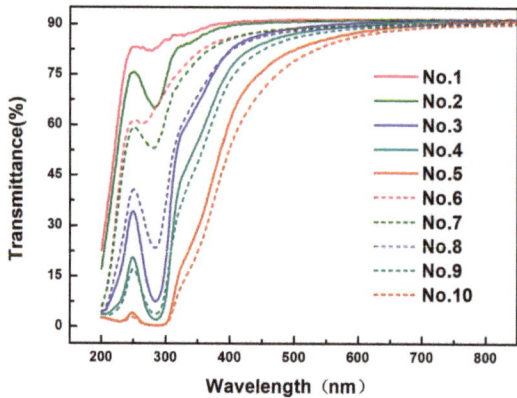

Figure 8. Transmittance of the PVA, PVA/CQDs, PVA/CNF, and PVA/CNF/CQDs films.

Compared with the pure PVA films and PVA/CQDs films, the transmittance of PVA/CNF films and PVA/CNF/CQDs films decreases with the addition of CNF (8 wt.%) into PVA-based film, showing that barrier property to light of PVA/CNF film improved by the introduction of CNF. For PVA/CQDs films and PVA/CNF/CQDs films, the transmittance of the films decreased with the increase of the dosage of CQDs. It indicates that barrier property to light of the films enhances after the addition of CQDs. The result coincides with Liu [68]. It is attributed to the increase of cluster size doped CQDs [69]. The pure PVA film has almost no barrier to UV light, while the transmittance of the films in the UV-light region reduces after adding CNF or CQDs; especially around 280 nm, the transmittance is very low, which shows that the PVA films with the added CNF or CQDs have excellent barrier to UV light. Furthermore, it is observed that, when the CQDs are added up to 4 mL (No. 5 and No. 10), the transmittance of PVA/CQDs and PVA/CNF/CQDs films is almost down to zero in UV wavelength range of 200–310 nm, demonstrating that the PVA/CQDs and PVA/CNF/CQDs films with CQDs dosage of 4 mL in the composite matrix have good barrier property of UV light.

In addition, for PVA film (No. 1 in Table 1) and PVA/CQDs films (No. 2 to No. 5 in Table 1), the peaks of the light transmittance curve shift to the right at UV wavelength range of 240 to 300 nm, with the increase of CQDs contents. One reason is that, although the spectra of PVA/CQDs films (No. 2 to No. 5 in Table 1) are similar to those of the parent compound PVA, the observed shift in the band positions may be attributed to the substituent chains attached. Another reason is the mutual interactions of the CQDs surface dangling bonds, the absorption peak with CQDs added happens to the red-shift in the UV-visible absorption spectrum, which generates the self-reabsorption. This red-shift overlaps with emission, causing re-absorption with the loss of energy [70]. Compared with the PVA/CNF films (No. 6 in Table 1), the PVA/CNF/CQDs films (No. 7–10 in Table 1) shift to right; this is consistent with our previous reasons. Consequently, the advantage of CQDs to PVA films is the reduction of the UV-transmittance [66].

As a result, it is shown that the modified PVA-based film can block almost all of the UV region, so the PVA/CNF/CQDs films can be used to block UV-rays' materials.

3.5. Tensile Properties of the Prepared PVA-Based Films

The tensile properties of the prepared PVA-based films are listed in Table 3, and the typical stress–strain curves are shown in Figure 9. Figure 10 gives SEM graphs of the fracture section of PVA, PVA/CQDs, PVA/CNF, and PVA/CNF/CQDs films. As shown in Figure 9, the preferential tensile orientation is observed in the fractured surface of all the PVA-based films, which results in the "drawing" at strains when the stress reaches a plateau before the film failure [71], as shown in Figure 9.

Table 3. Tensile properties of the prepared PVA-based films.

Film Number	Tensile Modulus (MPa)	Tensile Strength (MPa)	Elongation at Break (%)	Film Thickness (mm)
No. 1	580.76 (98.27 [a])	42.50 (2.42 [a])	345.02 (25.49 [a])	0.129 [b]
No. 2	824.84 (121.76 [a])	48.58 (6.68 [a])	302.83 (48.24 [a])	0.115 [b]
No. 4	881.28 (125.59 [a])	45.91 (3.42 [a])	287.69 (32.89 [a])	0.177 [b]
No. 6	1291.00 (245.55 [a])	51.68 (6.51 [a])	254.77 (54.01 [a])	0.137 [b]
No. 7	1185.25 (121.26 [a])	56.37 (4.64 [a])	264.39 (27.02 [a])	0.157 [b]
No. 9	1316.06 (164.72 [a])	59.71 (5.22 [a])	313.01 (10.43 [a])	0.163 [b]

[a] The standard deviation value of three samples. [b] The average value of three samples.

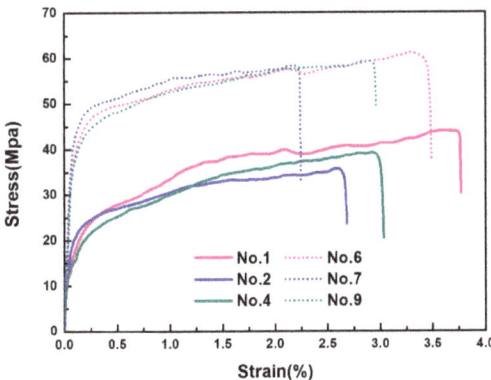

Figure 9. Stress–strain curves of the PVA-based films.

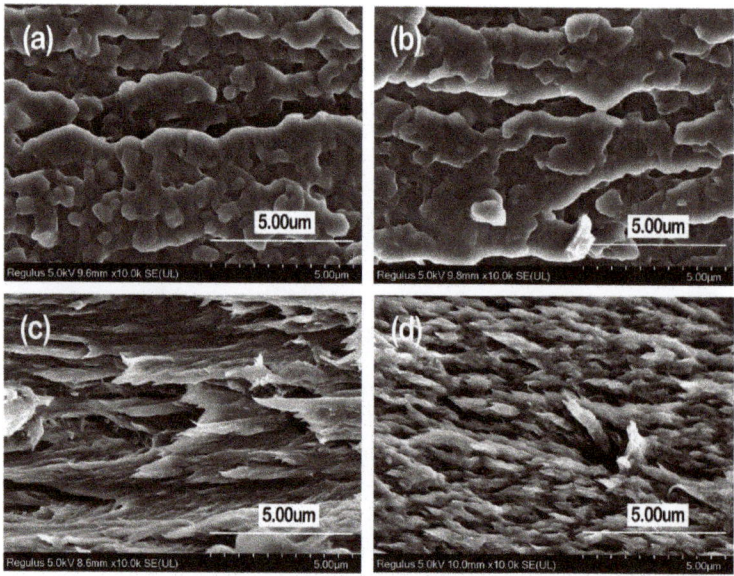

Figure 10. SEM graphs of the fracture section of PVA film, No. 1 (**a**); PVA/CQDs film, No. 4 (**b**); PVA/CNF, No. 6 (**c**); and PVA/CNF/CQDs film, No. 9 (**d**).

When adding an 8 wt.% CNF into the PVA matrix, the tensile modulus and tensile strength increases from 580.76 and 42.50 MPa (No. 1 in Table 1) to 1291.00 and 51.68 MPa (No. 6 in Table 1), respectively. It indicates that the introduction of CNF to the PVA-based film (No. 5 in Table 1) presents a higher tensile modulus and tensile strength, as well as a lower elongation at break, demonstrating that the CNF in PVA/CNF matrix can significantly improve its tensile performance. This is mainly attributed to the stable hydrogen bond between the CNF and PVA polymer matrix, which restricts propagation of the crack [72]. Compared with the smooth and flat tensile failure section of PVA film (Figure 10a), the failure section of PVA/CNF film (Figure 10c) is rough and uneven, which is evidence of a good combination of PVA and CNF interface. Yan et al. also found that the CNF could beneficially improve the ductility and tensile strength of PVA/CNF films [73].

As shown in Table 3 and Figure 9, introducing CQDs into PVA or PVA/CNF matrix leads to an increase of tensile modulus and tensile strength of the composite. The tensile modulus and tensile strength of the PVA/CQDs films with 0.2 mL CQDs (No. 2) were 42.03% and 14.3% higher than PVA (No. 1), respectively; and when the dosage of CQDs increases to 2 mL in the matrix, the tensile modulus and tensile strength of the PVA/CQDs film (No. 4) increases by 51.75% and 8.02%, respectively. Similarly, the tensile modulus and tensile strength of the PVA/CNF/CQDs film with 2 mL CQDs (No. 9) was 1.90% and 15.60% higher than the PVA/CNF film (No. 6), respectively. Although we cannot explain for the time being why CQDs can enhance the tensile modulus and tensile strength of PVA or PVA/CNF films, we hypothesize it may be the result of hydrogen bonds, van der Waals force bond, and other chemical bonds between the active groups on the surface of CQDs and PVA and CNF. The SEM graphs in Figure 10 support this hypothesis. Compared with the PVA/CNF film (Figure 10c), the PVA/CNF/CQDs film (Figure 10d) presents a denser failure section, which is evidence of a better combination of PVA, CNF, and CQDs interface.

As described in Section 3.4, the introduction of CQDs into PVA or PVA/CNF gives PVA-based composites a certain UV-light-barrier property; and the tensile test results here show that the CQDs in PVA or PVA/CNF matrix improve the tensile strength and modulus of the composite films. Compared with the proper films used in packaging [74], the prepared PVA/CQDs and PVA/CNF/CQDs films in

3.6. Water-Resistance of the Prepared PVA-Based Films

Water-barrier property of the prepared PVA-based films is shown in Figure 11a. The water absorption of films changes significantly within 0.5 h, which is due to the hydrophilic character of CNF and CQDs [75,76]. In addition, the average value of the water absorption of the PVA film (No. 1), PVA/CQDs film (No. 2 to No. 5), PVA/CNF film (No. 6), and PVA/CNF/CQDs film (No. 7 to No. 10) was 82.19%, 82.99%, 74.36%, and 77.15% after 0.5 h, reaching 89.04%, 89.25%, 84.62%, and 84.70% after 24 h, respectively. It shows that the water absorption of films is slightly unchanged after 0.5 h, which is due to saturation.

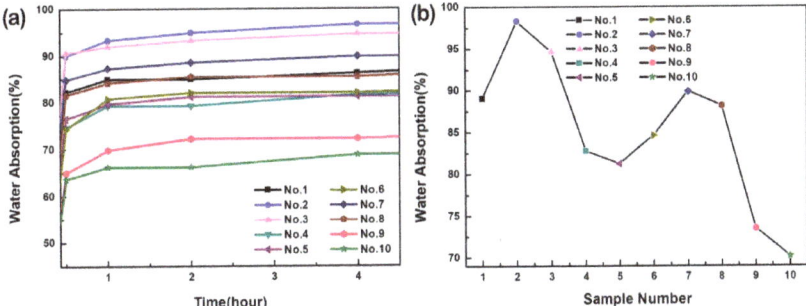

Figure 11. (a) Water absorption of the PVA, PVA/CQDs, PVA/CNF, and PVA/CNF/CQDs films with different contents of CQDs in 4 h. (b) Average water absorption of the PVA, PVA/CQDs, PVA/CNF, and PVA/CNF/CQDs films with different contents of CQDs within 24 h.

It shows the trend of the water absorption of the 10 composites is presented after 24 h in Figure 11b. The water absorption is a vital index to characterize the water-resistance of composites. With the decrease of water absorption, the water-resistance of the material increases. Compared with the PVA-based films, the water absorption of PVA/CQDs films (0.2 mL) is higher, which maybe be attributed to the introduction of CQDs. Meanwhile, the water absorption of PVA/CNF/CQDs films (0.2 mL) is higher than that of PVA/CNF films, which is consistent with the previous one; that is, the water absorption increase of the films is due to the introduction of CQDs. In addition, when the content of CQDs exceeds 0.2 mL, the water absorption of PVA/CQDs (No. 3 to No. 5) and PVA/CNF/CQDs (No. 8 to No. 10) films decreased significantly, which may be due to the reduction of porosity [77]. Another reason is, when CQDs with many carboxyl groups are introduced, the decrease of free hydroxyl radicals is due to the reaction between hydroxyl and carboxyl groups. This leads to the decrease of water absorption of the films [78]. This indicates that the change of CQDs content could affect the water absorption of the films. The water absorption decreases from 89% of the pure PVA films to 81.25% of the PVA/CQDs films (No. 5), and reduces to 70.13% of the PVA/CNF/CQDs films (No. 10). Compared with the water absorption of PVA/CQDs (No. 1–No. 5), the water absorption of the PVA/CNF/CQDs (No. 6 to No. 10) films reduces significantly. The result shows that the introduction of CNF reduces the water absorption of PVA/CQDs-based films [79]. When doping CNF in the PVA/CQDs films, the composites form more hydrogen bonds to reduce the amount of hydrophilic hydroxyl. Therefore, the surface polar groups of the PVA-based films have less contact with water molecules [80]. The results show that the doping of CNF and CQDs improves the water-resistance of PVA-based films.

3.7. Surface Wettability of PVA-Based Films

The contact angles of the 10 PVA-based composites are presented in Figure 12. The water contact angle is an important index to evaluate surface wettability of the prepared films. The contact angles of the films (No. 1 to No. 10) fall within a narrow range (18°–32°), due to the fact that the composites have a large amount of the hydrophilic groups. In contrast, the pure PVA films (No. 1) and the PVA/CNF films (No. 6) have contact angles of 23° and 23.5°, respectively, which are lower than those of the PVA-based films (28.2°) and PVA/CNF-based films (32.0°) loaded with 0.2 mL CQDs. This suggests that the wettability of PVA/CQDs films (No. 2) and PVA/CNF/CQDs films (No. 7) is worse than that of the pure PVA films and the PVA/CNF films, respectively. The change of contact angle is mainly related to the surface smoothness, porosity, pore size, and distribution of the films. This is probably due to the slight increase of surface smoothness of the prepared films after addition of CQDs (<0.2 mL), which is shown in Figure 10 [81], and is in good agreement with existing work [82]. Compared with the pure PVA films, the PVA/CQDs films (No. 3 to No. 5) show lower contact angles, which is attributed to the large amount of hydroxyl formed by high content of CQDs in the films. This happens for the same reason that the PVA/CNF/CQDs films (No. 8 to No. 10) have lower contact angles. Furthermore, the contact angles of PVA/CQDs films (No. 3 to No. 5) and PVA/CNF/CQDs films (No. 8 to No. 10) are small, which is due to no new chemical functional groups generated. This indicates that the hydrophilicity of the films is not changed with the introduction of CQDs. With the increase of CQDs content, the water contact angle decreases, which leads to the wettability of the prepared films improves, due to a change in surface properties of the blend films. Compared with the PVA/CQDs films (No. 1 to No. 5), respectively, the PVA/CNF/CQDs films (No. 6 to No. 10) present lightly larger water contact angle (<32°), due to the slightly higher surface roughness of the films (No. 6 to No. 10) after the introduction of CNF [83], which can be observed from the SEM graphs of the PVA/CNF/CQDs films in Figure 10. The results show the films after the introduction of CNF show a worse wettability.

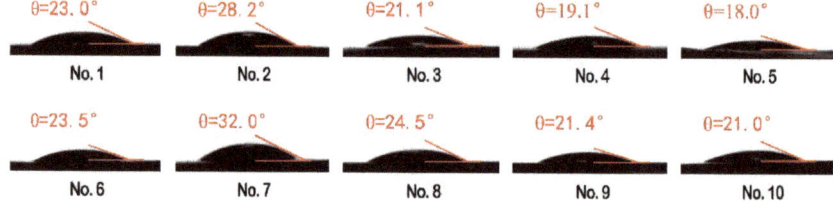

Figure 12. Water contact angles of the PVA/CQDs films and the PVA/CNF/CQDs films with different contents of CQDs.

4. Conclusions

In summary, a one-step hydrothermal method was adopted to prepare CQDs from Radiata pine at a constant temperature of 200 °C for 8 h for the first time. The synthesized Radiata-pine-derived CQDs exhibited excitation-dependent PL emissions with a QY of 1.60%.

When the excitation wavelength was 365 nm, the CQDs presented the strongest blue fluorescence intensity (563 a.u.). Introducing our CQDs into PVA films can endow PVA-based composites with certain photoluminescence; the fluorescence of films increases at first and then descends, reaching self-quenching with the increase of CQDs dosage in the composites. Meanwhile, CNF contributes to the fluorescence of the composite film, which creates the red-shift phenomenon at optimum emission peak. In addition, doping CNF and CQDs in the PVA-blend films is also confirmed by FT-IR, which indicates that the PVA matrix's introduction to CNF and CQDs went well, due to no new chemical bonds being produced. The pure PVA film has almost no barrier to UV light, while the transmittance of the films in the UV-light region reduces after adding CNF or CQDs; especially around 280 nm, the transmittance is very low, which shows that the PVA films adding CNF or CQDs have excellent barrier to UV light.

Dramatically, introducing CQDs and CNF into PVA matrix leads to an increase of tensile modulus and tensile strength of the composite. Demonstrating the CNF in PVA/CNF matrix can significantly improve its tensile performance. The introduction of CQDs affects the water absorption of PVA/CQDs based films. When doping CNF in the PVA/CQDs films, the composites form more hydrogen bonds to reduce the amount of hydrophilic hydroxyl. Therefore, the doping of CNF and CQDs improves the water-resistance of PVA-based films. A significant reduction in the contact angle of the PVA-modified film is observed, suggesting that the functionalized PVA improved the wettability of the film's surface. With the increase of CQDs content, the water contact angle decreases, which leads to an improvement in the wettability of the prepared films.

Overall, the improvement in water-resistance property, enhancement of the physical properties (structural and mechanical), and UV barrier of these the modified PVA-based films makes a variety of uses in the multifunctional applications. For example, the composites have broad prospects in the application of fruit and vegetable packaging. Therefore, we believe that the composite films can be used to make transparent, anti-counterfeiting, anti-UV, good-mechanical-strength materials, which have better application prospects in future.

Author Contributions: Conceptualization, L.X. and C.X.; methodology, L.X.; software, L.X., J.C., H.P., and H.M.; formal analysis, L.X. and Y.Z.; investigation, L.X. and H.P.; project administration, N.X. and H.M.; supervision, C.X. and C.M.; writing—original draft preparation, L.X., C.X., and Y.Z.; funding acquisition, C.X., C.M., and W.Z. All authors have read and agreed to the published version of the manuscript.

Funding: This research was funded by Youth Fund for Humanities and Social Sciences Research of the Ministry of Education (grant number 19YJC760132) and Jiangsu Provincial Key Lab of Environmental Engineering for Open Project Foundation (grant number ZX2017009).

Acknowledgments: The authors would like to thank First-Class Discipline Construction, International Innovation Highland of Forest Product Chemistry, Taizhou Efficient Processing Engineering Technology Research Center for Radiata Pine, Advanced Analysis and Testing Center of Nanjing Forestry University for supporting the work.

Conflicts of Interest: The authors declare no conflicts of interest.

References

1. Zhu, Y.K.; Chen, D.J. Clay-based nanofibrous membranes reinforced by multi-walled carbon nanotubes. *Ceram. Int.* **2018**, *44*, 15873–15879. [CrossRef]
2. Wu, M.C.; Chan, S.H.; Lin, T.H. Fabrication and photocatalytic performance of electrospun PVA/silk/TiO2 nanocomposite textile. *Funct. Mater. Lett.* **2015**, *8*, 1540013. [CrossRef]
3. Hassannejad, H.; Nouri, A.; Soltani, S.; Molavi, F.K. Study of corrosion behavior of the biodegradable chitosan-polyvinyl alcohol coatings on AA8011 aluminum alloy. *Mater. Res. Express* **2019**, *6*, 055312. [CrossRef]
4. Zhu, C.; Xiang, Q.; Liu, X.Y.; Dong, L.M. Study on antiseptic property of water soluble polyvinyl alcohol building adhesive. In Proceedings of the 2016 International Conference on Advances in Energy, Environment and Chemical Science (AEECS 2016), Changsha, China, 23–24 April 2016; Volume 76, pp. 4–8.
5. Abdullah, Z.W.; Dong, Y.; Davies, I.J.; Barbhuiya, S. PVA, PVA blends, and their nanocomposites for biodegradable packaging application. *Polym. Plast. Technol.* **2017**, *56*, 1307–1344. [CrossRef]
6. Gaaz, T.S.; Sulong, A.B.; Akhtar, M.N.; Kadhum, A.A.H.; Mohamad, A.B.; Al-Amiery, A.A. Properties and applications of polyvinyl Alcohol, halloysite nanotubes and their nanocomposites. *Molecules* **2015**, *20*, 22833–22847. [CrossRef] [PubMed]
7. Arain, M.F.; Wang, M.X.; Chen, J.Y.; Zhang, H.P. Study on PVA fiber surface modification for strain-hardening cementitious composites (PVA-SHCC). *Constr. Build. Mater.* **2019**, *197*, 107–116. [CrossRef]
8. Lelifajri; Nawi, M.A.; Sabar, S.; Supriatno; Nawawi, W.I. Preparation of immobilized activated carbon-polyvinyl alcohol composite for the adsorptive removal of 2,4-dichlorophenoxyacetic acid. *J. Water Process Eng.* **2018**, *25*, 269–277. [CrossRef]

9. Baker, M.I.; Walsh, S.P.; Schwartz, Z.; Boyan, B.D. A review of polyvinyl alcohol and its uses in cartilage and orthopedic applications. *J. Biomed. Mater. Res. Part B* **2012**, *100*, 1451–1457. [CrossRef]
10. Kim, S.J.; Park, S.J.; Kim, S.I. Swelling behavior of interpenetrating polymer network hydrogels composed of poly (vinyl alcohol) and chitosan. *React. Funct. Polym.* **2003**, *55*, 53–59. [CrossRef]
11. Heuschmid, F.F.; Schneider, S.; Schuster, P.; Lauer, B.; van Ravenzwaay, B. Polyethylene glycol-g-polyvinyl alcohol grafted copolymer: Reproductive toxicity study in wistar rats. *Food. Chem. Toxicol.* **2013**, *51* (Suppl. 1), S24–S35. [CrossRef]
12. Limpan, N.; Prodpran, T.; Benjakul, S.; Prasarpran, S. Influences of degree of hydrolysis and molecular weight of poly(vinyl alcohol) (PVA) on properties of fish myofibrillar protein/PVA blend films. *Food Hydrocoll.* **2012**, *29*, 226–233. [CrossRef]
13. Chen, C.W.; Xie, J.; Yang, F.X.; Zhang, H.L.; Xu, Z.W.; Liu, J.L.; Chen, Y.J. Development of moisture-absorbing and antioxidant active packaging film based on poly (vinyl alcohol) incorporated with green tea extract and its effect on the quality of dried eel. *J. Food Process. Preserv.* **2018**, *42*, e13374. [CrossRef]
14. Hong, H.Q.; Liao, H.Y.; Chen, S.J.; Zhang, H.Y. Facile method to prepare self-healable PVA hydrogels with high water stability. *Mater. Lett.* **2014**, *122*, 227–229. [CrossRef]
15. Toyoda, N.; Yamamoto, T. Dispersion of carbon nanofibers modified with polymer colloids to enhance mechanical properties of PVA nanocomposite film. *Colloid Surf. A* **2018**, *556*, 248–252. [CrossRef]
16. Yu, Z.; Li, B.Q.; Chu, J.Y.; Zhang, P.F. Silica in situ enhanced PVA/chitosan biodegradable films for food packages. *Carbohydr. Polym.* **2018**, *184*, 214–220. [CrossRef]
17. Lan, W.J.; Zhang, R.; Ahmed, S.; Qin, W.; Liu, Y.W. Effects of various antimicrobial polyvinyl alcohol/tea polyphenol composite films on the shelf life of packaged strawberries. *LWT Food. Sci. Technol.* **2019**, *113*, 108297. [CrossRef]
18. Rowe, A.A.; Tajvidi, M.; Gardner, D.J. Thermal stability of cellulose nanomaterials and their composites with polyvinyl alcohol (PVA). *J. Therm. Anal. Calorim.* **2016**, *126*, 1371–1386. [CrossRef]
19. Han, J.Q.; Yue, Y.Y.; Wu, Q.L.; Huang, C.B.; Pan, H.; Zhan, X.X.; Mei, C.T.; Xu, X.W. Effects of nanocellulose on the structure and properties of poly (vinyl alcohol)-borax hybrid foams. *Cellulose* **2017**, *24*, 4433–4448. [CrossRef]
20. Dufresne, A. Nanocellulose: A new ageless bionanomaterial. *Mater. Today* **2013**, *16*, 220–227. [CrossRef]
21. Klemm, D.; Kramer, F.; Moritz, S.; Lindstrom, T.; Ankerfors, M.; Gray, D.; Dorris, A. Nanocelluloses: A new family of nature-based materials. *Angew. Chem. Int. Ed.* **2011**, *50*, 5438–5466. [CrossRef]
22. Wang, H.Y.; Chen, C.C.; Fang, L.; Li, S.Y.; Chen, N.; Pang, J.W.; Li, D.G. Effect of delignification technique on the ease of fibrillation of cellulose II nanofibers from wood. *Cellulose* **2018**, *25*, 7003–7015. [CrossRef]
23. Phanthong, P.; Reubroycharoen, P.; Hao, X.G.; Xu, G.W.; Abudula, A.; Guan, G.Q. Nanocellulose: Extraction and application. *Carbon Resour. Convers.* **2018**, *1*, 32–43. [CrossRef]
24. Wang, H.Y.; Wu, T.T.; Wang, X.X.; Cheng, X.D.; Chen, N.; Li, D.G. Effect of ethylenediamine treatment on cellulose nanofibers and the formation of high-strength hydrogels. *Bioresources* **2019**, *14*, 1141–1156.
25. Chen, C.C.; Li, D.G.; Abe, K.; Yano, H. Formation of high strength double-network gels from cellulose nanofiber/polyacrylamide via NaOH gelation treatment. *Cellulose* **2018**, *25*, 5089–5097. [CrossRef]
26. Ching, Y.C.; Rahman, A.; Ching, K.Y.; Sukiman, N.L.; Cheng, H.C. Preparation and characterization of polyvinyl alcohol-based composite reinforced with nanocellulose and nanosilica. *Bioresources* **2015**, *10*, 3364–3377. [CrossRef]
27. Hietala, M.; Sain, S.; Oksman, K. Highly redispersible sugar beet nanofibrils as reinforcement in bionanocomposites. *Cellulose* **2017**, *24*, 2177–2189. [CrossRef]
28. Acharya, A. Luminescent magnetic quantum dots for in vitro/in vivo imaging and applications in therapeutics. *J. Nanosci. Nanotechnol.* **2013**, *13*, 3753–3768. [CrossRef]
29. Shi, Y.X.; Liu, X.; Wang, M.; Huang, J.B.; Jiang, X.Q.; Pang, J.H.; Xu, F.; Zhang, X.M. Synthesis of N-doped carbon quantum dots from bio-waste lignin for selective irons detection and cellular imaging. *Int. J. Biol. Macromol.* **2019**, *128*, 537–545. [CrossRef]
30. Lei, C.W.; Hsieh, M.L.; Liu, W.R. A facile approach to synthesize carbon quantum dots with pH-dependent properties. *Dyes Pigments* **2019**, *169*, 73–80. [CrossRef]
31. Zhao, J.X.; Liu, C.; Li, Y.C.; Liang, J.Y.; Liu, J.Y.; Qian, T.H.; Ding, J.J.; Cao, Y.C. Preparation of carbon quantum dots based high photostability luminescent membranes. *Luminescence* **2017**, *32*, 625–630. [CrossRef]

32. Zhang, L.G.; Wang, Y.; Liu, W.; Ni, Y.H.; Hou, Q.X. Corncob residues as carbon quantum dots sources and their application in detection of metal ions. *Ind. Crops Prod.* **2019**, *133*, 18–25. [CrossRef]
33. Yan, J.Y.; Hou, S.L.; Yu, Y.Z.; Qiao, Y.; Xiao, T.Q.; Mei, Y.; Zhang, Z.J.; Wang, B.; Huang, C.C.; Liu, C.H.; et al. The effect of surface charge on the cytotoxicity and uptake of carbon quantum dots in human umbilical cord derived mesenchymal stem cells. *Colloid Surf. B* **2018**, *171*, 241–249. [CrossRef] [PubMed]
34. Atchudan, R.; Edison, T.N.J.I.; Aseer, K.R.; Perumal, S.; Karthik, N.; Lee, Y.R. Highly fluorescent nitrogen-doped carbon dots derived from Phyllanthus acidus utilized as a fluorescent probe for label-free selective detection of Fe^{3+} ions, live cell imaging and fluorescent ink. *Biosens. Bioelectron.* **2018**, *99*, 303–311. [CrossRef] [PubMed]
35. Chandra, S.; Singh, V.K.; Yadav, P.K.; Bano, D.; Kumar, V.; Pandey, V.K.; Talat, M.; Hasan, S.H. Mustard seeds derived fluorescent carbon quantum dots and their peroxidase-like activity for colorimetric detection of H_2O_2 and ascorbic acid in a real sample. *Anal. Chim. Acta* **2018**, *1054*, 145–156. [CrossRef] [PubMed]
36. Arul, V.; Sethuraman, M.G. Facile green synthesis of fluorescent N-doped carbon dots from Actinidia deliciosa and their catalytic activity and cytotoxicity applications. *Opt. Mater.* **2018**, *78*, 181–190. [CrossRef]
37. Atchudan, R.; Edison, T.N.J.I.; Lee, Y.R. Nitrogen-doped carbon dots originating from unripe peach for fluorescent bioimaging and electrocatalytic oxygen reduction reaction. *J. Colloid Interface Sci.* **2016**, *482*, 8–18. [CrossRef] [PubMed]
38. Wang, X.; Yang, P.; Feng, Q.; Meng, T.T.; Wei, J.; Xu, C.Y.; Han, J.Q. Green Preparation of Fluorescent Carbon Quantum Dots from Cyanobacteria for Biological Imaging. *Polymers* **2019**, *11*, 616. [CrossRef]
39. Dong, L.; Xiong, Z.R.; Liu, X.D.; Sheng, D.K.; Zhou, Y.; Yang, Y.M. Synthesis of carbon quantum dots to fabricate ultraviolet-shielding poly(vinylidene fluoride) films. *J. Appl. Polym. Sci.* **2019**, *136*, 47555. [CrossRef]
40. El-Shamy, A.G. New free-standing and flexible PVA/Carbon quantum dots (CQDs) nanocomposite films with promising power factor and thermoelectric power applications. *Mater. Sci. Semicond. Proc.* **2019**, *100*, 245–254. [CrossRef]
41. Vandarkuzhali, S.A.A.; Jeyalakshmi, V.; Sivaraman, G.; Singaravadivel, S.; Krishnamurthy, K.R.; Viswanathan, B. Highly fluorescent carbon dots from Pseudo-stem of banana plant: Applications as nanosensor and bio-imaging agents. *Sens. Actuator B Chem.* **2017**, *252*, 894–900. [CrossRef]
42. Bao, R.Q.; Chen, Z.Y.; Zhao, Z.W.; Sun, X.; Zhang, J.Y.; Hou, L.R.; Yuan, C.Z. Green and facile synthesis of nitrogen and phosphorus co-doped carbon quantum dots towards fluorescent ink and sensing applications. *J. Nanomater.* **2018**, *8*, 386. [CrossRef] [PubMed]
43. Xue, B.L.; Yang, Y.; Sun, Y.C.; Fan, J.S.; Li, X.P.; Zhang, Z. Photoluminescent lignin hybridized carbon quantum dots composites for bioimaging applications. *Int. J. Biol. Macromol.* **2019**, *122*, 954–961. [CrossRef] [PubMed]
44. Temerov, F.; Belyaev, A.; Ankudze, B.; Pakkanen, T.T. Preparation and photoluminescence properties of graphene quantum dots by decomposition of graphene-encapsulated metal nanoparticles derived from Kraft lignin and transition metal salts. *J. Lumin.* **2019**, *206*, 403–411. [CrossRef]
45. Zhou, L.F.; Qiao, M.; Zhang, L.; Sun, L.; Zhang, Y.; Liu, W.W. Green and efficient synthesis of carbon quantum dots and their luminescent properties. *J. Lumin.* **2019**, *206*, 158–163. [CrossRef]
46. Eda, G.; Lin, Y.Y.; Mattevi, C.; Yamaguchi, H.; Chen, H.A.; Chen, I.S.; Chen, C.W.; Chhowalla, M. Blue photoluminescence from chemically derived graphene oxide. *Adv. Mater.* **2010**, *22*, 505–509. [CrossRef] [PubMed]
47. Sun, Y.P.; Zhou, B.; Lin, Y.; Wang, W.; Fernando, K.A.S.; Pathak, P.; Meziani, M.J.; Harruff, B.A.; Wang, X.; Wang, H.F.; et al. Quantum-sized carbon dots for bright and colorful photoluminescence. *J. Am. Chem. Soc.* **2006**, *128*, 7756–7757. [CrossRef]
48. Pan, D.Y.; Zhang, J.C.; Li, Z.; Wu, M.H. Hydrothermal route for cutting graphene sheets into blue-luminescent graphene quantum dots. *Adv. Mater.* **2010**, *22*, 734–738. [CrossRef]
49. Wang, C.J.; Wang, Y.B.; Shi, H.X.; Yan, Y.J.; Liu, E.Z.; Hu, X.Y.; Fan, J. A strong blue fluorescent nanoprobe for highly sensitive and selective detection of mercury (II) based on sulfur doped carbon quantum dots. *Mater. Chem. Phys.* **2019**, *232*, 145–151. [CrossRef]
50. Hoang, Q.B.; Mai, V.T.; Nguyen, D.K.; Truong, D.Q.; Mai, X.D. Crosslinking induced photoluminescence quenching in polyvinyl alcohol-carbon quantum dot composite. *Mater. Today Chem.* **2019**, *12*, 166–172. [CrossRef]

51. Dulkeith, E.; Morteani, A.C.; Niedereichholz, T.; Klar, T.A.; Feldmann, J.; Levi, S.A.; van Veggel, F.C.J.M.; Reinhoudt, D.N.; Moller, M.; Gittins, D.I. Fluorescence quenching of dye molecules near gold nanoparticles: Radiative and nonradiative effects. *Phys. Rev. Lett.* **2002**, *89*, 203002. [CrossRef]
52. Liu, T.; Li, N.; Dong, J.X.; Luo, H.Q.; Li, N.B. Fluorescence detection of mercury ions and cysteine based on magnesium and nitrogen co-doped carbon quantum dots and IMPLICATION logic gate operation. *Sens. Actuators B Chem.* **2016**, *231*, 147–153. [CrossRef]
53. Zhao, Q.L.; Zhang, Z.L.; Huang, B.H.; Peng, J.; Zhang, M.; Pang, D.W. Facile preparation of low cytotoxicity fluorescent carbon nanocrystals by electrooxidation of graphite. *Chem. Commun.* **2008**, *41*, 5116–5118. [CrossRef] [PubMed]
54. Htun, M.T. Characterization of high-density polyethylene using laser-induced fluorescence (LIF). *J. Polym. Res.* **2012**, *19*, 9823. [CrossRef]
55. Yang, L.; Huang, B.Q.; Wei, X.F.; Zhang, W.; Wang, D.D. The Research on Fluorescence Intensity Attenuation of UV Fluorescent Inkjet Ink. In Proceedings of the 29th International Conference on Digital Printing Technologies (NIP29)/Digital Fabrication 2013, Seattle, WA, USA, 29 September–3 October 2013.
56. Wang, Q.Q.; Zhu, J.Y.; Gleisner, R.; Kuster, T.A.; Baxa, U.; McNeil, S.E. Morphological development of cellulose fibrils of a bleached eucalyptus pulp by mechanical fibrillation. *Cellulose* **2014**, *19*, 1631–1643. [CrossRef]
57. Jing, X.; Li, H.; Mi, H.Y.; Liu, Y.J.; Feng, P.Y.; Tan, Y.M.; Turng, L.S. Highly transparent, stretchable, and rapid self-healing polyvinyl alcohol/cellulose nanofibril hydrogel sensors for sensitive pressure sensing and human motion detection. *Sens. Actuators B Chem.* **2019**, *252*, 159–167. [CrossRef]
58. Wang, Y.Q.; Xue, Y.A.; Wang, J.H.; Zhu, Y.P.; Wang, X.; Zhang, X.H.; Zhu, Y.; Liao, J.W.; Li, X.N.; Wu, X.G.; et al. Biocompatible and photoluminescent carbon dots/hydroxyapatite/PVA dual-network composite hydrogel scaffold and their properties. *J. Polym. Res.* **2019**, *26*, 248. [CrossRef]
59. Yan, Z.D.; Sun, L.D.; Hu, C.G.; Hu, X.T.; Zeppenfeld, P. Factors influencing the ability of fluorescence emission and fluorescence quenching experimental research. *Spectrosc. Spect. Anal.* **2012**, *32*, 2718–2721.
60. Ma, X.T.; Li, S.R.; Hessel, V.; Lin, L.L.; Meskers, S.; Gallucci, F. Synthesis of luminescent carbon quantum dots by microplasma process. *Chem. Eng. Process. Process Intensif.* **2019**, *140*, 29–35. [CrossRef]
61. Mahmud, H.N.M.E.; Kassim, A.; Zainal, Z.; Yunus, W.M.M. Fourier transform infrared study of polypyrrole-poly (vinyl alcohol) conducting polymer composite films: Evidence of film formation and characterization. *J. Appl. Polym. Sci.* **2006**, *100*, 4107–4113. [CrossRef]
62. Saikia, M.; Hower, J.C.; Das, T.; Dutta, T.; Saikia, B.K. Feasibility study of preparation of carbon quantum dots from Pennsylvania anthracite and Kentucky bituminous coals. *Fuel* **2019**, *243*, 433–440. [CrossRef]
63. Li, D.P.; Wu, Z.J.; Hang, C.H.; Chen, L.J.; Zhang, Y.L.; Ne, Y.X. Analysis of the Character of Film Decomposition of Methyl Methacrylate (MMA) Coated Urea by Infrared Spectrum. *Spectrosc. Spect. Anal.* **2012**, *32*, 635–641.
64. Behera, B.; Das, P.K. Blue- and Red-Shifting Hydrogen Bonding: A Gas Phase FTIR and Ab Initio Study of RR' CO center dot center dot center dot DCCl3 and RR' S center dot center dot center dot DCCl3 Complexes. *J. Phys. Chem. A* **2018**, *122*, 4481–4489. [CrossRef] [PubMed]
65. El-Shamy, A.G. Novel conducting PVA/Carbon quantum dots (CQDs) nanocomposite for high anti-electromagnetic wave performance. *J. Alloy Compd.* **2019**, *810*, 151940. [CrossRef]
66. Hertmanowski, R.; Biadasz, A.; Martynski, T.; Bauman, D. Optical spectroscopy study of some 3,4,9,10-tetra-(n-alkoxy-carbonyl)-perylenes in Langmuir-Blodgett films. *J. Mol. Struct.* **2003**, *646*, 25–33. [CrossRef]
67. Baraker, B.M.; Lobo, B. Spectroscopic Analysis of CdCl2 doped PVA-PVP Blend Films. *Can. J. Phys.* **2017**, *95*, 738–747. [CrossRef]
68. Lee, J.H.; Ko, K.H.; Park, B.O. Electrical and optical properties of ZnO transparent conducting films by the sol–gel method. *J. Cryst. Growth* **2003**, *247*, 119–125. [CrossRef]
69. Baraker, B.M.; Lobo, B. UV irradiation induced microstructural changes in CdCl2 doped PVA–PVP blend. *J. Mater. Sci. Mater. Electron.* **2018**, *29*, 4106–4121. [CrossRef]
70. Riaz, R.; Ali, M.; Maiyalagan, T.; Anjum, A.S.; Lee, S.; Ko, M.J.; Jeong, S.H. Dye-sensitized solar cell (DSSC) coated with energy down shift layer of nitrogen-doped carbon quantum dots (N-CQDs) for enhanced current density and stability. *Appl. Surf. Sci.* **2019**, *483*, 425–431. [CrossRef]
71. Liu, D.G.; Sun, X.; Tian, H.F.; Maiti, S.; Ma, Z.S. Effects of cellulose nanofibrils on the structure and properties on PVA nanocomposites. *Cellulose* **2013**, *20*, 2981–2989. [CrossRef]

72. Yang, X.M.; Shang, S.M.; Li, L.A. Layer-structured poly (vinyl alcohol)/graphene oxide nanocomposites with improved thermal and mechanical properties. *J. Appl. Polym. Sci.* **2011**, *120*, 1355–1360. [CrossRef]
73. Wu, Y.; Tang, Q.W.; Yang, F.; Xu, L.; Wang, X.H.; Zhang, J.L. Mechanical and thermal properties of rice straw cellulose nanofibrils-enhanced polyvinyl alcohol films using freezing-and-thawing cycle method. *Cellulose* **2019**, *26*, 3193–3204. [CrossRef]
74. Kumar, S.V.; George, J.; Sajeevkumar, V.A. PVA Based Ternary Nanocomposites with Enhanced Properties Prepared by Using a Combination of Rice Starch Nanocrystals and Silver Nanoparticles. *J. Polym. Environ.* **2018**, *26*, 3117–3127. [CrossRef]
75. Sehaqui, H.; Zimmermann, T.; Tingaut, P. Hydrophobic cellulose nanopaper through a mild esterification procedure. *Cellulose* **2014**, *21*, 367–382. [CrossRef]
76. Xu, M.H.; Zhang, W.; Yang, Z.; Yu, F.; Ma, Y.J.; Hu, N.T.; He, D.N.; Liang, Q.; Su, Y.J.; Zhang, Y.F. One-pot liquid-phase exfoliation from graphite to graphene with carbon quantum dots. *Nanoscale* **2015**, *7*, 10527–10537. [CrossRef]
77. Yin, J.; Deng, B.L. Polymer-matrix nanocomposite membranes for water treatment. *J. Membr. Sci.* **2015**, *479*, 256–275. [CrossRef]
78. Oza, G.; Ravichandran, M.; Merupo, V.I.; Shinde, S.; Mewada, A.; Ramirez, J.T.; Velumani, S.; Sharon, M.; Sharon, M. Camphor-mediated synthesis of carbon nanoparticles, graphitic shell encapsulated carbon nanocubes and carbon dots for bioimaging. *Sci. Rep. U. K.* **2016**, *6*, 21286. [CrossRef]
79. Wang, Z.; Zhao, S.J.; Zhang, W.; Qi, C.S.; Zhang, S.F.; Li, J.Z. Bio-inspired cellulose nanofiber-reinforced soy protein resin adhesives with dopamine-induced codeposition of "water-resistant" interphases. *Appl. Surf. Sci.* **2019**, *478*, 441–450. [CrossRef]
80. Shahbazi, M.; Rajabzadeh, G.; Rafe, A.; Ettelaie, R.; Ahmadi, S.J. Physico-mechanical and structural characteristics of blend film of poly (vinyl alcohol) with biodegradable polymers as affected by disorderto-order conformational transition. *Food Hydrocoll.* **2017**, *71*, 259–269. [CrossRef]
81. Gai, W.X.; Zhao, D.L.; Chung, T.S. Thin film nanocomposite hollow fiber membranes comprising Na+-functionalized carbon quantum dots for brackish water desalination. *Water Res.* **2019**, *154*, 54–61. [CrossRef]
82. Tang, C.Y.Y.; Kwon, Y.N.; Leckie, J.O. Effect of membrane chemistry and coating layer on physiochemical properties of thin film composite polyamide RO and NF membranes I. FTIR and XPS characterization of polyamide and coating layer chemistry. *Desalination* **2009**, *242*, 149–167. [CrossRef]
83. Priezjev, N.V.; Darhuber, A.A.; Troian, S.M. Slip behavior in liquid films on surfaces of patterned wettability: Comparison between continuum and molecular dynamics simulations. *Funct. Mater. Lett.* **2005**, *8*, 1–11. [CrossRef] [PubMed]

© 2019 by the authors. Licensee MDPI, Basel, Switzerland. This article is an open access article distributed under the terms and conditions of the Creative Commons Attribution (CC BY) license (http://creativecommons.org/licenses/by/4.0/).

Article

Preparation and Characterization of Dyed Corn Straw by Acid Red GR and Active Brilliant X-3B Dyes

Yanchen Li [1,2], Beibei Wang [1,2], Yingni Yang [1,2], Yi Liu [1,2,*] and Hongwu Guo [1,*]

1. Key Laboratory of Wooden Material Science and Application, College of Material science And Technology, Beijing Forestry University, Beijing 100083, China; lyc100083@163.com (Y.L.); SyliaWong@163.com (B.W.); yyingni_0922@163.com (Y.Y.)
2. Key Laboratory of Wood Science and Engineering, College of Material science And Technology, Beijing Forestry University, Beijing 100083, China
* Correspondence: liuyi.zhongguo@163.com (Y.L.); ghw5052@163.com (H.G.)

Received: 2 September 2019; Accepted: 21 October 2019; Published: 24 October 2019

Abstract: Corn straw is a kind of biomass material with huge reserves, which can be used in plate processing, handicraft manufacturing, indoor decoration, and other fields. To investigate the dyeing mechanism of corn straw with different dyes, corn straw was pretreated and dyed with Acid Red GR and Brilliant Red X-3B. The dyeing properties and light resistance of the two dyes were analyzed by dyeing rate, photochromaticity, FTIR, SEM, and water-washing firmness. The results showed that the structure and stability of the dyes were the main factors which influenced fading. A bleaching pretreatment could remove the waxiness of the corn straw epidermis and increase the porosity on the surface of the straw, which accelerated the photochromic coloring of the corn straw skin. The corn straw dyed with both dyes had good light resistance, but the straw dyed with Reactive Brilliant Red X-3B had higher dyeing rate, brighter color, and higher photochromaticity than the straw dyed with Acid Red GR. FTIR and water-washing firmness showed that Acid Red GR mainly bound to lignin, while Reactive Brilliant Red X-3B mainly bound to cellulose, hemicellulose, and lignin in corn straw through covalent bonds, which increased the coloring rate.

Keywords: corn straw; pretreatment; dyeing; chemical structure

---

1. Introduction

With changes in the quantity and structure of the global forest, the available timber resources for human production have been significantly reduced, and crop straws with a fast green growth cycle have gradually received attention [1–4]. Corn straw is a natural and renewable biomass resource, mainly composed of cellulose, hemicellulose, and lignin [5]. It has excellent physical and mechanical properties and can even be used as a substitute for wood through some innovative technologies for furniture, interior decoration, and artistic creation. The color of straw is an important indicator of its value [6,7]. However, the waxy layer on the surface of corn straw makes it difficult to be dyed. There is not enough research on how dye molecules bind to corn straw. Furthermore, dyed straw is an excellent light absorber due to the presence of chromophores as phenolic hydroxyl groups, aromatic skeleton, double bonds, and carbonyl groups in the lignin and extractives molecules it contains. These colored unsaturated compounds may cause changes in the surface color and produce new aromatic and other chromophores after irradiation [8–12]. The dyes are also predisposed to fade through either oxidation or reduction reactions when stimulated by light radiation [13,14].

At present, research of dyed corn straw and photochromism is not systematic. In order to compare how different dyes bind to corn straw and how the surface color changes after irradiation, this study drew lessons from the theory and process of wood dyeing to dye corn straw, selecting acid dyes and reactive dyes which are commonly used for wood dyeing. Acid Red GR and Active Brilliant

Red X-3B were selected as dyeing agents to systematically explore the dyeing of corn straw after pretreatment. Dyeing rate, photochromism, FTIR, SEM, and water-washing firmness were used to analyze the dyeing, light resistance, chemical structure, microstructure, and water-washing firmness of corn straw. This study has a great practical significance to optimize the decorative color of corn straw and obtain a rich color system.

2. Materials and Methods

2.1. Materials

Corn straw epidermis: the straw was selected in Fengning, Hebei, as shown in Figure 1. Corn straw with uniform texture and no defect on the surface was cut into 50 mm × 10 mm × 0.1 mm pieces, which were put in black plastic bags for later utilization.

Figure 1. Corn straw epidermis.

Dyes: Acid Red GR and Reactive Brilliant Red X-3B were provided by the Second Plant Dye Chemical Company (Tianjin, China). Their molecular structure formulas are shown in Figure 2.

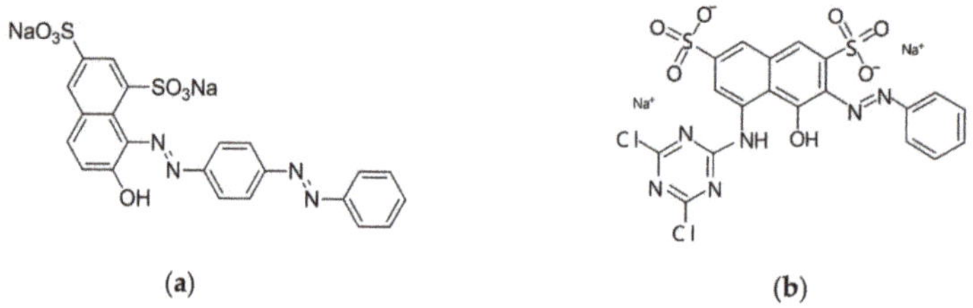

(a)   (b)

Figure 2. Molecular structures of Acid Red GR (**a**) and Reactive Brilliant Red X-3B (**b**).

Reagents: sodium chloride (NaCl), 30% dilute sulfuric acid (H_2SO_4), soda ash (Na_2CO_3), sodium hydroxide (NaOH), hydrogen peroxide (H_2O_2).

Instruments: constant-temperature water bath (HH-4, Kai hang, Shanghai, China); UV–visible spectrophotometer (Genesys10 UV-335903, Thermo Scientific Spectronic, Waltham, MA, USA); automatic spectrophotometer (MF-FS97Pro, Med Future, Shandong, China); solar weather tester (MQ-UV-2, ZKMQ, Tianjin, China); Fourier infrared spectrometer (Nicolet Nexus 670, Waltham, MA, USA); colorimeter (WR10, FRU, China); fourier transform infrared spectrometry(Avatar-380, Thermo-Nicolet, Waltham, MA, USA); scanning electron microscope (Sigma500, ZEISS, Jena, Germany).

2.2. Methods

2.2.1. Pretreatment

The specimens were immersed in H_2O_2 with a solute mass fraction of 27%, avoiding superposition between the specimens. The solution was heated to 50 °C, and the samples were taken out after 30 min. The surface residue was washed with deionized water at room temperature. After air-drying, the specimens were immersed in NaOH solution with solute mass fraction of 0.1% at room temperature, avoiding superposition among the specimens. After 30 min of treatment, the specimens were air-dried to a moisture content of 8%.

2.2.2. Dyeing Method

Acid Red GR is a cationic dye existing in the form of sodium sulfonate in water, while Reactive Brilliant Red X-3B exists in the form of an anion. The cellulose of corn straw is also charged in aqueous solution as an anion, so there is a repulsive force between Reactive Brilliant Red X-3B and straw cellulose, resulting in low dye utilization. By adding an inorganic salt, NaCl, a large amount of Na^+ can promote the activity of the dye molecules in water to achieve the purpose of dyeing. Furthermore, NaCl is also good at increasing the coloring rate of the dye molecules. In order to ensure accurate test results, repeated each dyeing test three times.

(1) Acid Red GR staining

Step 1: An atmospheric-pressure dip-dye method was used for wood dyeing. A 0.5% Acid Red GR solution in distilled water (w/v) of at 25 °C was prepared. The pH value of the solution was adjusted to 4.0 by addition of 10% H_2SO_4 (w/w). Gaps were left between the samples to avoid overlapping and to ensure dye homogeneity.

Step 2: A bath ratio of 1:20 ($V_{straw}/V_{dye\ solution}$) was used, and the straw was placed into an 50 °C electrically heated thermostatic water bath dye vat for 1 h. After 15 min, a cup of 1% NaCl solution was added to assist the dyeing process. After 30 min, another cup of 1% NaCl solution was added. After dyeing, all dyed straw was washed with distilled water and then air-dried to a moisture content of 12%.

(2) Reactive Brilliant Red X-3B staining

Step 1: Refer to Acid Red GR staining, but the pH value of the solution was adjusted to 10 by addition of 10% Na_2CO_3 (w/w).

Step 2: Refer to Acid Red GR staining, but after 30 min, another cup of 1% NaCl and 1% Na_2CO_3 solution was added.

The dyeing effects of Acid Red GR and Reactive Brilliant Red X-3B on corn straw are shown in Figure 3.

Figure 3. Dyeing effects Acid Red GR (**a**) and Reactive Brilliant Red X-3B on corn straw (**b**).

2.2.3. Dyeing Rate Calculation

The maximum absorption spectrum curves of Acid Red GR and Reactive Brilliant Red X-3B were measured by a ultraviolet spectrophotometer, and the absorbance of the dye solution before and after dyeing was measured at the maximum absorption wavelength. The dye uptake rate C_t was calculated as follows:

$$C_t (\%) = (A_0 - A_1) \times 100\%/A_0 \tag{1}$$

In the formula, C_t is the dye uptake, A_0 is the absorbance before dyeing, A_1 is the absorbance after dyeing.

2.2.4. Chromaticity Value Calculation

The 1976 International Illumination Commission CIE (L*a*b*) standard colorimetric characterization system is used for color measurement in the case of small chromatic aberrations. By comparing the chromaticity values of Acid Red GR and Reactive Brilliant Red X-3B, the light resistance level of the dyed straw could be analyzed and calculated as follows:

$$\Delta E^* = [(\Delta L^*)^2 + (\Delta a^*)^2 + (\Delta b^*)^2]^{1/2} \tag{2}$$

where ΔE^* indicates the degree of total color change, L^* stands for lightness from 0 (for black) to 100 (for white), a^* represents the red-green chromaticity coordinates (+a^* is for red, −a^* for green), b^* denotes the yellow-blue chromaticity index (+b^* is for yellow,−b^* for blue), ΔL^*, Δa^*, and Δb^* are the differences of the values of L^*, a^*, and b^* before and after treatment, respectively. The unit of chromaticity value is represented by "NBS". When the value of ΔE^* is 1NBS, it is equivalent about 5 times to the visual recognition threshold. There is approximately such a correspondence between color difference and visual perception: 0.0~0.50, the feel of visual perception is gentle; 0.5~1.50, the feel of visual perception is slight; 1.50~3.0, the feel of visual perception is noticeable; 3.0~6.0, the feel of visual perception is appreciable; 6.0 or more, the feel of visual perception is violent.

The test samples were prepared by laminating the straw which was dyed by Acid Red GR and Reactive Brilliant Red X-3B. Each test piece was placed in a colorimeter (WR10, FRU, China) using a D65 standard illuminant and 10° standard observer for 50 h ultraviolet-lamp irradiation test. The chromatic value ΔE^* of the samples were measured after irradiation at each wavelengths of light source for 0, 1, 2, 5, 10, 20, 30, and 50 h, respectively.

2.2.5. Surface Chemical Structure Analysis

The chemical composition and structure of corn stalks are complex. During the dyeing process, the dye molecules have a large volume, which makes it difficult for them to pass through the microporous structure of the straw, so they are easily adsorbed by the straw components. Therefore, the presence of dyes in straw can be utilized. Attenuated total reflectance–Fourier-transform infrared spectrometry (ATR–FTIR) was performed for the characterization of the surface chemical structure of the dyed straw. All samples were placed on the diamond crystal of the ATR–FTIR spectrometer (Avatar-380, Thermo-Nicolet, Waltham, MA, USA), and the spectra were collected in transmittance mode by 64 scans in the range of 4000–400 cm^{-1} at a resolution of 8 cm^{-1}.

2.2.6. Microstructure Analysis

The untreated corn straw, the straw treated by the pretreatment process, and the straw dyed with the two dyes were cut out transversely to make slices, and after gold plating by an ion-coating machine, the images were observed by SEM (Sigma 500, ZEISS, Jena, Germany).

2.2.7. Water-Washing Firmness

Straw dyed by Acid Red GR and Reactive Brilliant Red X-3B was made into straw panels, and the surface chromaticity value ΔE^* was measured by an automatic spectrophotometer. Then, the straw was placed in a 1000 ml beaker, and the beaker was placed in an electrically heated thermostatic water bath and soaked at 65 °C for 2 h. After soaking, the straw was taken out and air-dried. Then, the surface chromaticity value ΔE^* was measured.

3. Results

3.1. Dyeing Rate

The average dyeing rates of Acid Red GR and Reactive Brilliant Red X-3B were calculated in three tests, as shown in Table 1.

Table 1. Dyeing rate of Acid Red GR and Reactive Brilliant Red X-3B. A_0: absorbance before dyeing, A_1: absorbance after dyeing.

	Acid Red GR				
	Test 1	Test 2	Test 3	AVG	Dyeing Rate/%
A0	0.716	0.715	0.721	0.717	29
A1	0.497	0.512	0.516	0.508	
	Reactive Brilliant Red X-3B				
	Test 1	Test 2	Test 3	AVG	Dyeing Rate/%
A0	0.789	0.785	0.781	0.785	46
A1	0.420	0.432	0.424	0.425	

The dyeing rate of Reactive Brilliant Red X-3B was 46%, and that of Acid Red GR was 29%. The dyeing rate of Reactive Brilliant Red X-3B was much higher than that of Acid Red GR, and both dyes all had good dyeing ability acting on corn straw. According to studies on wood dyeing with acid dyes and reactive dyes, acid dyes mainly act on lignin, and the binding mode of dyes to corn stalks is physical adsorption [15,16]. However, the epidermis of corn straw contains a very high amount of cellulose, so the combination of acid dye and straw is limited. Reactive dyes mainly act on cellulose and hemicellulose and react with them to form covalently bound reactive groups [17,18]. When NaCl was added to the reactive dye solution, the negative-charge repulsion between the straw fiber surface and the dye anions in the solution was weakened, greatly improving the opportunity for dye molecules to approach the straw fiber. However, when NaCl was added into the acid dye solution, it only improved the fastness of dye molecules adsorption on the straw epidermis but did not promote the covalent binding between dye molecules and straw components. Therefore, Reactive Brilliant Red X-3B had higher dyeing rate of corn straw epidermis.

3.2. Chromatic Value

In the accelerated aging process of light irradiation, the brightness index L^* curve of the two dyes dyeing corn straw is shown in Figure 4a, the curve of the chromaticity index a^*, changing with the illumination time, is shown in Figure 4b, the curve of the chromaticity index b^*, changing with the illumination time, is shown in Figure 4c, and the curve of value of chromatism ΔE^* is shown in Figure 4d.

In Figure 4a, the measured data showed that the change of the brightness index L^* of the two dyed straw was basically the same, and both curves showed an increasing trend. The brightness index L^* of Reactive Brilliant Red X-3B-dyed corn straw showed a slower increase than that of corn straw dyed with Acid Red GR, indicating that the brightness change of corn stalk dyed with Reactive Brilliant Red X-3B was slighter after irradiation with ultraviolet light. The curves of dyed straw chromaticity index a^* showed a decreasing trend, while the curves of the chromaticity index b^* adopted first and then rose, shown in Figure 4b,c, corresponding to the incremental reflection of whiteness, indicating the discoloration of the surface of straw.

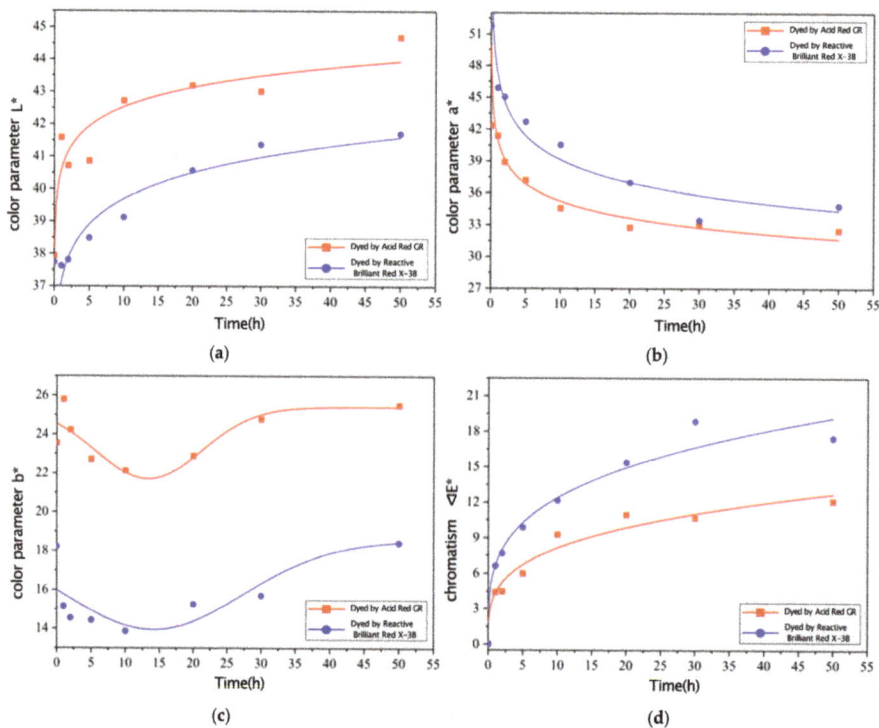

Figure 4. Color comparison of corn straw dyed with the two dyes during light irradiation.

It can be seen from Figure 4d that the value of chromatism ΔE^* of each group increased with the extension of the irradiation time. The dyed straw underwent significant fading. After 50 h of light irradiation, the value of chromatism ΔE^* of the two test pieces exceeded to12NBS, and the maximum was 17.43NBS. The chromophore system of the dyed straw consisted of the dye, lignin, and extractives present in the straw, which combined the chromophore group and the auxiliary color group in many forms and absorbed the visible light [19,20]. In the early stage of light irradiation (the first 30 h), the main reason for the fading of dyed straw is that the dyed material absorbs light and exposes the unsaturated groups in the dye and wood extract, causing photooxidation and photodegradation and bringing irreversible changes and destruction of the dyeing specimen. After 30 h of light irradiation, the chemical structure of the surface of the dyed straw and the active groups in the chromogenic system were basically deteriorated, and other stable groups were formed by the photooxidation and photodegradation. The dyed straw entered a slow deterioration stage, and the color gradually became lighter.

After 50 h of light irradiation, the value of chromatism ΔE^* (6.61NBS~17.43NBS) of the straw dyed with Reactive Brilliant Red was significantly larger than that of the straw dyed with Acid Red GR (4.39NBS~12.09NBS). This is because Acid Red GR combined with straw by physical adsorption. The dye molecules adsorbed on the cell wall of the straw were the first to undergo photo-discoloration when light was irradiated. The Acid Red GR is an azo dye with good light resistance [21]. In addition, acid dyes mainly dye lignin in straw components, and photodegradation of lignin is the main cause of wood color fading. The corn straw which was dyed with a reactive dye had the characteristics of high dyeing rate and poor light resistance, maybe because the surface wax of the straw was removed, and the fiber structure of the straw epidermis became weak after pretreatment. The structure of the straw is mainly composed of cellulose and hemicellulose, and the inner core is mainly composed of lignin.

After removing the surface wax, the reactive dye could easily bind to the surface fibers [19]. However, the pretreatment process promoted the combination of reactive dyes and surface fibers, and the fiber structure became fragile. With the prolongation of the light time, the fiber structure on the surface of the straw gradually broke, so the light resistance of the straw dyed with the reactive dye was not as good as that dyed with the acid dye. The above parameters are shown in Table 2.

Table 2. Photochromic parameters of Acid Red GR and Reactive Brilliant Red X-3B.

Acid Red GR							
Time(h)	L^*	a^*	b^*	ΔL^*	Δa^*	Δb^*	ΔE^*
0	37.92	42.28	23.52	0	0	0	0
1	41.57	41.34	25.77	3.65	−0.94	2.25	4.39
2	40.7	38.88	24.21	2.78	−3.4	0.69	4.45
5	40.85	37.16	22.69	2.93	−5.12	−0.83	5.96
10	42.7	34.51	22.11	4.78	−7.77	−1.41	9.24
20	43.18	32.73	22.87	5.26	−9.55	−0.65	10.92
30	43	32.97	24.77	5.08	−9.31	1.25	10.67
50	44.68	32.45	25.49	6.76	−9.83	1.97	12.09
Reactive Brilliant Red X-3B							
Time(h)	L^*	a^*	b^*	ΔL^*	Δa^*	Δb^*	ΔE^*
0	37.73	51.74	18.21	0	0	0	0
1	37.62	45.89	15.14	−0.11	−5.85	−3.07	6.61
2	37.8	45.02	14.54	0.07	−6.72	−3.67	7.66
5	38.48	42.69	14.42	0.75	−9.05	−3.79	9.84
10	39.11	40.51	13.85	1.38	−11.23	−4.36	12.13
20	40.56	36.95	15.24	2.83	−14.79	−2.97	15.35
30	41.36	33.42	15.7	3.63	−18.32	−2.51	18.84
50	41.7	34.77	18.4	3.97	−16.97	0.19	17.43

3.3. Surface Chemical Structure Changes

A Fourier-transform infrared spectrometer was used to calculate the infrared spectrum data of corn straw dyed with Acid Red GR and Reactive Brilliant Red X-3B before and after illumination. The results of the analysis are shown in Figures 5 and 6.

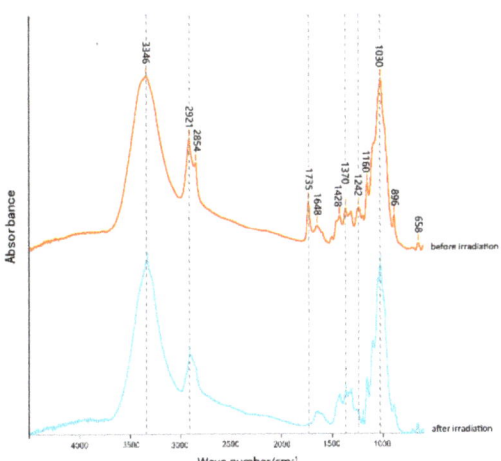

Figure 5. FTIR of corn straw dyed with Acid Red GR under irradiation.

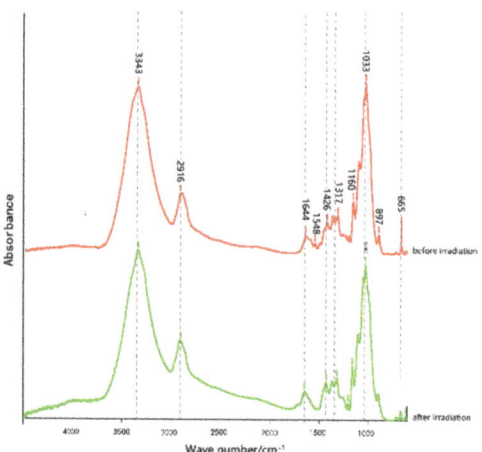

Figure 6. FTIR of corn straw dyed with Reactive Brilliant Red X-3B.

The FTIR spectra of the corn straw dyed with Acid Red GR before and after ultraviolet radiation are shown in Figure 5. Considering the peak of 1300 cm^{-1} as a reference, obvious transmittance changes can be observed in the fingerprint region from 800 cm^{-1} to 1800 cm^{-1}. The light sensitivities of the carbonyl structure (C=O) and the aromatic ring (C=C) are particularly noticeable. Under irradiation with artificial light, some unsaturated functional structures in lignin and/or dyes were degraded by photooxidation, confirmed by the considerable disappearance of C=O bonds at 1242 cm^{-1} and 1735 cm^{-1}. The absorption peaks at 2921 cm^{-1} and 2854 cm^{-1} were significantly reduced, reflecting the C–H stretching vibration in lignin [21–23]. These changes indicate that the original chromophoric system of the dyed straw was reduced, inducing an undesirable discoloration. The narrow change of the spectral band at 3346 cm^{-1} reduced the phenolic hydroxyl group –OH, which was gradually broken after 50 h of irradiation. This was probably due to the reaction of the methoxy group on the benzene ring and/or to hydrogen bond cleavage caused by the evaporation of water molecules.

Lignin is an excellent light absorber whose molecular structure can undergo free-radical reactions after irradiation and form new carbonyl and carboxyl chromophoric groups. These molecular structures include carbonyl groups, phenolic hydroxyl, methoxy, and other functional groups [8,15,20]. In this process, lignin degrades through free-radical reactions. Meanwhile, the spectral band of cellulose and hemicellulose at 1426 cm^{-1}, 2916 cm^{-1}, 1370 cm^{-1}, 896 cm^{-1}, and 1160 cm^{-1} were almost unaffected by UV irradiation. It can be deduced that Acid Red GR mainly combined with lignin. After UV irradiation, lignin (including aromatic extractives) was reduced and formed new chromophoric groups which caused surface chromatism of the straw.

The FTIR spectra of the corn straw dyed with Reactive Brilliant Red X-3B before and after ultraviolet radiation are shown in Figure 6. Compared with the straw dyed with Acid Red GR, the spectra are similar, especially, the peaks of cellulose and hemicellulose at 1426 cm^{-1}, 2916 cm^{-1}, 897 cm^{-1}, and 1160 cm^{-1} were almost unaffected by UV irradiation. The absorption peak at 1644 cm^{-1} was enhanced, indicating that the hydroxyl group –OH in cellulose could be oxidized to a ketone carbonyl group after irradiation and to an aldehyde, carbonyl group, or carboxyl group by free-radical reactions, and these chromophore groups were unstable under the applied illumination conditions. The narrow change of the spectral band at 3343 cm^{-1} represents the phenolic hydroxyl group (–OH) which was gradually broken after 50 h of irradiation. This was probably due to the reaction of the methoxy group in the benzene ring and/or to hydrogen bond cleavage caused by the evaporation of water molecules.

3.4. Surface Microstructure Changes

The appearance of untreated, pretreated, and dyed corn straw slices is shown in Figure 7.

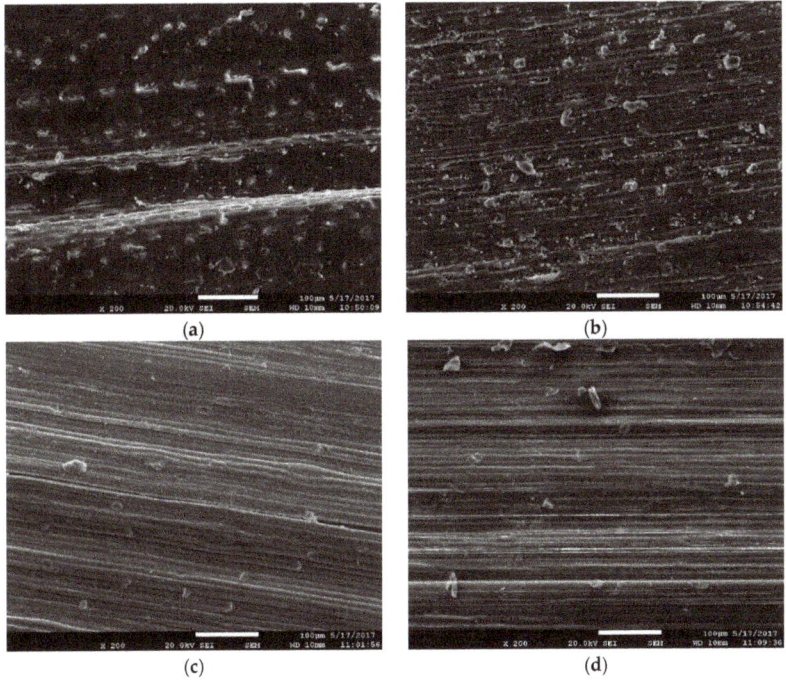

Figure 7. (a) Surface of untreated straw; (b) Surface of pretreated straw; (c) Straw surface dyed with Acid Red GR; (d) Straw surface dyed with Reactive Brilliant Red X-3B.

It can be seen from Figure 7a that the surface of the corn straw was covered by a waxy layer, whose pore distribution was relatively uniform. After pretreatment, the surface of the straw became rough and many breaks appeared, meanwhile the fiber texture was clear, as shown in Figure 7b. This is because the wax on the surface of the straw was an ester composed of fatty alcohols and higher fatty acids, which was soluble in alkali. In the presence of sodium hydroxide (NaOH), the wax and pectin were dissolved. After the barrier was removed, penetration and diffusion of the dyes were possible. It can be seen from Figure 7c,d that the surface of the straw became smoother, and the texture was clear after dyeing. It can be inferred that there was a physical or organic combination between dye's molecules and straw microregions. As for Acid Red GR, it exists in aqueous solution as ions (D− or D+), ionic micelles (nDn− or nDn+), and micelles $\{[(HD)_m nD]n-$ or $[(DA)_m n\ D]n+\}$. The electrostatic repulsion between the dye SO_3H-dissociated sulfonic acid group SO_3^- and the negatively charged region in the fiber caused aggregation and blockage of the dye molecules. As for Reactive Brilliant Red X-3B, it can chemically react with cellulose to form ester and ether bonds after the dissociation of the hydroxyl groups in cellulose and forms hydroxyl anions. By combining FTIR and SEM, the penetration or organic binding of Brilliant Red X-3B could be demonstrated, whose mechanism of reaction as shown in Figure 8. In the first step, the ionization of cellulose under alkaline condition and form cellulose anions. In the second step, the cellulose negative ions attack the carbon atoms with the lowest electron cloud distributed around the active group, causing a nucleophilic substitution reaction.

Step 1:

$$CellOH \xrightarrow{OH^-} Cell\text{—}O^-$$

Step 2: Figure 8

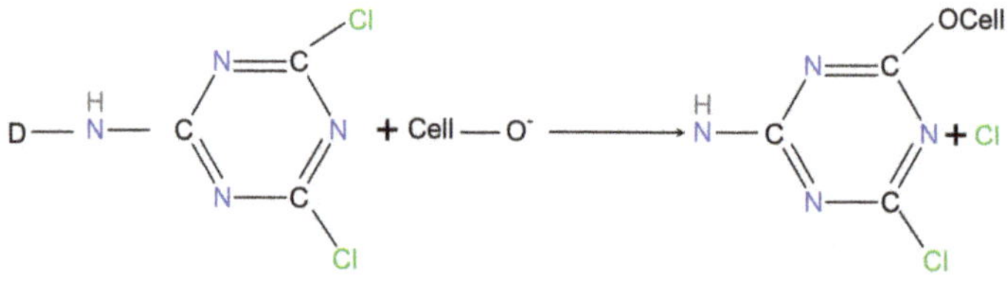

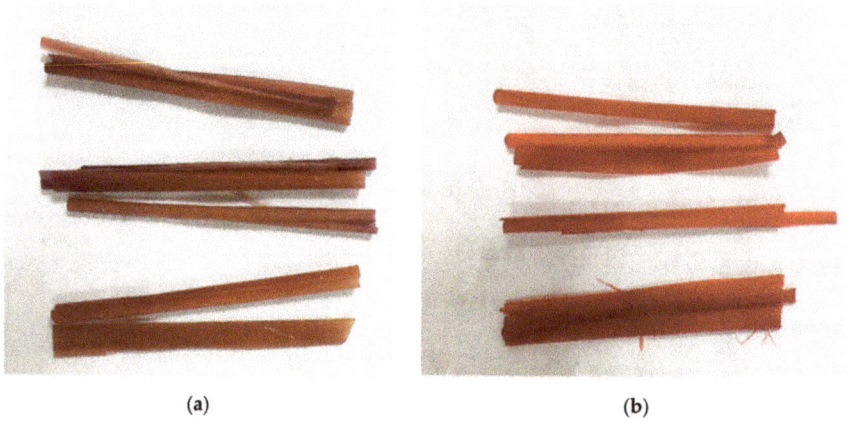

Figure 8. Mechanism of reaction between Brilliant Red X-3B and straw.

3.5. Water-Washing Firmness

After the washing treatment, the surface color of the straw dyed with Acid Red GR became obviously lighter, and the surface color of the straw dyed with Reactive Brilliant Red X-3B remained bright red, as shown in Figure 9.

Figure 9. (a) Water washing effect of the straw dyed with Acid Red GR (b) Water washing effect of the straw dyed with Reactive Brilliant Red X-3B.

It can be seen from Table 3 that the chromaticity value ΔE^* of the corn straw dyed with Acid Red GR was greatly reduced, and its color was lighter after water washing, while the chromaticity value ΔE^* of the corn straw dyed with Reactive Brilliant Red X-3B changed little, and the color did not change obviously at the naked eyes. This is because the dye molecules of Acid Red GR cellulose only physically combined with the lignin of corn straw, so the bonding strength was fragile [19]. When the dyed straw was immersed in hot water, since the concentration of the dye in the water was zero, a large concentration gradient was generated between the inside of the straw and the external environment. With the prolongation of the soaking time, the dye molecules in the straw desorbed and diffused into the water, which produced a great change of the chromaticity value. The combination of Reactive Brilliant Red X-3B and straw was mainly based on covalent bonds, and Reactive Brilliant Red X-3B mainly acted on cellulose, becoming a part of the components of the straw, so Reactive Brilliant Red X-3B did not desorb, and the color changed little.

Table 3. Chromaticity values of straw before and after washing.

Dyeing Materials	ΔE* before Washing	ΔE* after Washing	Change of ΔE*
Acid Red GR	4.39	1.29	−3.1
Reactive Brilliant Red X-3B	6.61	6.24	−0.37

3.6. Utilization of Dyed Corn

Dyed corn straw can be utilized in decorating furniture as pasteup. Corn straw dyed with various reactive dyes is used for decoration elements to improve their color system and aesthetics, as shown in Figure 10.

Figure 10. Utilization of dyed straw

4. Conclusions

1. Corn straw was bleached with 27% hydrogen peroxide (H_2O_2) and 0.1% NaOH, which had a good effect of decolorization. The dilute solution of NaOH could significantly improve the permeability of straw and opened the partially blocked pores of the straw, facilitating dyeing.

2. Under the condition of 1 h dyeing time, the dyeing rate of corn straw dyed with Reactive Brilliant Red X-3B was 46%, while the dyeing rate of corn straw dyed with Acid Red GR was 29%, indicating that Reactive Brilliant Red X-3B has a high dyeing rate for corn straw. The most content of cellulose in corn straw is high, indicating that Reactive Brilliant Red X-3B mainly binds cellulose, whereas Acid Red GR mainly binds lignin.

3. After 50 h of irradiation, the brightness change of corn straw dyed with Reactive Brilliant Red X-3B was small. The values of chromatism ΔE^* of the corn straw dyed with Acid Red GR and Reactive Brilliant Red X-3B were 12.09NBS and 17.43NBS, respectively, indicating that the straw dyed with Acid Red GR had a good light resistance.

4. The results of FTIR showed that Acid Red GR mainly combined with lignin, and the lignin (including aromatic extractives) distributed on the surface of the straw was seriously damaged after ultraviolet irradiation, when new chromophoric groups were generated, resulting in surface discoloration of the straw. Reactive Brilliant Red X-3B mainly combined with cellulose and hemicellulose in the corn straw specimens and formed covalent bonds. After the straw was irradiated with ultraviolet radiation for a long time, the fiber inside the straw broke, so that the binding rate between dye and fiber was reduced.

5. SEM showed that when 27% H_2O_2 and 0.1% NaOH were used to bleach the corn straw, removing the silica (SiO_2) covering the surface of the straw, obvious damage on the surface of the straw occurred, which was conducive to the organic combination of dye molecules with the internal components of the straw. After dyeing, the surface of the straw became smoother, and its texture was clear, which proved the infiltration or organic combination of the dye.

6. The analysis of water-washing firmness showed that the value of chromatism ΔE^* of corn straw dyed with Reactive Brilliant Red X-3B changed little after soaking at 65 °C for 2 h, and the color did not change obviously at the naked eyes, confirming that the combination of Reactive Brilliant Red X-3B with straw was based mainly on covalent bonds, while Reactive Brilliant Red X-3B mainly acted on cellulose and became a component of the straw.

5. Patents

Patent of invention: A method of pretreatment and dyeing of corn straw, grant number ZL201710873966.X.

Author Contributions: Conceptualization, Y.L. and B.W.; methodology, Y.L.; software, B.W.; validation, Y.L.; formal analysis, Y.L.; investigation, Y.L. and B.W.; resources, Y.L. and H.G.; data curation, Y.L; writing—original draft preparation, Y.L.; writing—review and editing, Y.L., H.G., and Y.L.; visualization, Y.Y. and H.G.; supervision, H.G.; project administration, Y.L.; funding acquisition, H.G. and Y.L.

Funding: This research was funded by the National Key R & D Program of China (2017YFD0601104), the Fundamental Research Funds for the Central Universities (2018ZY12), and the Natural Science Foundation of Beijing Municipality (6184045).

Acknowledgments: Thanks to the Key Laboratory of Wooden Material Science and Application, Beijing Forestry University; Key Laboratory of Wood Science and Engineering, Beijing Forestry University; MOE Engineering Research Center of Forestry Biomass Materials and Bioenergy, Beijing Forestry University.

Conflicts of Interest: The authors declare no conflict of interest.

References

1. Xu, B.T. Analysis report of corn market in Heilongjiang province in the year of 2014. *HL Grain* **2015**, *5*, 31–33.
2. Xu, Y.G.; Ma, Q.; Zhou, H.; Jiang, C.M.; Yu, W.T. Effect of straw returning and deep loosening on soil physical and chemical properties and maize yields. *Chin. J. Soil Sci.* **2015**, *46*, 428–432.
3. Ashori, A.; Nourbakhsh, A. Bio-based composites from waste agricultural residues. *Waste Manag.* **2010**, *30*, 680–684. [CrossRef] [PubMed]
4. Zhao, N.; Yang, X.; Zhang, J.; Zhu, L.; Lv, Y. Adsorption Mechanisms of Dodecylbenzene Sulfonic Acid by Corn Straw and Poplar Leaf Biochars. *Materials* **2017**, *10*, 1119. [CrossRef]
5. Yang, F.; Miao, Y.; Wang, Y.; Zhang, L.-M.; Lin, X. Electrospun Zein/Gelatin Scaffold-Enhanced Cell Attachment and Growth of Human Periodontal Ligament Stem Cells. *Materials* **2017**, *10*, 1168. [CrossRef]
6. Thangalazhy-Gopakumar, S.; Adhikari, S.; Ravindran, H.; Gupta, R.B.; Fasina, O.; Tu, M.; Fernando, S.D. Physiochemical properties of bio-oil produced at various temperatures from pine wood using an auger reactor. *Bioresour. Technol.* **2010**, *101*, 8389–8395. [CrossRef]

7. Forero, J.C.; Roa, E.; Reyes, J.G.; Acevedo, C.; Osses, N. Development of Useful Biomaterial for Bone Tissue Engineering by Incorporating Nano-Copper-Zinc Alloy (nCuZn) in Chitosan/Gelatin/Nano-Hydroxyapatite (Ch/G/nHAp) Scaffold. *Materials* **2017**, *10*, 1177. [CrossRef]
8. Pandey, K.K.; Vuorinen, T. Comparative study of photodegradation of wood by a UV laser and a xenon light source. *Polym. Degrad. Stab.* **2008**, *93*, 2138–2146. [CrossRef]
9. Chang, T.C.; Chang, H.T.; Wu, C.L.; Chang, S.T. Influences of extractives on the photodegradation of wood. *Polym. Degrad. Stab.* **2010**, *95*, 516–521. [CrossRef]
10. Pandey, K.K. Study of the effect of photo-irradiation on the surface chemistry of wood. *Polym. Degrad. Stab.* **2005**, *90*, 9–20. [CrossRef]
11. Živković, V.; Arnold, M.; Pandey, K.K.; Richter, K.; Turkulin, H. Spectral sensitivity in the photodegradation of fir wood (Abies alba Mill.) surfaces: Correspondence of physical and chemical changes in natural weathering. *Wood Sci. Technol.* **2016**, *50*, 989–1002. [CrossRef]
12. Mitsui, K.; Murata, A.; Tolvaj, L. Changes in the properties of light-irradiated wood with heat treatment: Part 3. Monitoring by DRIFT spectroscopy. *Holz als Roh-und Werkst* **2004**, *62*, 164–168. [CrossRef]
13. Campbell, B.; Rushworth, J. Relation between Structure of Chrome Dyes and Effect on Wood Fibre Properties in Dyeing. *J. Soc. Dye. Colour.* **1962**, *78*, 491–495. [CrossRef]
14. Liu, Y.; Xing, F.R.; Hu, J.H.; Guo, H.W. Photochromic inhibition of dyed veneer covered with modified transparent film. *Adv. Mater. Res.* **2014**, *933*, 38–42. [CrossRef]
15. Wang, L.; Liu, Y.; Zhan, X.; Luo, D.; Sun, X. Photochromic transparent wood for photo-switchable smart window applications. *J. Mater. Chem. C* **2019**, *7*, 8649–8654. [CrossRef]
16. Liu, Y.; Guo, H.; Gao, J.; Zhang, F.; Shao, L.; Via, B.K. Effect of Bleach Pretreatment on Surface Discoloration of Dyed Wood Veneer Exposed to Artificial Light Irradiation. *Bioresources* **2015**, *10*, 5607–5619. [CrossRef]
17. Liu, Y.; Shao, L.; Gao, J.; Guo, H.; Chen, Y.; Cheng, Q.; Via, B.K. Surface photo-discoloration and degradation of dyed wood veneer exposed to different wavelengths of artificial light. *Appl. Surf. Sci.* **2015**, *331*, 353–361. [CrossRef]
18. Liu, Y.; Hu, J.; Gao, J.; Guo, H.; Chen, Y.; Cheng, Q.; Via, B.K. Wood veneer dyeing enhancement by ultrasonic-assisted treatment. *Bioresources* **2015**, *10*, 1198–1212. [CrossRef]
19. Pandey, K.K. A note on the influence of extractives on the photo-discoloration and photo-degradation of wood. *Polym. Degrad. Stab.* **2005**, *87*, 375–379. [CrossRef]
20. Oltean, L.; Teischinger, A.; Hansmann, C. Verfärbung von Holzoberflächen Aufgrund von Simulierter Sonneneinstrahlung im Innenraum. *Holz als Roh-und Werkst* **2008**, *66*, 51–56. [CrossRef]
21. Rosu, D.; Teaca, C.A.; Bodirlau, R.; Rosu, L. FTIR and color change of the modified wood as a result of artificial light irradiation. *J. Photochem. Photobiol. B Biol.* **2010**, *99*, 144–149. [CrossRef] [PubMed]
22. Kitao, M.; Yamada, S.; Yoshida, S.; Akram, H.; Urabe, K. Preparation conditions of sputtered electrochromic WO_3 films and their infrared absorption spectra. *Sol. Energy Mater. Sol. Cells* **1992**, *25*, 241–255. [CrossRef]
23. Kataoka, Y.; Kiguchi, M.; Williams, R.S.; Evans, P.D. Violet light causes photodegradation of wood beyond the zone affected by ultraviolet radiation. *Holzforschung* **2007**, *61*, 23–27. [CrossRef]

© 2019 by the authors. Licensee MDPI, Basel, Switzerland. This article is an open access article distributed under the terms and conditions of the Creative Commons Attribution (CC BY) license (http://creativecommons.org/licenses/by/4.0/).

Article

Combined Effects of Color and Elastic Modulus on Antifouling Performance: A Study of Graphene Oxide/Silicone Rubber Composite Membranes

Huichao Jin [1,2,3], Wei Bing [1,4], Limei Tian [1,*], Peng Wang [5] and Jie Zhao [1]

1. Key Laboratory of Bionic Engineering, Ministry of Education, Jilin University, Changchun 130022, China
2. School of Mechanical Engineering and Automation, Beihang University, Beijing 100191, China
3. College of Physics, Jilin University, Changchun 130012, China
4. Advanced Institute of Materials Science, Changchun University of Technology, Changchun 130012, China
5. State Key Laboratory of Superhard Materials, Jilin University, Changchun 130012, China
* Correspondence: lmtian@jlu.edu.cn; Tel.: +86-13944850095; Fax: +0431-85095253

Received: 27 June 2019; Accepted: 14 August 2019; Published: 16 August 2019

Abstract: Biofouling is a significant maritime problem because the growth of fouling organisms on the hulls of ships leads to very high economic losses every year. Inspired by the soft skins of dolphins, we prepared graphene oxide/silicone rubber composite membranes in this study. These membranes have low surface free energies and adjustable elastic moduli, which are beneficial for preventing biofouling. Diatom attachment studies under static conditions revealed that color has no effect on antifouling behavior, whereas the studies under hydrodynamic conditions revealed that the combined effects of color and elastic modulus determine the antifouling performance. The experimental results are in accordance with the "harmonic motion effect" theory proposed by us, and we also provide a supplement to the theory in this paper. On the basis of the diatom attachment test results, the membrane with 0.36 wt % of graphene oxide showed excellent antifouling performance, and is promising in practical applications. The results confirmed that the graphene oxide and graphene have similar effect to enhance silicone rubber antifouling performance. This study provides important insight for the design of new antifouling coatings; specifically, it indicates that lighter colors and low Young's moduli provide superior performance. In addition, this study provides a reference for the application of graphene oxide as fillers to enhance the composite antifouling performance.

Keywords: graphene oxide; silicone rubber; composite materials; antifouling; harmonic motion

1. Introduction

Biofouling is a significant maritime problem because fouling organisms that grow on the hulls of ships promote their deterioration and increase drag, thereby increasing fuel costs [1,2]. Another impact of biofouling is bioinvasion [3]. In the coast of California, more than 60% of invasive species arrived by clinging to the surface of ships [4]. To solve these problems, the California government announced regular biological inspection of hulls starting from January 2018 to reduce the invasion of alien species [5]. These negative impacts have caused enormous economic losses worldwide [6]. Intense methods have been developed to combat biofouling, including the use of copper coatings [7] and tributyltin self-polishing copolymer (TBT-SPC) [8]. However, in the 1980s, a series of studies reported the high toxicity of TBT-SPC to marine organisms [9,10]. Since then, the use of toxic antifouling coatings has gradually been banned in many countries [11,12]. Therefore, the development of new eco-friendly antifouling coatings is urgent.

In recent years, composite coatings have aroused considerable interest as economic and eco-friendly solutions for preventing marine biofouling [13]. Composite coatings based on polydimethylsiloxane, silicone, and polyurethane-acrylate, among others, have been shown to prevent biofouling effectively [14–16]. Recently, considerable effort has been devoted to developing new composite antifouling coatings with graphene/graphene oxide (GO) as fillers [17,18], which promote the antifouling properties of the coatings [19]. However, several challenges still need to be overcome before these methods can be applied, such as high cost and low durability [20]. Hence, low cost and efficient antifouling coatings need to be designed and developed.

Dolphins, soft corals (*Sarcophyton trocheliophorum*), and seals were found to have antifouling capacity. These organisms have elastic skin, enabling the creation of a dynamic surface around the flow to resist biofouling, which is called the "harmonic motion effect." [21,22] Inspired by this antifouling strategy, six GO/silicone rubber (GOSR) composite membranes were prepared in the present study. The preparation method is simple, highly efficient, and low-cost, and the resulting GOSR membranes are environmentally friendly. They are characterized by adjustable Young's moduli, low surface free energies, and smooth surfaces, which are conducive to preventing biofouling. In this study, diatoms were selected as fouling organisms. The diatom attachment was examined under both static and hydrodynamic conditions, which revealed that color and Young's modulus play major roles in diatom attachment. We previously reported a new understanding of the effect of elastic modulus on antifouling performance, which we referred to as the "harmonic motion effect" [21]; the experimental results of the present study can be explained on the basis of this effect. This study provides insight that will be useful for the design and fabrication of new antifouling coatings.

2. Materials and Methods

2.1. Materials

Silicone rubber (SR) was purchased from Guangdong Bo Rui Co., Ltd, Shenzhen, China. GO was purchased from Yuhuang New Energy Technology Co., Ltd, Heze, China. Acetone, tetrahydrofuran, and anhydrous ethanol were supplied by Beijing Chemical Works (Beijing, China). The diatom *Triceratium* sp. was obtained from Nanhuaqianmu Biotechnology Co. Ltd, Zhengzhou, China. Algal broth medium (028820) was purchased from Huankai Microbial Sci. & Tech. Co., Ltd, Guangzhou, China.

2.2. Membrane Preparation

Scheme 1 illustrates the preparation of the GOSR composite membranes. (a) First, GO was added to acetone under mechanical agitation for 3 h, then remove the mixture into an ultrasonic cleaner for 3 h, which yielded GO dispersions, and the weight ratio of GO/acetone was 1/9. (b) SR was added to tetrahydrofuran under magnetic stirrer for 2 h, then SR dispersions were obtained, and the weight ratio of SR/tetrahydrofuran was 1/4. (c) According to the following Equation,

$$\text{Weight percent (wt\%)} = \frac{1/10 M_{\text{Go dispersions}}}{1/10 M_{\text{Go dispersions}} + 1/5 M_{\text{SR dispersions}}} \quad (1)$$

Five different GO/SR composite materials were prepared, the GO in composite materials were expressed as 0.16, 0.36, 0.64, 1.28, and 2.56 wt %, respectively. The mixtures were then stirred for 1 h under mechanical agitation, after which they were placed in a vacuum chamber for 1 h at 60 °C until acetone and tetrahydrofuran were removed. (d–e) Finally, the mixtures were poured into an acrylic mold and cured at room temperature for 24 h. A pristine SR (PSR) membrane with 0 wt % GO was prepared as a control.

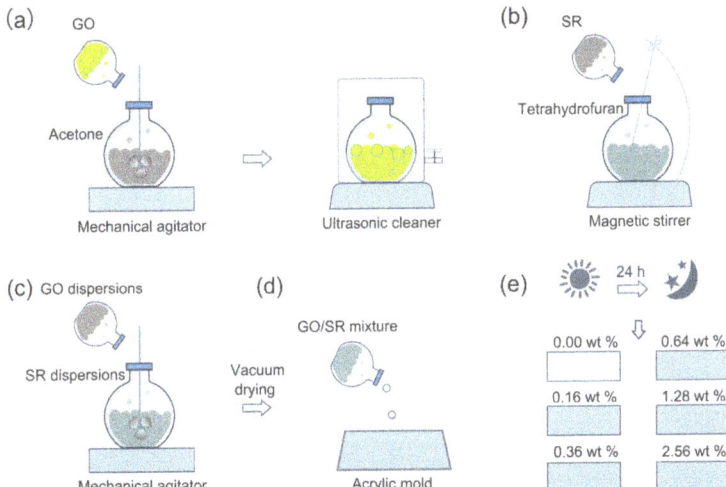

Scheme 1. Preparation schematic of graphene/graphene oxide (GO)/silicone rubber (GOSR) composite membranes: (**a**) Preparation of a GO dispersion, (**b**) preparation of an silicone rubber (SR) dispersion, (**c**) the two dispersions are mixed to produce GOSR composite material, and (**d,e**) the mixture is poured into an acrylic mold and cured.

2.3. Diatom Attachment Testing

First, 3.8 g of the algal broth medium was added to 1 L of deionized water, and the mixture was heated in an autoclave for 15 min at 121 °C. After cooling, 100 g of diatoms was added, and the mixture was cultured for 2 days at room temperature. As diatoms are photosynthetic organisms, a light-emitting diode (LED) plant lamp was used with a light/dark time-cycle of 14/10 h per day. After culturing for 2 days (see Figure S1, Supporting Information), the diatom suspension was added to containers for static and hydrodynamic diatom attachment tests. The containers were filled with 50 L of water and 50 g of the algal broth medium. The size of the specimens was approximately 5 cm × 2 cm × 2 mm (length × width × thickness). The specimens were submerged in a static container for 8 days, after which they were examined and measured. In the hydrodynamic test, the specimens were fixed to a hexagonal prism, and the speed of the electric motor was set as 500 rad/min. Therefore, the linear velocity (V) near the specimens was 3.4 m/s (Figure S2 and Table S1). Some studies reported the negative relevant relations between the amount of fouling organisms adhered on ship hulls and the flow velocity [23,24]. Therefore, a low flow velocity can reveal the antifouling capability of coatings under hydrodynamic conditions. The specimens were examined after 10 days.

2.4. Characterization

Raman spectra was obtained using a Raman spectrometer (T64000, HORIBA, Paris, France) which combined an Olympus microscope lens with a 532.05 nm CW He-Ne laser. The composition of GO nanosheets were analyzed using a Fourier transform infrared spectroscopy (FTIR, Thermo Fisher Scientific, Nicolet 6700, Waltham, MA, USA). The morphologies of GO nanosheets were obtained using a transmission electron microscope (TEM, JEOL, JEM1200EX, Mitaka-shi, Japan). The elemental composition was analyzed using an energy dispersive spectrometer (EDS, X-Max50, Oxford, UK) installed in a scanning electron microscope (JSM-7610, JEOL, Mitaka-shi, Japan). The water contact angles were measured using a surface tension meter (WCA, XG-CAM, Xuanyichuangxi, Shanghai, China) at room temperature with deionized water as testing liquid, and three specimens were measured for each membrane. The Young's moduli were tested using a rubber testing machine (UTM5305, YOUHONG, Shanghai, China), and three specimens were measured for each membrane.

The surface topographies of the specimens were analyzed by scanning probe microscopy (SPM, ICON, BRUKER, Karlsruhe, Germany). Specimens under static and hydrodynamic conditions were observed by scanning electron microscopy (SEM; MAGELLAN 400, FEI, Hillsboro, Oregon, USA, and JSM-7610, JEOL, Mitaka-shi, Japan). After the diatom attachment test, the six specimens were removed and placed into the same amounts of normal saline (4 ml), after which they were placed in an ultrasonic cleaner for 1 h to separate the diatoms from the specimens, and to produce diatom solutions. The optical densities of these solutions were measured at 440 nm (OD_{440}) using an ultraviolet-visible spectrophotometer (UV-5500/PC, METASH, Shanghai, China), which revealed the amounts of diatoms on the different specimen surfaces. Three specimens were measured for each membrane to minimize the experimental error.

3. Results and Discussions

3.1. Membrane Composition Analysis

The morphologies of GO nanosheets were observed by SEM (Figure S3a) and TEM (Figure S3b). As can be seen, the GO nanosheets resemble crumpled silk veil waves, which indicates that they are multi-layered. The GO nanosheets were confirmed by Raman spectroscopy (Figure S3c), and it is evident that GO has typical D (1359 cm^{-1}) and G (1589 cm^{-1}) peaks and a 2D region [25,26]. The functional groups of GO nanosheets were analyzed by FTIR (Figure S3d). The band at 3199.55 cm^{-1} is assigned to the O–H stretching vibration. The peak at 1722.31 cm^{-1} is attributed to C=O stretching vibration. The band at 1617.94 cm^{-1} is assigned to C=C skeletal vibration. The three peaks at 1401.60 cm^{-1}, 1221.08 cm^{-1}, and 1049.12 cm^{-1} correspond to the C–O stretching vibration. The Raman spectra of the GOSR membrane (red curve in Figure 1a) exhibits a peak at 1594 cm^{-1}, which is the G peak of GO. The elemental compositions of the membranes were determined using the EDS (Figure 1b), and the results in Table S2 reveal that the carbon content increases with increasing GO content. The presence of GO in the membranes is evident (Figure S4). The PSR membrane appears milky white and semi-transparent, whereas the GOSR membranes are brown because of the imbedded GO, and the color deepens with increasing GO content.

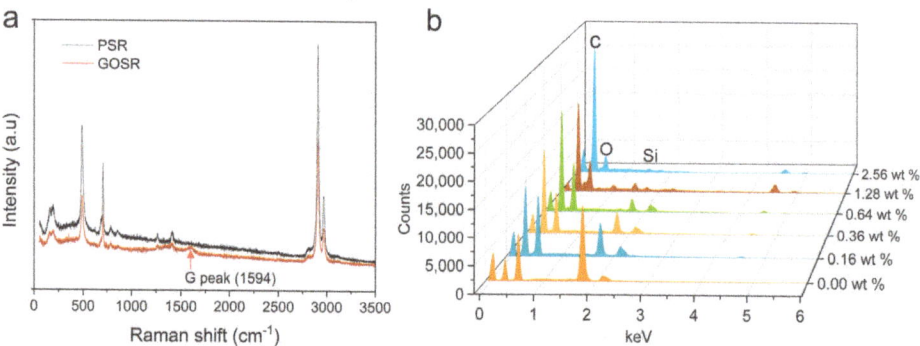

Figure 1. Characterization of specimens with different GO contents: (**a**) Raman spectra, (**b**) energy dispersive spectrometer (EDS) results.

3.2. Membrane Properties

The surface free energies, Young's moduli, and surface topographies of these membranes were also examined, as they are vital for bio-adhesion. The contact angles of GOSR were measured (Figure 2a, red curve), and the correlation between contact angle and surface energy is given by [27]:

$$\cos\theta = -1 + 2\sqrt{\frac{\gamma_S}{\gamma_L}}[1 - \beta(\gamma_L - \gamma_S)^2] \tag{2}$$

where θ stands for contact angle, γ_S and γ_L represents surface energy of solid and liquid, respectively. β is a constant with the value 1.057×10^{-4} m^2/mJ, and surface energy of deionized water is 72.8 mJ/m^2.

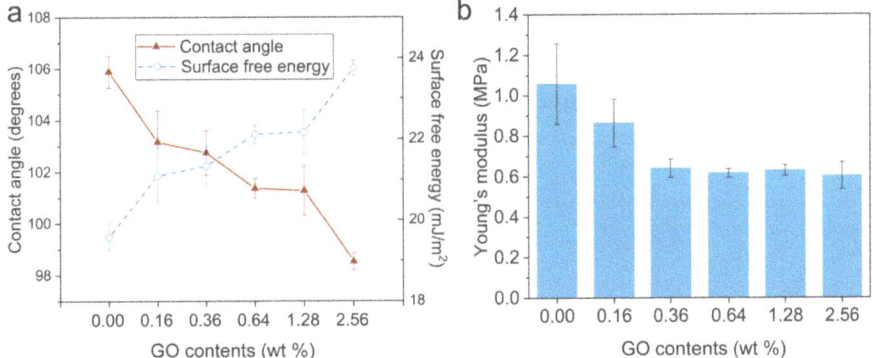

Figure 2. Characterization of specimens with different GO contents: (**a**) Water contact angles and surface free energies, (**b**) Young's moduli.

The surface free energies of the membranes with varying GO contents range from 19.57 mJ·m^{-2} to 23.74 mJ·m^{-2} (Figure 2a, blue curve), which are conducive to reducing marine biofouling [28]. It is clear that GO significantly affects the Young's modulus (Figure 2b). Some studies have reported that Young's modulus increases with GO contents [29]. However, the Young's modulus results from this test contradict reported data as well as our previous studies [21,30], which is ascribable to the GO not being pre-treated with the coupling agent (KH-550) in this experiment. The GO in the membrane can aggregate, resulting in points of converging stress in the membrane [17]; consequently, the Young's modulus decreases with increasing GO contents. Another explanation is that an excess of the GO filler will weaken the interaction between the polymer chain segments and decrease the tensile strength [31]. It is generally assumed that a low Young's modulus is beneficial for mitigating the adhesion of fouling organisms [21,32]. The surface topographies were also examined. Figure 3a reveals that the membranes have smooth surfaces at the micron level, and this low surface roughness is important for combating biofouling [33]. In addition, SPM results (Figure 3b and Figure S5) show that nanostructures are present on the surfaces, and some studies reported that nanoscale roughness is conducive to combating biofouling [34].

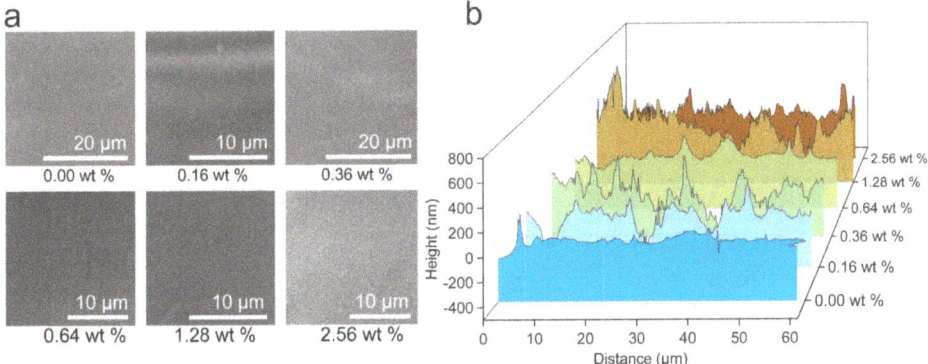

Figure 3. (**a**) SEM images of pristine SR (PSR)/GOSR membrane surfaces. (**b**) Scanning probe microscopy (SPM) height profiles of PSR/GOSR membranes.

3.3. Diatom Adhesion Testing Under Static Conditions

The specimens were examined after soaking for 8 days. In summary, these specimens show similar results (Figure 4b,c), and no biofilms were observed (Figure 4d), and only scattered diatoms were present on the surfaces. It is commonly assumed that the Young's modulus does not influence static attachment [35]. Since the surface free energies of the specimens are similar (Figure 2a), they show similar diatom attachment results which conform to the prediction of "Baier curve" [28]. Previous studies confirmed that diatoms prefer to adhere to dark surfaces [36,37]. However, their experiments were performed under hydrodynamic conditions. The results in Figure 4 show that colors have no effect on diatom adhesion under static conditions.

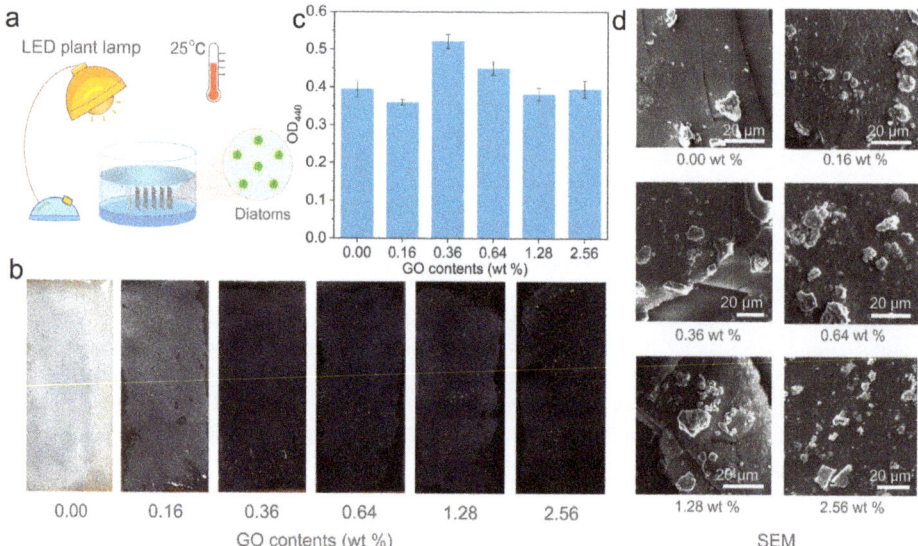

Figure 4. (a) Schematic illustration of diatom adhesion testing under static conditions. (b) Optical images of sample surfaces at 0 days and 8 days. (c) OD_{440} results of the membranes after 8 days. (d) SEM images of specimens after 8 days.

3.4. Diatom Adhesion Testing Under Hydrodynamic Conditions

The specimens were examined after 10 days. Biofilms were observed on the 0 wt %, 0.16 wt %, and 1.28 wt % membranes (Figure 5d). The 0.36 wt % membrane showed the cleanest surface (Figure 5c,d), and scattered diatoms were present on the 0.64 wt % and 2.56 wt % membranes. A low Young's modulus is beneficial to combat adhesion of fouling organisms [21,38]. According to Figures 2b and 5, the 0.64 wt %, 1.28 wt %, and 2.56 wt % GO-containing membranes have the lowest Young's moduli; however, they show poor antifouling performances, due to their dark colors. On the other hand, the 0 wt % and 0.16 wt % membranes perform poorly due to high Young's moduli. The 0.36 wt % membrane with appropriate color and Young's moduli exhibits the best antifouling performance. We conclude that the combined effects of color and Young's modulus determine the antifouling performance of GOSR membranes.

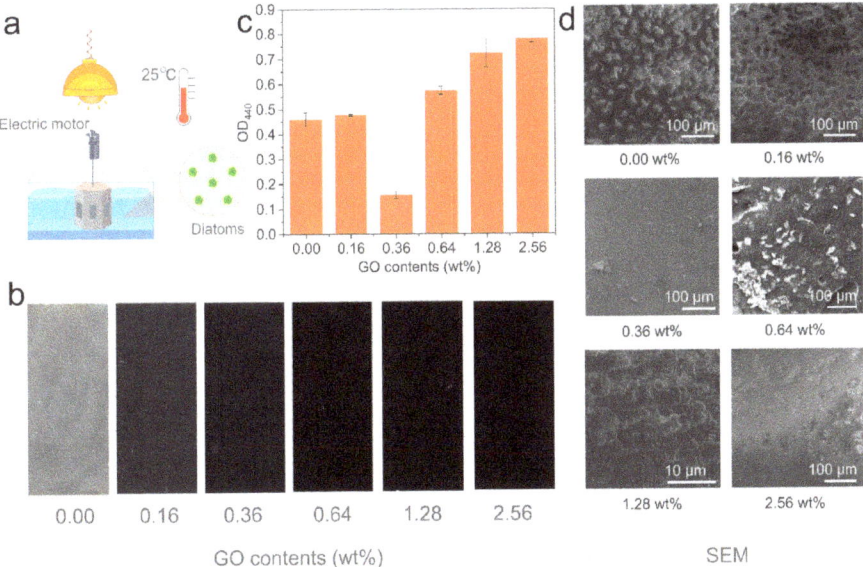

Figure 5. (a) Schematic illustration of diatom adhesion testing under hydrodynamic conditions. (b) Optical images of sample surfaces at 0 days and 10 days. (c) OD$_{440}$ results of the membranes after 10 days. (d) SEM images of specimens after 10 days.

In our previous studies, we proposed the "harmonic motion effect" to explain the antifouling behavior of elastic membranes [21], in which the deformation of an elastic membrane under turbulent flow imparts biofouling resistance. Biofouling resistance involves three components: (i) Fouling organisms have difficulty identifying dynamic surfaces (Figure 6a); (ii) when these organisms approach an elastic membrane, its dynamic surface sweeps them away (Figure 6b); (iii) if some organisms adhere to the elastic membrane (Figure 6c), they cannot adhere tightly because of the micro-flaws [21] produced at the interface.

To reveal the mechanism via which a low elastic modulus aids in combating fouling organisms in the present study, a theory by Kulik is introduced below. Studies by Kulik revealed that the deformation amplitude of the elastic coatings in turbulent flow is approximately 0 μm to a few microns [39,40]. Three equations [41] (from Kulik) were employed to calculate the deformation amplitude of a surface:

$$C_n^* = \frac{C_n}{H/E} = \frac{\lambda}{H}\left(\frac{V}{C_t^0}\right)2 \frac{(1+\sigma)aF}{2\pi(1-i\mu)^2[2 - \frac{(V/C_t^0)^2}{1-i\mu} - 2S]} \tag{3}$$

$$\eta_{rms} = \sqrt{\overline{\eta^2}} = \left[\int_0^\infty |C_n(\omega)|^2 P(\omega)d\omega\right]^{1/2}, \tag{4}$$

where C_n^* is the vertical (normal to the surface) compliance; $P(\omega)$ stands for the energy spectrum of pressure pulsations; η_{rms} denotes the vertical displacement; η and H are deformation and elastic-membrane thickness, respectively; and V and C_t^0 are the flow rate and shear wave rate for an ideal material, respectively.

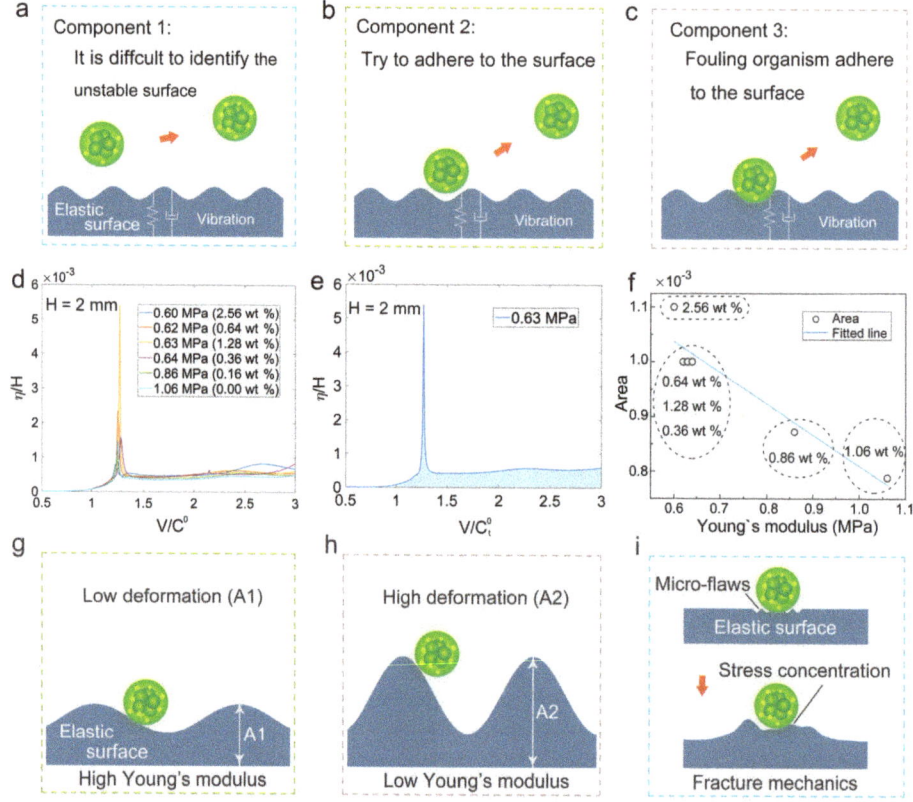

Figure 6. (**a–c**) Contributions of "harmonic motion" to biofouling resistance. (**d**) Deformations (η/H) of elastic membranes as functions of V/C_t^0 for different Young's moduli. (**e**) $E = 0.63$ MPa curve. (**f**) Area as a function of Young's modulus. (**g**) Elastic surface with high Young's modulus. (**h**) Elastic surface with low Young's modulus. (**i**) Schematic illustration of the fracture mechanics on antifouling.

η/H is plotted as a function of V/C_t^0 for various Young's moduli in Figure 6d. In a practical situation, the flow velocity is not constant; thus, various velocities need to be studied. Figure 6e shows the curve of $E = 0.63$ MPa; the area between the curve and the x-axis (blue area in Figure 6e) characterizes the total deformation of membrane surface. Figure 6f shows the areas for $E = 0.60$–1.06 MPa (Figure 2b), which reveal that a lower Young's modulus leads to a bigger area, i.e., a bigger deformation. According to contact mechanics [20,42], this higher deformation is conducive to combating the adhesion of fouling organisms (Figure 6g,h). Thus, 0 wt % and 0.16 wt % membranes show poor antifouling performances, due to their high Young's modulus. On the basis of this theory, the 0.64 wt %, 1.28 wt %, and 2.56 wt % GO-containing membranes have a low Young's moduli which are conducive to resisting biofouling; however, they show poor antifouling performances, due to their dark colors. In the past, fracture mechanics [38,43] was employed to reveal the mechanisms of fouling organisms and elastic surfaces. Figure 6i shows a schematic illustration of the fracture mechanics, which indicates that the contact between the fouling organism and elastic surface is not strong enough; consequently, there are some microcracks at the contact area. Fouling organism tend to detach from the surface when the microcracks grow increasing large due to the stress concentration. The Griffith theory of brittle fracture is as follows [43]:

$$F = \sqrt{\frac{2E\gamma}{A\pi}} \qquad (5)$$

where F represents the stress; A stands for the crack length; E represents for the Young's modulus of the surface; γ stands for the surface energy density.

On the basis of the Griffith equation, A is assumed to be a constant. Hence, $F\sqrt{A\pi}$ is a function of $E\gamma$. Here, $F\sqrt{A\pi}$ denotes a separating force (the force required to separate the fouling organism from the surface), and a low Young's modulus (E) leads to a low separating force ($F\sqrt{A\pi}$). The low Young's modulus is beneficial for fouling organisms to isolate from the surface due to low separating force. The experimental results in the present study also are in accord with the fracture mechanics. The harmonic motion effect sheds new light on the effects of Young's modulus on antifouling behavior.

4. Conclusions

SR composite membranes containing 0.36 wt % GO exhibited excellent antifouling performance in this study, while they are low-cost with inexpensive silicone rubber as the main material. Hence, these membranes may be promising for applications in antifouling. The combined effects of color and Young's modulus determine the antifouling performance of GOSR membranes. The findings provide insight that could facilitate the fabrication of new antifouling coatings; specifically, lighter colors and low Young's moduli provide superior performance. In addition, we provide a novel insight into the role of the elastic modulus toward the antifouling performance.

Supplementary Materials: The following are available online at http://www.mdpi.com/1996-1944/12/16/2608/s1, Figure S1: SEM images of diatoms (Triceratium sp.) on silicon slice after culturing for 2 days, Figure S2: Specimens were fixed to a hexagonal prism, Figure S3a: SEM image of GO nanosheets, Figure S3b: TEM image of GO nanosheets, Figure S3c: Raman spectra of GO nanosheets, Figure S3d: FTIR spectra of GO nanosheets, Figure S4: Colors of the membranes with different GO contents, and Figure S5: Roughness or GOSR membranes with different GO contents, Table S1: Nomenclature and values, Table S2: Elemental compositions of membranes.

Author Contributions: Conceptualization and methodology, L.T. and H.J.; validation, H.J., W.B., L.T., P.W., J.Z.; formal analysis and investigation, H.J., W.B. and P.W. writing—original draft preparation, H.J.; review and editing, H.J., W.B., L.T., P.W., J.Z.; visualization, H.J.; supervision, L.T.

Funding: This work was supported by the National Natural Science Foundation of China General Program (Grant No. 51875240), the Fund Project in Equipment Pre-research Field (Grant No. 61400040403), the China Postdoctoral Science Foundation funded project (Grant No. 2018M630324), the Special Industrialization Demonstration project "New Material Special Project" of Jilin province and school co-construction plan (Grant No. SXGJSF2017-3), the Major Scientific and Technological Project of Changchun (Grant No.17SS023), and the Department of Science and Technology of Jilin Province (Grant No. 20190103114JH).

Conflicts of Interest: The authors declare no conflict of interest.

References

1. Selim, M.S.; Shenashen, M.A.; El-Safty, S.A.; Higazy, S.A.; Isago, H.; Elmarakbi, A. Recent progress in marine foul-release polymeric nanocomposite coatings. *Prog. Mater. Sci.* **2017**, *87*, 1–32. [CrossRef]
2. Sullivan, T.; Regan, F. Marine diatom settlement on microtextured materials in static field trials. *J. Mater. Sci.* **2017**, *52*, 5846–5856. [CrossRef]
3. Lacoursière-Roussel, A.; Dan, G.B.; Cristescu, M.E.; Guichard, F.; Mckindsey, C.W. Effect of shipping traffic on biofouling invasion success at population and community levels. *Biol. Invasions* **2016**, *18*, 3681–3695. [CrossRef]
4. Scianni, C.; Brown, C.; Nedelcheva, R.; Dobroski, N. Hull husbandry practices and biofouling management of vessels operating in California. In Proceedings of the Oceans IEEE, San Diego, CA, USA, 23–27 September 2013.
5. CLSC. Marine Invasive Species Program. Available online: http://www.slc.ca.gov/Programs/MISP.html (accessed on 3 February 2019).

6. Xu, Y.; He, H.; Schulz, S.; Liu, X.; Fusetani, N.; Xiong, H.; Qian, P.Y. Potent antifouling compounds produced by marine Streptomyces. *Bioresour. Technol.* **2009**, *101*, 1331–1336. [CrossRef] [PubMed]
7. Chen, C.L.; Maki, J.S.; Dan, R.; Teo, L.M. Early marine bacterial biofilm on a copper-based antifouling paint. *Int. Biodeter. Biodegr.* **2013**, *83*, 71–76. [CrossRef]
8. Jelic-Mrcelic, G.; Sliskovic, M.; Antolic, B. Biofouling communities on test panels coated with TBT and TBT-free copper based antifouling paints. *Biofouling* **2006**, *22*, 293–302. [CrossRef] [PubMed]
9. Alzieu, C. Environmental problems caused by TBT in France: Assessment, regulations, prospects. *Mar. Environ. Res.* **1991**, *32*, 7–17. [CrossRef]
10. Antizar-Ladislao, B. Environmental levels, toxicity and human exposure to tributyltin (TBT)-contaminated marine environment. A review. *Environ. Int.* **2008**, *34*, 292–308. [CrossRef] [PubMed]
11. Minchin, D.; Oehlmann, J.; Duggan, C.B.; Stroben, E.; Keatinge, M. Marine TBT antifouling contamination in Ireland, following legislation in 1987. *Mar. Pollut. Bull.* **1995**, *30*, 633–639. [CrossRef]
12. Blanck, H.; Dahl, B. Recovery of marine periphyton communities around a swedish marina after the ban of TBT use in antifouling paint. *Mar Pollut Bull* **1998**, *36*, 437–442. [CrossRef]
13. Selim, M.S.; Elmarakbi, A.; Azzam, A.M.; Shenashen, M.A.; El-Saeed, A.M.; El-Safty, S.A. Eco-friendly design of superhydrophobic nano-magnetite/silicone composites for marine foul-release paints. *Prog. Org. Coat.* **2018**, *116*, 21–34. [CrossRef]
14. Selim, M.S.; El-Safty, S.A.; Fatthallah, N.A.; Shenashen, M.A. Silicone/graphene oxide sheet-alumina nanorod ternary composite for superhydrophobic antifouling coating. *Prog. Org. Coat.* **2018**, *121*, 160–172. [CrossRef]
15. Wang, B.; Wu, Z.; Zhang, D.; Wang, R.M.; Song, P.; Xiong, Y.; He, Y. Antibacterial silicylacrylate copolymer emulsion for antifouling coatings. *Prog. Org. Coat.* **2018**, *118*, 122–128. [CrossRef]
16. Mo, F.; Ren, H.; Chen, S.; Ge, Z. Novel zwitterionic polyurethanes with good biocompatibility and antibacterial activity. *Mater. Lett.* **2015**, *145*, 174–176. [CrossRef]
17. Lee, J.; Chae, H.R.; Won, Y.J.; Lee, K.; Lee, C.H.; Lee, H.H.; Kim, I.C.; Lee, J.M. Graphene oxide nanoplatelets composite membrane with hydrophilic and antifouling properties for wastewater treatment. *J. Membr. Sci.* **2013**, *448*, 223–230. [CrossRef]
18. Huang, Y.; Li, H.; Wang, L.; Qiao, Y.; Tang, C.; Jung, C.; Yoon, Y.; Li, S.; Yu, M. Ultrafiltration membranes with structure-optimized graphene-oxide coatings for antifouling Oil/Water separation. *Adv. Mater. Interfaces* **2015**, *2*, 1400433. [CrossRef]
19. Shi, Y.; Chang, L.; He, D.; Shen, L.; Bao, N. Preparation of graphene oxide-cellulose acetate nanocomposite membrane for high-flux desalination. *J. Mater. Sci.* **2017**, *52*, 13296–13306. [CrossRef]
20. Fu, J.; Zhang, H.; Guo, Z.; Feng, D.Q.; Thiyagarajan, V.; Yao, H. Combat biofouling with microscopic ridge-like surface morphology: a bioinspired study. *J. R. Soc. Interface* **2018**, *15*, 1–8. [CrossRef]
21. Jin, H.; Zhang, T.; Bing, W.; Dong, S.; Tian, L. Antifouling performance and mechanism of elastic graphene–silicone rubber composite membranes. *J. Mater. Chem. B* **2019**, *7*, 488–497. [CrossRef]
22. Bing, W.; Tian, L.; Wang, Y.; Jin, H.; Ren, L.; Dong, S. Bio-Inspired non-bactericidal coating used for antibiofouling. *Adv. Mater. Technol.* **2019**, *4*, 1800480. [CrossRef]
23. Qian, P.Y.; Rittschof, D.; Sreedhar, B.; Chia, F.S. Macrofouling in unidirectional flow: Miniature pipes as experimental models for studying the effects of hydrodynamics on invertebrate larval settlement. *Mar. Ecol. Prog. Ser.* **1999**, *191*, 141–151. [CrossRef]
24. Nishizaki, M.T.; Carrington, E. The effect of water temperature and flow on respiration in barnacles: patterns of mass transfer versus kinetic limitation. *J. Exp. Biol.* **2014**, *217*, 2101–2109. [CrossRef] [PubMed]
25. Kaniyoor, A.; Ramaprabhu, S. A Raman spectroscopic investigation of graphite oxide derived graphene. *Aip Adv.* **2012**, *2*, 241. [CrossRef]
26. Kudin, K.N.; Ozbas, B.; Prud'Homme, R.K.; Aksay, I.A.; Car, R. Raman spectra of graphite oxide and functionalized graphene sheets. *Nano Lett.* **2008**, *8*, 36. [CrossRef] [PubMed]
27. Kwok, D.Y.; Neumann, A.W. Contact angle measurement and contact angle interpretation. *Adv. Colloid Int. Sci.* **1999**, *81*, 167–249. [CrossRef]
28. Brady, R.F.; Singer, I.L. Mechanical factors favoring release from fouling release coatings. *Biofouling* **2000**, *15*, 73–81. [CrossRef] [PubMed]
29. Compton, O.C.; Nguyen, S.T. Graphene oxide, highly reduced graphene oxide, and graphene: versatile building blocks for carbon-based materials. *Small* **2010**, *6*, 711–723. [CrossRef] [PubMed]

30. Tian, L.; Jin, E.; Mei, H.; Ke, Q.; Li, Z.; Kui, H. Bio-inspired Graphene-enhanced Thermally Conductive Elastic Silicone Rubber as Drag Reduction Material. *J. Bionic Eng.* **2017**, *14*, 130–140. [CrossRef]
31. Yan, X.; Gao, Q.; Liang, H.; Zheng, K. Effects of functional graphene oxide on the properties of phenyl silicone rubber composites. *Polym. Test.* **2016**, *54*, 168–175.
32. Chaudhury, M.K.; Finlay, J.A.; Chung, J.Y.; Callow, M.E.; Callow, J.A. The influence of elastic modulus and thickness on the release of the soft-fouling green alga Ulva linza (syn. Enteromorpha linza) from poly(dimethylsiloxane) (PDMS) model networks. *Bioflouling* **2005**, *21*, 41–48. [CrossRef]
33. Jia, H.; Wu, Z.; Liu, N. Effect of nano-ZnO with different particle size on the performance of PVDF composite membrane. *Plast. Rubber Compos.* **2017**, *46*, 1–7. [CrossRef]
34. Scardino, A.J.; Zhang, H.; Cookson, D.J.; Lamb, R.N.; de Nys, R. The role of nano-roughness in antifouling. *Biofouling* **2009**, *25*, 757–767. [CrossRef] [PubMed]
35. Molena, E.; Credi, C.; Marco, C.D.; Levi, M.; Turri, S.; Simeone, G. Protein antifouling and fouling-release in perfluoropolyether surfaces. *Appl. Surf. Sci.* **2014**, *309*, 160–167. [CrossRef]
36. Shan, C.; Wang, J.; Yan, Z.; Chen, D. The effectiveness of an antifouling compound coating based on a silicone elastomer and colored phosphor powder against Navicula species diatom. *J. Coat. Technol. Res.* **2013**, *10*, 397–406.
37. Geoffrey, S.; Herpe, S.; Emily, R.; Melissa, T. Short-term testing of antifouling surfaces: The importance of colour. *Biofouling* **2006**, *22*, 425–429.
38. Brady, R.F. A fracture mechanical analysis of fouling release from nontoxic antifouling coatings. *Prog. Org. Coat.* **2001**, *43*, 188–192. [CrossRef]
39. Kulik, V.M. Boundary conditions on a compliant wall in the turbulent flow. *Thermophys. Aeromech.* **2013**, *20*, 435–439. [CrossRef]
40. Kulik, V.M.; Rodyakin, S.V.; Lee, I.; Chun, H.H. Deformation of a viscoelastic coating under the action of convective pressure pulsation. *Thermophys. Aeromech.* **2004**, *11*, 1–12.
41. Kulik, V.M. Action of a turbulent flow on a hard compliant coating. *Int. J. Heat Fluid Flow* **2012**, *33*, 232–241. [CrossRef]
42. Chaudhury, M.K.; Weaver, T.; Hui, C.Y.; Kramer, E.J. Adhesive contact of cylindrical lens and a flat sheet. *J. Appl. Phys.* **1996**, *80*, 30–37. [CrossRef]
43. Griffith, A.A. The phenomena of rupture and flow in solids. *Philos. Trans. R. Soc. London* **1921**, *221*, 163–198. [CrossRef]

© 2019 by the authors. Licensee MDPI, Basel, Switzerland. This article is an open access article distributed under the terms and conditions of the Creative Commons Attribution (CC BY) license (http://creativecommons.org/licenses/by/4.0/).

Article

Investigating the Mechanical Properties of ZrO$_2$-Impregnated PMMA Nanocomposite for Denture-Based Applications

Saleh Zidan [1,*], Nikolaos Silikas [1], Abdulaziz Alhotan [1], Julfikar Haider [2] and Julian Yates [1]

1. Dentistry, School of Medical Sciences, University of Manchester, Manchester M13 9PL, UK; Nikolaos.Silikas@manchester.ac.uk (N.S.); abdulaziz.alhotan@postgrad.manchester.ac.uk (A.A.); Julian.yates@manchester.ac.uk (J.Y.)
2. School of Engineering, Manchester Metropolitan University, Manchester M1 5GD UK; j.haider@mmu.ac.uk
* Correspondence: saleh.zidan@postgrad.manchester.ac.uk; Tel.: 44-79-3309-6536

Received: 28 March 2019; Accepted: 20 April 2019; Published: 25 April 2019

Abstract: Acrylic resin PMMA (poly-methyl methacrylate) is used in the manufacture of denture bases but its mechanical properties can be deficient in this role. This study investigated the mechanical properties (flexural strength, fracture toughness, impact strength, and hardness) and fracture behavior of a commercial, high impact (HI), heat-cured denture base acrylic resin impregnated with different concentrations of yttria-stabilized zirconia (ZrO$_2$) nanoparticles. Six groups were prepared having different wt% concentrations of ZrO$_2$ nanoparticles: 0% (control), 1.5%, 3%, 5%, 7%, and 10%, respectively. Flexural strength and flexural modulus were measured using a three-point bending test and surface hardness was evaluated using the Vickers hardness test. Fracture toughness and impact strength were evaluated using a single edge bending test and Charpy impact instrument. The fractured surfaces of impact test specimens were also observed using a scanning electron microscope (SEM). Statistical analyses were conducted on the data obtained from the experiments. The mean flexural strength of ZrO$_2$/PMMA nanocomposites (84 ± 6 MPa) at 3 wt% zirconia was significantly greater than that of the control group (72 ± 9 MPa) ($p < 0.05$). The mean flexural modulus was also significantly improved with different concentrations of zirconia when compared to the control group, with 5 wt% zirconia demonstrating the largest (23%) improvement. The mean fracture toughness increased in the group containing 5 wt% zirconia compared to the control group, but it was not significant. However, the median impact strength for all groups containing zirconia generally decreased when compared to the control group. Vickers hardness (HV) values significantly increased with an increase in ZrO$_2$ content, with the highest values obtained at 10 wt%, at 0 day (22.9 HV$_{0.05}$) in dry conditions when compared to the values obtained after immersing the specimens for seven days (18.4 HV$_{0.05}$) and 45 days (16.3 HV$_{0.05}$) in distilled water. Incorporation of ZrO$_2$ nanoparticles into high impact PMMA resin significantly improved flexural strength, flexural modulus, fracture toughness and surface hardness, with an optimum concentration of 3–5 wt% zirconia. However, the impact strength of the nanocomposites decreased, apart from the 5 wt% zirconia group.

Keywords: PMMA; zirconia (ZrO$_2$); nanocomposite; denture base; flexural strength; impact strength; fracture toughness; hardness

1. Introduction

In practical applications, denture base materials experiences different types of stresses, such as compressive, tensile and shear, which can lead to premature failure. Intra-orally, repeated mastication over a period of time can lead to denture base fatigue failure. Extra-orally, denture bases can also experience high impact forces when dropped by accident [1,2]. Impact fractures occur extra-orally as a result of inadvertent denture damage [1,3]. The incidence of denture fracture is relatively high: 68% of dentures fail within three years of fabrication and the incidence in partial denture is greater than that of complete dentures [4,5]. Studies have also reported that 33% of the repairs in dental laboratories are as a result of de-bonded teeth, and 29% percent of fractures occur in the midline of the denture base, being seen more frequently in the upper than in the lower prosthesis [6,7]. The remaining 38% of fractures are caused by other types of failure [6,7].

High impact (HI) denture base resins are widely used in prosthetic dentistry. These materials are provided in either powder or liquid forms and are processed in the same manner as other heat-cured, poly-methyl methacrylate (PMMA) resins. HI resins are reinforced with butadiene-styrene rubber, with the rubber particles grafted to the poly-methyl methacrylate so that the particles are covalently bonded into the polymerized acrylic matrix in order to better absorb mechanical loads [4,8–10]. Incorporation of butadiene-styrene rubber into PMMA resins improves impact strength and dimensional stability [8,11,12]. However, such reinforcement can result in the reduction of mechanical properties, including flexural strength, fatigue strength and stiffness [8,11,13].

Many attempts have been made to improve the strength of denture base resins, including the addition of metal wires and plates made of either Co-Cr alloy or stainless steel. However, these materials present limitations contrary to the standard requirements, including poor adhesion between the acrylic resin and reinforcing metal. This separation can result in a reduction in overall mechanical strength within the prosthesis, as well as poor aesthetics. Additionally, metal-reinforced denture bases can become noticeably heavier [13,14]. Other attempts to improve denture base mechanical properties include fibre reinforcement to enhance fracture toughness, flexural and impact strength, and fatigue properties [13,15]. Different fibre types, such as ultra-high modulus polyethylene fibre (UHMPE), aramid fibre, nylon fibre, carbon fibre and glass fibre, have all been investigated [13,15,16]. UHMPE fibre does not demonstrate good adhesion to PMMA, and therefore, no significant increase in flexural properties has been demonstrated [17]. Carbon and aramid fibres are not practical materials because of difficulties in polishing the final prostheses, and resultant poor aesthetics [18]. However, nylon reinforcement enhances fracture resistance and structural elasticity of acrylic resins [15]. A study undertaken by Vallittu et al., on the flexural and transverse strength of heat-cured PMMA denture bases reinforced with a high concentration of continuous glass fibre demonstrated an improvement in these properties [19]. Additionally, silane coupling agents have been added to enhance adhesion between the polymer resin and glass fibres to improve mechanical strength, resulting in enhanced flexural and fatigue strength [19,20]. However, fibre orientation in the resin matrix is technically difficult to control and a random distribution could result in defects within the finished product [21].

In recent years, several investigations have focused on improving the mechanical properties of PMMA acrylic resins by adding nanomaterials, such as bio-ceramic nanoparticles, due to their special characteristics [22]. Zirconia (ZrO_2) is a bio-ceramic material that has been widely used for various dental applications, such as crowns and bridges, implant fixture "screws" and abutments, and orthodontic brackets [23]. Zirconia has a high flexural strength (900 to 1200 MPa), hardness (1200 HV), and fracture toughness (9–10 MPa $m^{1/2}$) [24]. Furthermore, zirconia shows excellent biocompatibility compared to other ceramic materials, such as alumina [22,24]. A number of studies found that reinforcement of conventional, heat-cured denture base resins with zirconia nanoparticles significantly improved mechanical properties such as flexural and impact strength, as well as surface hardness [22,25]. However, no systematic study on the effect of zirconia addition in the high impact (HI) heat-cured PMMA denture base material has been reported in the literature. Therefore, research is needed to identify an optimum amount of zirconia suitable for improving performance and life of HI PMMA denture bases.

The purpose of this study is to evaluate the effects of zirconia nanoparticle addition at low concentrations (up to 10%) to a commercially available, high-impact, PMMA denture base resin on selected mechanical properties such as flexural strength, impact strength, fracture toughness, hardness and fracture behaviour.

2. Material and Methods

2.1. Materials

A commercially available, Metrocryl HI denture base powder, (PMMA, poly-methyl methacrylate) and Metrocryl HI (X-linked) denture base liquid (MMA, methyl methacrylate) (Metrodent Limited, Huddersfield, UK) were selected as the denture base material. Yttria-stabilized zirconia (ZrO_2) nanoparticles (94% purity; Sky Spring Nano materials, Inc., Houston, TX, USA) were chosen as the inorganic filler agent for fabricating the nanocomposite denture base specimens.

2.2. Specimen Preparation

2.2.1. Silane Functionalization of Zirconia Nanoparticle Surfaces

Fifteen grams of zirconia nanoparticles and 70 mL of toluene solvent were deposited into a plastic container, which was then placed in a speed mixer (DAC 150.1 FVZK, High Wycombe, UK), and mixed at 1500 rpm for 20 min. Following the initial mixing, 7 wt% silane coupling agent (3-trimethoxysilyl propyl methacrylate; product no. 440159, Sigma Aldrich, Gillingham, UK) was added slowly over a period of 20 s. The mixture was then placed in the speed mixer at 1500 rpm for 10 min and divided equally into two tubes and spun in a centrifuge at 23 °C at 4000 rpm for 20 min. The supernatant (separated toluene) was removed, and the remaining silanized nanoparticles were transferred into a personal solvent evaporator (EZ-2 Elite, Genevac Ltd., SP Scientific Company, Ipswich, UK) for 3 h of drying at 60 °C.

2.2.2. Selection of Appropriate Percentages of Zirconia Nanoparticles

To determine the most appropriate weight percentages of zirconia nanoparticles for the current study, preliminary investigations were undertaken using 1.5 wt%, 10 wt% and 15 wt% mixtures. Based on these results and knowledge from relevant literature, a decision was made to utilize the following weight percentages of silanized zirconia nanoparticles in the denture base formulation: 0.0% (control), 1.5 wt%, 3.0 wt%, 5.0 wt%, 7.0 wt%, and 10.0 wt%. The composition details of the specimen groups used in this study are described in Table 1 (all used an acrylic resin powder:monomer ratio of 21 g:10 mL, in accordance with manufacturer's instructions).

Table 1. Weight percent zirconia in combination with acrylic resin powder as well as monomer content of the specimen groups. HI: High impact; PMMA: Poly-methyl methacrylate; MMA: methyl methacrylate.

Experimental Groups	Zirconia (wt%)	Zirconia (g)	HI PMMA Powder (g)	HI MMA Monomer (mL)
Control	0.0	0.000	21.000	10.0
1.5	1.5	0.315	20.685	10.0
3.0	3.0	0.630	20.370	10.0
5.0	5.0	1.050	19.950	10.0
7.0	7.0	1.470	19.530	10.0
10.0	10.0	2.100	18.900	10.0

2.2.3. Mixing of Zirconia with PMMA

The silane-treated zirconia and acrylic resin powders were weighed according to Table 1 using an electronic balance (Ohaus Analytical with accuracy up to 3 decimal points). The zirconia powder was added to the acrylic resin monomer and mixed by hand using a stainless-steel spatula to make sure all the powder was uniformly distributed within the resin monomer. The HI acrylic resin powder was then added to the solution, and mixing continued until a consistent mixture was obtained, according to the manufacturer's instruction. The mixing continued for approximately 20 min until the mixture reached a dough-like stage, which was suitable for handling. When the mixture reached a consistent dough-like stage (working stage), it was packed into a mould by hand. The moulds were made from aluminium alloy, which contained five cavities with a dimension of 65 mm (l) × 10 mm (w) × 2.50 mm (d) for producing flexural strength and hardness test samples. However, the cavity dimensions for the impact test was as follows: 80 mm (l) × 10 mm (w) × 4 mm (d) and fracture toughness was 40 mm (l) × 8 mm (w) × 4 mm (d). Before pouring the mixture into the mould, sodium alginate as a separating medium (John Winter, Germany) was applied to the surfaces of the mould for easy removal of the specimens. The mould was then closed and placed in a hydraulic press (Sirio P400/13045) under a pressure of 15 MPa in the first cycle, and then the pressure was released. Excess mixture was removed from the mould periphery, which was then re-pressed at room temperature for 15 min under the same pressure. The mould was then immersed in a temperature-controlled curing water bath for 6 h to allow polymerization. The curing cycle involved increasing the temperature to 60 °C over 1 h and maintained this temperature for 3 h. After this time, the temperature was increased to 95 °C over an additional 2 h to complete the heat polymerization cycle. The mould was removed from the curing bath and cooled slowly for 30 min at room temperature. The mould was then opened and the specimens were removed. The specimens were then trimmed using a tungsten carbide bur, ground with an emery paper and polished with pumice powder in a polishing machine (Tavom, Wigan, UK) in accordance with British International Standard Organization (BS EN ISO 20795-1:2008) and British Standard Specification for Denture Base Polymers (BS 2487: 1989 ISO 1567; 1988) [26,27].

2.3. Mechanical Characterization of the Nanocomposite

2.3.1. Flexural Strength Test

Flexural strength of the nanocomposite specimens was evaluated using a 3-point bend test in a universal testing machine (Zwick/Roell Z020 Leominster, UK) in accordance with British International Standard for Denture Base Polymers (2487: 1989) [27]. The dimensions of the specimens were 65 mm length × 10 ± 0.01 mm width × 2.50 ± 0.01 mm thickness. All specimens were stored in distilled water at a temperature of 37 ± 1 °C for 50 ± 2 h in an incubator before testing. The specimens were then removed from the distilled water and placed on a support jig. The loading plunger (diameter 7.0 mm) was fixed at the center of the specimen midway between two supports, which were parallel and separated by 50 ± 0.1 mm, and the diameter of the load supports were 3.20 mm. A 500 N load cell was used to record force and the load was applied using a cross-head speed of 5 mm/min. The maximum force (F) was recorded in newtons, and flexural strength was calculated in MPa for all specimens using the following equation [28]:

$$\sigma = \frac{3Fl}{2bh^2} \tag{1}$$

where F is the maximum force applied in N, l is the distance between the supports in mm, b is the width of the specimen in mm, and h is the height of the specimen in mm. The flexural modulus was determined as the slope of the linear portion of the stress/strain curve for each test run.

2.3.2. Fracture Toughness Test

Fracture toughness tests were conducted using a single edge span notch bending test on the Zwick universal testing machine in accordance with the British International Standard Organization (BS EN ISO 20795-1:2008) [26,29]. The dimensions of the specimens were 40 mm (l) × 8 mm (w) × 4 mm (h), and a notch was created in the middle of the specimens with a diamond blade and a saw to a depth of 3.0 ± 0.2 mm along a marked centre line. All specimens were then stored in distilled water and placed in an incubator at 37 ± 1 °C for 168 ± 2 h before testing. The specimens were removed from the water, dried by a towel and placed edgewise on the supports of the testing rig. The notch of the specimen was placed directly opposite to the load plunger (diameter 7 mm) and in the middle of the span between the two supports (32.0 ± 0.1 mm). The load cell was 500 N, and the cross-head speed was 1.0 mm/min. Fracture toughness was determined by increasing the force from zero to a maximum value in order to propagate a crack from the opposite side of the specimen to the impact point. The maximum force (P) in newtons to fracture was recorded in order to calculate the fracture toughness (K_{IC}) in MPa m$^{1/2}$ according to Equation (2) [29]:

$$K_{IC} = \frac{3PL}{2BW^{\frac{3}{2}}} \times Y \tag{2}$$

where W is the height of the specimen in mm, B is the width of the specimen in mm, L is the distance between the supports in mm, and Y is a geometrical function calculated by Equation (3).

$$Y = 1.93 \times \left(\frac{a}{w}\right)^{1/2} - 3.07 \times \left(\frac{a}{w}\right)^{\frac{3}{2}} + 14.53 \times \left(\frac{a}{w}\right)^{\frac{5}{2}} - 25.11 \times \left(\frac{a}{w}\right)^{\frac{7}{2}} + 25.80 \times \left(\frac{a}{w}\right)^{\frac{9}{2}} \tag{3}$$

where a is the depth of the notch.

2.3.3. Impact Test

The Charpy V-notch impact test (kJ/m^2) utilized a universal pendulum impact testing machine (Zwick/Roell Z020 Leominster). Specimen dimensions were 80 mm (l) × 10 ± 0.01 mm (w) × 4 ± 0.01 mm (h), in accordance with the European International Standard Organization (EN ISO 179-1:2000) [30]. The specimens were notched in the middle to a depth of 2.0 ± 0.2 mm, a notch angle of 45 ° and a notch radius of 1.0 ± 0.05 mm and were then stored in distilled water at 37 ± 1 °C for 168 ± 2 h in an incubator before testing. The specimens were then removed from the water and dried with a towel. Each specimen was placed in the machine and were supported horizontally at its ends (40 ± 0.2 mm), and the centre of the specimen (the un-notched surface) was hit by a free-swinging pendulum that was released from a fixed height. The pendulum load cell was 0.5 J and directly faced the centre of the specimen, as shown in Figure 1. When the test was started, the pendulum was released to strike the specimen, and the impact energy absorbed was recorded in joules (J). The Charpy impact strength (a_{iN}) (kJ/m^2) was calculated using Equation (4) [10,30]:

$$a_{iN} = \frac{E_c}{h * b_N} \times 10^3 \tag{4}$$

where E_c is the breaking energy in joules absorbed by breaking, h is the thickness in mm, and b_N is the remaining width in mm after notching.

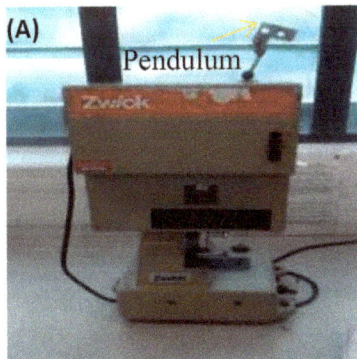
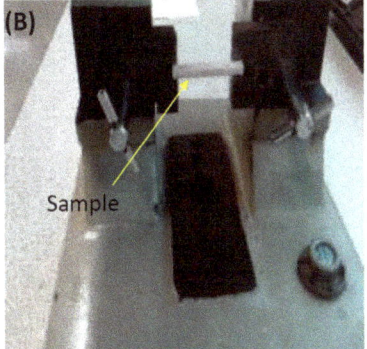

Figure 1. (**A**) Impact test machine and (**B**) position of sample in the machine before the test.

2.3.4. Hardness Test

The Vickers hardness ($HV_{0.05}$) of the specimens was measured using a micro-hardness testing machine (FM-700, Future Tech Corp, Tokyo, Japan). Specimens were 65 mm length × 10 mm width × 2.50 mm thickness, and the test load was fixed at 50 g for 30 s. The Vickers hardness was calculated by measuring the diagonals of the pyramid-shaped indentation impressed on the specimen. A total of three indentations were taken at different points in each specimen one side, and then a mean value was calculated. The mean hardness values for all the specimens were determined demonstrative of the materials in the dry condition at day 0. The specimens were then stored individually in 37 ± 1 °C distilled water for 7 d ± 2 h, and were then re-immersed for a total of 45 d ± 2 h. From the raw data, the mean hardness values for each sample group were calculated [31,32].

2.4. Scanning Electron Microscopy (SEM) Examination

The size and shape distribution of the PMMA powder and zirconia nanoparticles was analysed using a scanning electron microscope (SEM) (Carl Zeiss Ltd, 40 VP, Smart SEM, Cambridge, UK). The fractured surface was also studied to identify failure mechanism. Specimens were mounted onto aluminium stubs and sputter-coated with gold after which SEM visualization was performed using a secondary electron detector at an acceleration voltage of 2.0 kV.

2.5. Statistical Analyses

Flexural strength, modulus, impact strength, fracture toughness and Vickers hardness data were analysed using a statistical software (SPSS statistics version 23, IBM, New York, NY, USA). Non-significant Shapiro–Wilk and Levene tests showed that the data of flexural and fracture toughness were normally distributed and there was homogeneity of variance. The flexural and fracture toughness data were analysed using a one-way analysis of variance (ANOVA) with the Tukey honestly significant difference post-hoc test at a pre-set alpha of 0.05. Impact strength and hardness data demonstrated nonparametric distributions as evidenced by significant Shapiro–Wilk test results for two groups, and therefore the Kruskal–Wallis test was used to analyse the results as well as to compare the differences among the test groups at a pre-set alpha of 0.05. In addition, the Friedman's two-way analysis test was applied to identify any significant difference between the three immersion time groups ($p < 0.05$).

3. Results

3.1. Visual Analysis

SEM analysis revealed that the average particle size of the PMMA powder was approximately 50 μm with a range from 10 μm to 100 μm, as shown in Figure 2A. The rubber particles were also visible within the powder, with an average size of approximately 50 μm. The as-received, yttria-stabilized zirconia nanoparticles demonstrated an average size ranging between 30 nm and 60 nm for individual particles and 200 nm to 300 nm for clusters, as shown in Figure 2B.

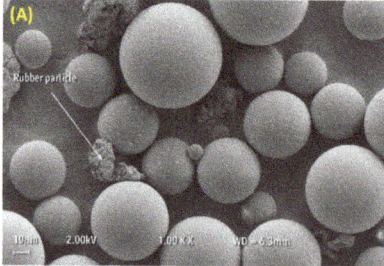

Figure 2. Particle size and shape distributions of (**A**) PMMA powder and (**B**) zirconia nanoparticles.

3.2. Mechanical Tests

3.2.1. Flexural Strength and Flexural Modulus

One-way analysis of variance (ANOVA) of flexural strength values presented in Table 2 show a significant difference ($p < 0.05$) for the specimen group containing 3 wt% zirconia. However, the mean values of flexural modulus showed a significant increase ($p < 0.05$) for all specimens, except that containing 7 wt% zirconia, which was not significantly different ($p > 0.05$) from the control group. The flexural strength data in the table demonstrates that an addition of zirconia nanoparticles to the HI PMMA gradually increased the strength up to 3 wt% and then gradually decreased for other compositions when compared to the control group (0 wt%). The highest value of flexural strength was recorded for the group containing 3 wt% zirconia (83.5 MPa) in comparison with the control group (72.4 MPa), representing a 15% increase in the flexural strength. However, a higher percentage of zirconia nanoparticles (7 wt% to 10 wt%) in the specimens reduced the strength, which was comparable to the control group. A similar behaviour was also found for the flexural modulus of the nanocomposites with increasing zirconia content (Table 2). However, a maximum value of the flexural modulus was reached at a zirconia content of 5 wt% (2419 MPa) when compared to the control group (1971 MPa), meaning an increase of 22.7%. Furthermore, even though at high zirconia content the modulus values decreased, they were still higher than those of the control group.

Table 2. Mean (MPa) Standard deviation (SD) values of flexural strength, flexural modulus and fracture toughness as well as median of impact strength (kJ/m^2) Interquartile range (IQR) for the test groups.

Zirconia Content (wt%)	Flexural Strength and SD (MPa)	Flexural Modulus and SD (MPa)	Impact Strength and (IQR) (kJ/m^2)	Fracture Toughness and (SD) (MPa m$^{1/2}$)
Control (0%)	72.4 (8.6) A	1971 (235) A	10.0 (2.69) A	2.12 (0.1) A
1.5	78.7 (6.9) A	2237 (117) B	7.03 (4.45) A	1.91 (0.2) A
3.0	83.5 (6.2) B	2313 (161) B	7.38 (4.50) A	1.97 (0.2) A
5.0	78.7 (7.2) A	2419 (147) B	9.05 (3.50) A	2.14 (0.1) A
7.0	72.2 (7.0) A	2144 (85) A	7.12 (1.50) A	1.86 (0.1) A
10.0	71.5 (5.7) A	2204 (91) B	5.89 (2.33) B	1.76 (0.8) B

Within a column, cells having similar (upper case) letters are not significantly different from the control (0% zirconia content) value. N = 10 specimens per group.

3.2.2. Fracture Toughness and Impact Strength

The mean values of the fracture toughness (Table 2) of the nanocomposites decreased significantly compared to that of the control group at the zirconia concentrations of 7% and 10% ($p < 0.05$). Furthermore, after the initial decrease of fracture toughness at 1.5 wt% zirconia, the values slightly increased in the groups containing 3 wt% and 5 wt% zirconia, but they were not statistically significant increases ($p > 0.05$). Table 2 shows that the best fracture toughness could be achieved at 5 wt% zirconia.

The values of the impact strength for all nanocomposite groups were not statistically significant ($p > 0.05$), as shown in Table 2. The median impact strength gradually decreased with the increase in zirconia content, except in the group containing 5 wt% zirconia, which showed the best impact strength (only 10% reduction compared to the control group). However, all measured impact strength values for the nanocomposites were lower than that for the control group.

3.2.3. Hardness

The median values of Vickers hardness in Table 3 show significant differences ($p < 0.05$) for the specimen groups containing 7 wt% and 10 wt% zirconia in both dry (0 day) and wet (7 days) conditions. From the graphical presentation of the hardness results (Figure 3), it is interesting to note that at lower zirconia contents (1.5–5.0%), the difference in hardness between dry and wet conditions was much lower than that at higher zirconia contents (7.0–10.0%). Furthermore, no significant difference was found between the hardness of the specimens stored in water for seven days and 45 days at all zirconia contents. This finding indicates that the hardness of the nanocomposites does not degrade over time in the wet condition at lower zirconia contents, particularly up to 3% zirconia.

Table 3. Vickers hardness (kg/mm^2) (median and interquartile range) after 0, 7 and 45 days of water immersion.

Weight Percent Zirconia	Day Zero (Dry) Vickers Hardness (kg/mm^2) Median (IQR)	7-Days Water-Immersion Vickers Hardness (kg/mm^2) Median (IQR)	45 Days Water-Immersion Vickers Hardness (kg/mm^2) Median (IQR)
Control (0.0%)	17.6 (1.7) ^{Aa}	15.2 (2.0) ^{Ab}	15.5 (3.3) ^{Ab*}
1.5%	18.9 (3.2) ^{Ab}	17.7 (1.1) ^{Ab}	17.0 (1.8) ^{Ab*}
3.0%	19.6 (4.0) ^{Ac}	17.8 (1.2) ^{Ac}	17.3 (2.8) ^{Ac}
5.0%	21.1 (3.1) ^{Ad}	17.9 (2.9) ^{Ad}	17.1 (2.2) ^{Ad*}
7.0%	21.7 (3.0) ^{Be}	19.4 (0.9) ^{Be}	16.8 (2.3) ^{Ae*}
10.0%	22.9 (2.9) ^{Bf}	18.4 (3.3) ^{Bf}	16.3 (1.2) ^{Af*}

Within a column, values identified using similar upper-case letters are not significantly different from the control group value; within rows values identified using the same lower-case letters are not significantly different; asterisks indicate significant differences between day 0 and 45 days; N = 5 specimens per experimental group.

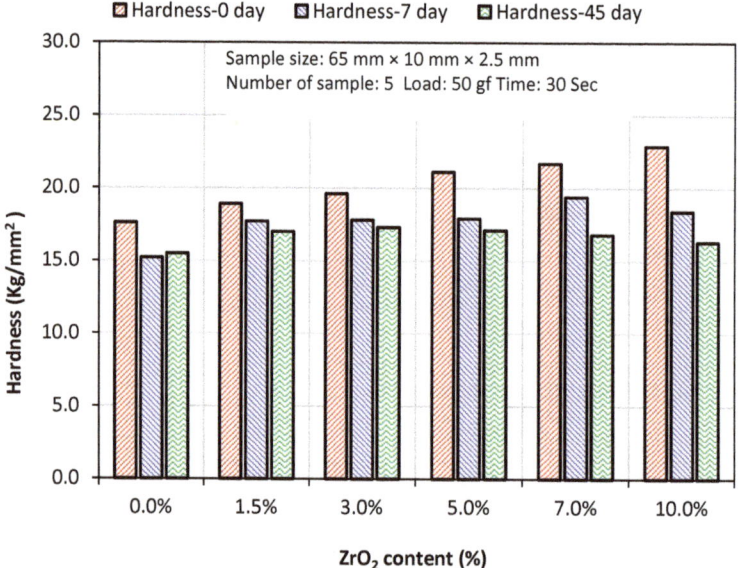

Figure 3. Vickers hardness median (kg/mm^2) after 0, 7, and 45 days of water immersion.

3.3. Microstructural Characteristics

The fractured surface of pure PMMA specimens displayed a smooth surface in small areas and revealed a ductile type failure behaviour exhibiting irregular and rough surface as is shown in Figure 4A. The composite fractured surface showed signs of cracks and particle clustering with small voids (Figure 4B). Figure 4C presents more clear fracture features and shows that the distribution of the nanoparticles was not uniform. The image highlights particle clustering in several places and voids on the fractured surface.

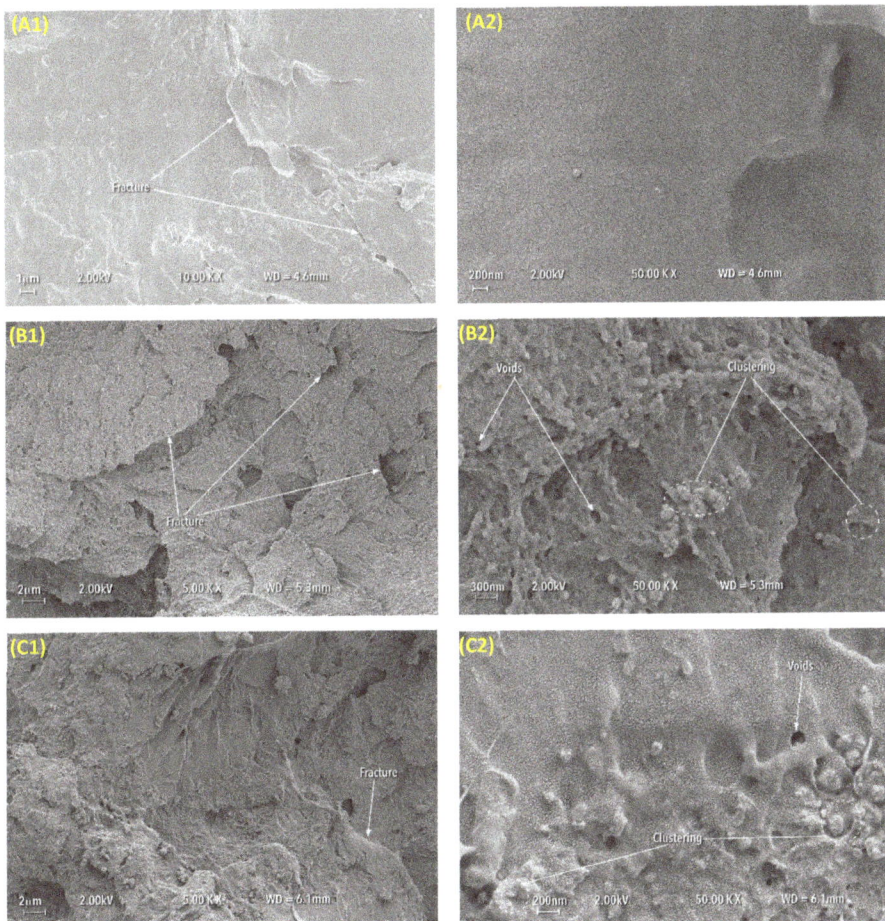

Figure 4. Representative SEM images of the fractured surfaces of impact strength test specimens at two different magnifications (1 at 10K and 2 at 50K for the control group (**A**) 0 wt%, (**B**) 5 wt% and (**C**) 10 wt% added zirconia, respectively).

4. Discussion

In this study, it was shown that combining zirconia nanoparticles to HI acrylic resin improved flexural strength and flexural modulus, which can lead to a reduction in different types of stresses encountered during the mastication process, including compressive, tensile and shear stresses [33]. However, the reinforced HI acrylic resin with lower concentration of zirconia (5%) did not show any significant difference from the control group on fracture toughness and impact strength.

The inorganic reinforcing nano-fillers have a large surface area that provides high surface energy, and this produces nanoparticles with a strong tendency to aggregate. This characteristic may decrease the chemical interaction between the nanoparticles and the base PMMA [22]. In this study, to enhance the chemical adhesion between the ZrO_2 nanoparticles and ZrO_2-PMMA, the surface of the ZrO_2 particles was treated with 7 wt% silane coupling agent (3-MPS) to create reactive functional groups. This could be responsible for improving the flexural properties of the nanocomposites at lower concentrations of zirconia nanoparticles. Moreover, the improvement in flexural strength and flexural modulus could be a result of the improved dispersion of the ZrO_2 nanoparticles when mixing with the speed mixer machine during the preparation stage. This improvement would decrease the agglomeration tendency in the composites. Additionally, the large interfacial area of the nanoparticles contributes to more contact points between the ZrO_2 and PMMA, thus enhancing mechanical interlocking and offers additional flexibility in the nanocomposites [34].

Only a few studies on the effect of adding ZrO_2 nanoparticles in HI heat-cured denture base acrylic resin are available in the literature. In contrast, investigators have worked on improving the mechanical properties of conventional heat-cured denture base acrylic resin by incorporating different types of fillers [35]. Alhareb et al. [36] showed a 16% increase in flexural strength value compared to control samples when PMMA was reinforced with Al_2O_3 and ZrO_2 with a filler concentration of 5 wt%. Moreover, the flexural modulus increased with an increase in Al_2O_3/ZrO_2 nanoparticle concentration [36]. The greater value of the modulus indicates a stiffer material [16], and this improvement can be explained by a homogenous distribution of the fillers within the polymer matrix. Vojdani et al. [3] evaluated the effect of adding Al_2O_3 particles to PMMA denture bases on flexural strength. They found that a 6% increase in flexural strength value with 2.5 wt% Al_2O_3 compared to a control group could be obtained. Zhang et al. [22] investigated the effect of hybrid ZrO_2 nanoparticles and micro-particles of aluminium borate whiskers (ABWS) at concentrations of 1 wt%, 2 wt%, 3 wt%, and 4 wt% on the flexural strength of PMMA denture base resin. They found that 2 wt% nano-ZrO_2 with a ZrO_2/ABWS ratio of 1:2 improved flexural strength by 32% when compared to a control group. These previous studies in the literature were in agreement with the results obtained in this study, which revealed that zirconia positively influenced the flexural properties of HI PMMA with an optimum zirconia concentration between 3 wt% and 5 wt%.

Fracture toughness (K_{IC}) is a critical stress intensity factor that provides information on crack formation [29] and the ability of a material to resist crack propagation [37]. The reduction of fracture toughness in the PMMA/ZrO_2 nanocomposites with increasing filler content could be due to a number of reasons, such as particle distribution in the polymer matrix, the type and size of the particles, the concentration of the added particles, and chemical reactions between the particles and polymer [15,35,38]. A high filler concentration leads to more filler-to-filler interactions than filler-to-matrix interactions; therefore, agglomeration may act as a point of stress concentration that could lead to non-uniform stress distribution. When applying the load, the agglomeration restrains the movement of molecular deformation and reduces the fracture toughness [38]. Sodagar et al. [39] determined that the incorporation of TiO_2 nanoparticles to the PMMA matrix causes agglomeration, which acts as a stress raiser in the centre of the matrix and reduces the mechanical properties of the polymer material with increasing concentrations of the TiO_2 nanoparticles. Fangqiang et al. [40] investigated the distribution of ZrO_2 particles in PMMA matrix using two strategies during mixing: Physical method and chemical method. The physical method was conducted by melt blending, high-energy ball milling or ultrasonic vibration. In the chemical method, when mixing nanoparticles with an MMA monomer, the inorganic ZrO_2 nanoparticles acted as a core, and the monomer as a shell structure by in situ polymerization of the monomers, known as grafting. The chemically modified nanoparticle surfaces with MMA enhanced the dispersion stability of the nanoparticles in the polymer matrix. Owing to a combined physical and chemical preparation, it was observed that the dispersion of ZrO_2 nanoparticles in the polymer matrix was enhanced and particle aggregation and phase separation decreased to a demonstrable extent. The results of the present study on fracture toughness are consistent with those reported in the study of Alhareb et al. [36], where a PMMA denture base reinforced with 5 wt% fillers (80/20 Al_2O_3/ZrO_2) showed an improvement in fracture toughness but an increase in zirconia concentration decreased toughness.

The incorporation of hard ZrO_2 ceramic into PMMA can increase brittleness in the specimens, which would reduce the impact strength. Additionally, the lack of adhesion due to poor chemical reaction at the interface between the particles and PMMA or the inhomogeneous distribution of the nanoparticles with frequent clustering could affect the impact strength negatively [25,36]. A study conducted by Gad et al. [25] evaluated the effect of the incorporation of ZrO_2 nanoparticles with varying concentrations (2.5 wt%, 5 wt% and 7 wt%) to PMMA denture bases on impact strength. The results showed that the impact strength decreased with an increase in ZrO_2 nanoparticle concentration. The finding of the impact strength in the present study is in agreement with that of the previous study, with the exception of the 5 wt% ZrO_2/PMMA nanocomposite results. This result can be explained by the fact that a concentration of 5 wt% might be the optimum quantity to improve particle distribution and reduce amalgamation. Asar et al. [35] investigated the influence of metal oxides, ZrO_2, TiO_3, and Al_2O_3, with 1% and 2% by volume on the impact strength of the PMMA acrylic resin. In contrast to the current study, the findings showed a slight increase in the values of impact strength with 2% ZrO_2 addition.

Denture base materials should also have adequate abrasion resistance to prevent high wear of the material by abrasive denture cleansers, food or general functional forces [41]. Greater hardness in the denture base will reduce abrasive wear. The improvement of hardness in the nanocomposites might be related to the inclusion of hard yttria-stabilized zirconia nanoparticles with fine grains, which are known as tetragonal zirconia poly-crystals (TZP). The size of the grains is dependent on the metastable nature of the tetragonal phase and can be important for providing improved mechanical properties in the nanocomposites, and this zirconia-yttria phase increases surface hardness to resist indentation [42]. However, the increase in surface hardness with the increase of the concentration of zirconia also reduces the impact strength, as seen in Table 3. The reason for hardness decrease after water immersion was described in a previous study conducted on acrylic resin denture base materials, where residual monomers release and water absorption occurring simultaneously caused the surface to become softened [43].

The finding of the present study is in agreement with a study by Yiqing et al. [43], who evaluated the hardness of PMMA/ZrO_2 nanocomposites with different ZrO_2 concentrations (0.5 wt%, 1 wt%, 2 wt%, 3 wt%, 4 wt%, 5 wt%, 7 wt% and 15 wt%) using indentation and pendulum hardness tests. They found that the hardness values were increased with an increase in the ratio of ZrO_2 to PMMA, with the highest value being 15 wt%. Zhang et al. [22] investigated the effect of zirconia nanoparticles and aluminium borate whiskers (ABW) in PMMA denture bases on the surface hardness at concentrations of 1 wt%, 2 wt%, 3 wt% and 4 wt%. The results showed an increase in surface hardness with an increase in ZrO_2/ABW content, and the optimum hardness was achieved at 3 wt% ZrO_2 nanoparticles. They suggested that the decrease in surface hardness with higher filler loading was caused by poor adhesion of the particles to the resin matrix and filler clustering within the matrix. In another study, the incorporation of aluminium oxide (Al_2O_3) with percentages of 0.5 wt%, 1 wt%, 2.5 wt% and 5 wt% to PMMA acrylic resin exhibited an improvement in Vickers hardness with an increase of Al_2O_3 filler concentrations [3].

The lower impact strength in the nanocomposites can be related to the presence of voids and clustering of the nanoparticles [22,36]. At high magnification, the SEM images showed voids on the fractured surface, and these voids could lead to the generation of stress concentration under loading and initiate crack propagation by crossing the HI PMMA/ZrO_2 nanocomposite matrix. At low magnification, the fractured surfaces of the nanocomposite specimens exhibited less ductile fracture compared to the control group with a large amount of fragment crack deformation, which formed an irregular surface. Furthermore, the distribution of ZrO_2 nanoparticles in the polymer matrix was not homogeneous with evidence of agglomerations, which could reduce the impact strength, particularly at high ZrO_2 concentrations (10 wt%).

5. Conclusions

With consideration to the limitations of this study, the following conclusions can be drawn:

1. The flexural strength of the high impact (HI) heat-cured PMMA denture base was significantly enhanced by the addition of zirconia nanoparticles with 3 wt% when compared to the pure acrylic material (control group).

2. The flexural modulus of the high impact (HI) heat-cured PMMA denture base was significantly enhanced compared to the control group by addition of zirconia nanoparticles with 1.5 wt%, 3 wt%, 5 wt% and 10 wt%. The 7 wt% of zirconia showed a non-significant enhancement compared to the control group.
3. The fracture toughness of the zirconia-reinforced PMMA was significantly decreased, particularly at 10 wt% ZrO_2 concentration. The fracture toughness was slightly increased at 5 wt%, but this was not significantly different compared to the control group.
4. For all zirconia contents, the impact strength of the nanocomposites was significantly lower than that of the control group. However, at 5 wt% and 3 wt% zirconia content, the proportion of reduction in impact strength was not significantly different from that of the control group.
5. Surface hardness continuously increased with increase of zirconia content, in the dry condition at day 0. However, in the wet condition after seven days, and 45 days surface hardness was decreased with all groups.
6. Addition of zirconia in PMMA between 3 wt% and 5 wt% zirconia would provide the optimum mechanical properties suitable for denture base applications.

Author Contributions: Conceptualization, J.Y. and S.Z.; Methodology, J.Y., S.Z., N.S., A.A. and J.H.; Validation, S.Z.; Formal Analysis, S.Z., J.H., N.S., and J.Y.; Investigation, S.Z; Data Curation, S.Z.; Writing—Original Draft Preparation, S.Z. and J.H.; Writing—Review & Editing, N.S., J.Y., J.H., S.Z., and A.A.; Visualization, S.Z. and J.H.; Supervision, J.Y., N.S. and J.H.; Project Administration, J.Y.

Funding: This research received no external funding.

Acknowledgments: The authors would like to thank the ministry of higher education of Libya for providing financial support for PhD study; Brian Daber from Department of Dental Biomaterial, University of Manchester; Paul Murphy from University Dental Hospital of Manchester; Michael Green and Hayley Andrews from the Faculty of Science and Engineering, Manchester Metropolitan University and Gary Pickles from School of Materials, University of Manchester, for supporting the experimental work.

Conflicts of Interest: The authors declare no conflict of interest.

Abbreviations

PMMA	Poly-methyl methacrylate
MMA	Methyl methacrylate
HI	High impact heat cured acrylic resin
HV	Vickers hardness
SD	Standard deviation
IQR	Interquartile range
SEM	Scanning Electron Microscope

References

1. Zappini, G.; Kammann, A.; Wachter, W. Comparison of fracture tests of denture base materials. *J. Prosthet. Dent.* **2003**, *90*, 578–585. [CrossRef]
2. Kanie, T.; Fujii, K.; Arikawa, H.; Inoue, K. Flexural properties and impact strength of denture base polymer reinforced with woven glass fibers. *Dent. Mater.* **2000**, *16*, 150–158. [CrossRef]
3. Vojdani, M.; Bagheri, R.; Khaledi, A.A.R. Effects of aluminum oxide addition on the flexural strength, surface hardness, and roughness of heat-polymerized acrylic resin. *J. Dent. Sci.* **2012**, *7*, 238–244. [CrossRef]
4. Sasaki, H.; Hamanaka, I.; Takahashi, Y.; Kawaguchi, T. Effect of long-term water immersion or thermal shock on mechanical properties of high-impact acrylic denture base resins. *Dent. Mater. J.* **2016**, *35*, 204–209. [CrossRef]
5. Jagger, D.C.; Harrison, A.; Jandt, K. The reinforcement of dentures. *J. Oral Rehabil.* **1999**, *26*, 185–194. [CrossRef]
6. Nejatian, T.; Johnson, A.; van Noort, R. Reinforcement of denture base resin. *Adv. Sci. Technol.* **2006**, *49*, 124–129. [CrossRef]

7. Agha, H.; Flinton, R.; Vaidyanathan, T. Optimization of Fracture Resistance and Stiffness of Heat-Polymerized High. Impact Acrylic Resin with Localized E-Glass FiBER FORCE(R) Reinforcement at Different Stress Points. *J. Prosthodont.* **2016**, *25*, 647–655. [CrossRef] [PubMed]
8. Jagger, D.; Harrison, A.; Jagger, R.; Milward, P. The effect of the addition of poly(methyl methacrylate) fibres on some properties of high strength heat-cured acrylic resin denture base material. *J. Oral Rehabil.* **2003**, *30*, 231–235. [CrossRef]
9. Stafford, G.; Bates, J.; Huggett, R.; Handley, R. A review of the properties of some denture base polymers. *J. Dent.* **1980**, *8*, 292–306. [CrossRef]
10. Abdulwahhab, S.S. High-impact strength acrylic denture base material processed by autoclave. *J. Prosthodont. Res.* **2013**, *57*, 288–293. [CrossRef]
11. Jagger, D.C.; Jagger, R.G.; Allen, S.M.; Harrison, A. An investigation into the transverse and impact strength of 'high strength' denture base acrylic resins. *J. Oral Rehabil.* **2002**, *29*, 263–267. [CrossRef]
12. Zheng, J.; Wang, L.; Hu, Y.; Yao, K. Toughening effect of comonomer on acrylic denture base resin prepared via suspension copolymerization. *J. Appl. Polym. Sci.* **2012**, *123*, 2406–2413. [CrossRef]
13. Kim, S.-H.; Watts, D.C. The effect of reinforcement with woven E-glass fibers on the impact strength of complete dentures fabricated with high-impact acrylic resin. *J. Prosthet. Dent.* **2004**, *91*, 274–280. [CrossRef]
14. Yu, S.-H.; Cho, H.-W.; Oh, S.; Bae, J.-M. Effects of glass fiber mesh with different fiber content and structures on the compressive properties of complete dentures. *J. Prosthet. Dent.* **2015**, *113*, 636–644. [CrossRef]
15. Gad, M.M.; Fouda, S.M.; Al-Harbi, F.; Näpänkangas, R.; Raustia, A.; Al-Harbi, F. PMMA denture base material enhancement: A review of fiber, filler, and nanofiller addition. *Int. J. Nanomed.* **2017**, *12*, 3801–3812. [CrossRef]
16. Uzun, G.; Hersek, N.; Tincer, T. Effect of five woven fiber reinforcements on the impact and transverse strength of a denture base resin. *J. Prosthet. Dent.* **1999**, *81*, 616–620. [CrossRef]
17. Köroğlu, A.; Özdemir, T.; Usanmaz, A. Comparative study of the mechanical properties of fiber-reinforced denture base resin. *J. Appl. Polym. Sci.* **2009**, *113*, 716–720. [CrossRef]
18. Kim, H.-H.; Kim, M.-J.; Kwon, H.-B.; Lim, Y.J.; Kim, S.-K.; Koak, J.-Y. Strength and cytotoxicity in glass-fiber-reinforced denture base resin with changes in the monomer. *J. Appl. Polym. Sci.* **2012**, *126*, E260–E266. [CrossRef]
19. Vallittu, P.K.; Lassila, V.P.; Lappalainen, R. Transverse strength and fatigue of denture acrylic-glass fiber composite. *Dent. Mater.* **1994**, *10*, 116–121. [CrossRef]
20. Vallittu, P.K.; Lassila, V.P.; Lappalainen, R. Acrylic resin-fiber composite—Part I: The effect of fiber concentration on fracture resistance. *J. Prosthet. Dent.* **1994**, *71*, 607–612. [CrossRef]
21. Pan, Y.; Liu, F.; Xu, D.; Jiang, X.; Yu, H.; Zhu, M. Novel acrylic resin denture base with enhanced mechanical properties by the incorporation of PMMA-modified hydroxyapatite. *Prog. Nat. Sci. Mater. Int.* **2013**, *23*, 89–93. [CrossRef]
22. Zhang, X.-Y.; Zhang, X.-J.; Huang, Z.-L.; Zhu, B.-S.; Chen, R.-R. Hybrid effects of zirconia nanoparticles with aluminum borate whiskers on mechanical properties of denture base resin PMMA. *Dent. Mater. J.* **2014**, *33*, 141–146. [CrossRef] [PubMed]
23. Wang, T.; Tsoi, J.K.-H.; Matinlinna, J.P. A novel zirconia fibre-reinforced resin composite for dental use. *J. Mech. Behav. Biomed. Mater.* **2016**, *53* (Suppl. C), 151–160. [CrossRef]
24. Kawai, N.; Lin, J.; Youmaru, H.; Shinya, A.; Shinya, A. Effects of three luting agents and cyclic impact loading on shear bond strengths to zirconia with tribochemical treatment. *J. Dent. Sci.* **2012**, *7*, 118–124. [CrossRef]
25. Gad, M.M.; Rahoma, A.; Al-Thobity, A.M.; ArRejaie, A.S. Influence of incorporation of ZrO_2 nanoparticles on the repair strength of polymethyl methacrylate denture bases. *Int. J. Nanomed.* **2016**, *11*, 5633–5643. [CrossRef]
26. British Standards. *Dentistry-Base polymers BS EN ISO 20795-1:2008*; Biritish Standards Institution (BSI): London, UK, 2008; p. 36.
27. British Standards. *British Standard Specification for Denture base Polymers BS 2487:1989 ISO 1567:1988*; Biritish Standards Institution (BSI): London, UK, 1989; p. 10.
28. Jerolimov, V.; Brooks, S.; Huggett, R.; Bates, J. Rapid curing of acrylic denture-base materials. *Dent. Mater.* **1989**, *5*, 18–22. [CrossRef]

29. Al-Haddad, A.; Roudsari, R.V.; Satterthwaite, J.D. Fracture toughness of heat cured denture base acrylic resin modified with Chlorhexidine and Fluconazole as bioactive compounds. *J. Dent.* **2014**, *42*, 180–184. [CrossRef] [PubMed]
30. European International Standard Organization. *European International Standard Organization (EN ISO 179-1:2000)*; International Organization for Standardization: Geneva, Switzerland, 2000.
31. Neppelenbroek, K.H.; Pavarina, A.C.; Vergani, C.E.; Giampaolo, E.T. Hardness of heat-polymerized acrylic resins after disinfection and long-term water immersion. *J. Prosthet. Dent.* **2005**, *93*, 171–176. [CrossRef]
32. Farina, A.P.; Cecchin, D.; Soares, R.G.; Botelho, A.L.; Takahashi, J.M.F.K.; Mazzetto, M.O.; Mesquita, M.F. Evaluation of Vickers hardness of different types of acrylic denture base resins with and without glass fibre reinforcement. *Gerodontology* **2012**, *29*, e155–e160. [CrossRef]
33. Li, B.B.; Bin Xu, J.; Cui, H.Y.; Lin, Y.; Di, P. In vitro evaluation of the flexural properties of All-on-Four provisional fixed denture base resin partially reinforced with fibers. *Dent. Mater. J.* **2016**, *35*, 264–269. [CrossRef]
34. Gad, M.M.; Abualsaud, R.; Rahoma, A.; Al-Thobity, A.M.; Al-Abidi, K.S.; Akhtar, S. Effect of zirconium oxide nanoparticles addition on the optical and tensile properties of polymethyl methacrylate denture base material. *Int. J. Nanomed.* **2018**, *13*, 283. [CrossRef]
35. Asar, N.V.; Albayrak, H.; Korkmaz, T.; Turkyilmaz, I. Influence of various metal oxides on mechanical and physical properties of heat-cured polymethyl methacrylate denture base resins. *J. Adv. Prosthodont.* **2013**, *5*, 241–247. [CrossRef]
36. Alhareb, A.O.; Ahmad, Z.A. Effect of Al_2O_3/ZrO_2 reinforcement on the mechanical properties of PMMA denture base. *J. Reinf. Plast. Compos.* **2011**, *30*, 86–93. [CrossRef]
37. Hamza, T.A.; Rosenstiel, S.F.; Elhosary, M.M.; Ibraheem, R.M. The effect of fiber reinforcement on the fracture toughness and flexural strength of provisional restorative resins. *J. Prosthet. Dent.* **2004**, *91*, 258–264. [CrossRef]
38. Kundie, F.; Azhari, C.H.; Ahmad, Z.A. Effect of nano-and micro-alumina fillers on some properties of poly (methyl methacrylate) denture base composites. *J. Serb. Chem. Soc.* **2018**, *83*, 75–91. [CrossRef]
39. Sodagar, A.; Bahador, A.; Khalil, S.; Shahroudi, A.S.; Kassaee, M.Z. The effect of TiO_2 and SiO_2 nanoparticles on flexural strength of poly (methyl methacrylate) acrylic resins. *J. Prosthodont. Res.* **2013**, *57*, 15–19. [CrossRef]
40. Fan, F.; Xia, Z.; Li, Q.; Li, Z.; Chen, H. ZrO_2/PMMA nanocomposites: Preparation and its dispersion in polymer matrix. *Chin. J. Chem. Eng.* **2013**, *21*, 113–120. [CrossRef]
41. Ali, I.L.; Yunus, N.; Abu-Hassan, M.I. Hardness, Flexural Strength, and Flexural Modulus Comparisons of Three Differently Cured Denture Base Systems. *J. Prosthodont.* **2008**, *17*, 545–549. [CrossRef]
42. Piconi, C.; Maccauro, G. Zirconia as a ceramic biomaterial. *Biomaterials* **1999**, *20*, 1–25. [CrossRef]
43. Hu, Y.; Zhou, S.; Wu, L. Surface mechanical properties of transparent poly(methyl methacrylate)/zirconia nanocomposites prepared by in situ bulk polymerization. *Polymer* **2009**, *50*, 3609–3616. [CrossRef]

© 2019 by the authors. Licensee MDPI, Basel, Switzerland. This article is an open access article distributed under the terms and conditions of the Creative Commons Attribution (CC BY) license (http://creativecommons.org/licenses/by/4.0/).

Review

Cellulose Composites with Graphene for Tissue Engineering Applications

Madalina Oprea [1] and Stefan Ioan Voicu [1,2,*]

1. Faculty of Applied Chemistry and Materials Science, University Politehnica of Bucharest, Gheorghe Polizu 1-7, 011061 Bucharest, Romania; madalinna.calarasu@gmail.com
2. Advanced Polymer Materials Group, Faculty of Applied Chemistry and Material Science, University Politehnica of Bucharest, Gheorghe Polizu 1-7, 011061 Bucharest, Romania
* Correspondence: svoicu@gmail.com or stefan.voicu@upb.ro; Tel.: +40-721-165-757

Received: 25 October 2020; Accepted: 24 November 2020; Published: 25 November 2020

Abstract: Tissue engineering is an interdisciplinary field that combines principles of engineering and life sciences to obtain biomaterials capable of maintaining, improving, or substituting the function of various tissues or even an entire organ. In virtue of its high availability, biocompatibility and versatility, cellulose was considered a promising platform for such applications. The combination of cellulose with graphene or graphene derivatives leads to the obtainment of superior composites in terms of cellular attachment, growth and proliferation, integration into host tissue, and stem cell differentiation toward specific lineages. The current review provides an up-to-date summary of the status of the field of cellulose composites with graphene for tissue engineering applications. The preparation methods and the biological performance of cellulose paper, bacterial cellulose, and cellulose derivatives-based composites with graphene, graphene oxide and reduced graphene oxide were mainly discussed. The importance of the cellulose-based matrix and the contribution of graphene and graphene derivatives fillers as well as several key applications of these hybrid materials, particularly for the development of multifunctional scaffolds for cell culture, bone and neural tissue regeneration were also highlighted.

Keywords: scaffolds; membranes; hydrogels; cellulose; tissue engineering

---

1. Introduction

1.1. General Aspects Concerning Tissue Engineering

In recent decades, the rapidly aging population, environmental stressors, frequent cases of traumatic injuries and chronic diseases lead to a growing interest in the revolutionary domain of tissue engineering (TE) [1–3]. Tissue engineering evolved from the field of biomaterials; its purpose is to combine scaffolds, cells, and biologically active molecules to obtain multifunctional materials that restore, maintain or improve damaged tissues or an entire organ. Some examples of Food and Drug Administration (FDA)-approved engineered tissues include artificial skin and cartilage, but they have limited use in human medicine due to several yet unknown aspects regarding their long term biocompatibility [4–9]. Bioactive scaffolds, cell therapy, smart drug delivery systems, and wound healing mats are some representative examples of the research topics approached by TE. In addition to medical applications, non-therapeutic findings include the use of tissues as biosensors to detect chemical or biological threats or the development of organs-on-a-chip for toxicity screening of experimental medication [4,8,10].

Porous three-dimensional (3D) scaffolds are an important component of tissue engineering. These constructs are used to provide an appropriate environment for tissue and organs regeneration. Biological scaffolds (e.g., fibrin, amniotic membrane, and perfusion-decellularized organs) are an accessible option because they already contain a broad spectrum of signaling molecules with an important role in the processes of cellular morphogenesis and function development [11,12]. However, their composition is strongly related to their source of origin, therefore they have poor reproducibility. Biomaterials-based scaffolds have the advantage that they can be tailored to meet specific requirements, the result being a controllable environment in which stem cells and growth factors can be incorporated to recreate various tissues [4,5,13]. Considering the response of the body's immune system, it is recommended that scaffolds replicate the native extracellular matrix (ECM) of different tissues, in terms of physical structure, chemical composition and biological functionality [13–16]. Biocompatibility, non-immunogenicity, and non-toxicity are mandatory features of biomaterials-based scaffolds. Their design and mechanical properties are also important because they should have the ability to enhance cell migration, proliferation, and differentiation, by presenting appropriate biomechanical, biophysical, and biochemical signals, in vivo, while maintaining their shape and integrity [17,18].

1.2. Characteristics Recommending Cellulose and Graphene for Applications in Tissue Engineering

Cellulose remarks itself among the biomaterials used for scaffold production due to its high availability and renewability. Cellulose is mainly extracted from plant cell walls. The nano-scaled forms of plant cellulose—cellulose nanofibers (CNFs) and cellulose nanocrystals (CNCs), are obtained following specific mechanical and chemical treatments (e.g., ball milling, enzymatic or chemical hydrolysis, TEMPO-mediated oxidation) [3,19]. In virtue of its natural origin, cellulose has a native biocompatibility and negligible cytotoxicity [20]. Some issue were posed related to the inflammatory effect and oxidative stress caused by the cellular uptake of the nano-scaled forms of cellulose [21] but studies showed that both cellulose nanocrystals and cellulose nanofibers presented a non-immunogenic and non-cytotoxic character when different mammalian cell lines were exposed to CNCs suspensions [22] or CNFs membranes [23].

Cellulose can also be produced by certain microorganisms, in this case being called bacterial cellulose (BC). BC is considered to be the most biocompatible form of cellulose because it lacks biogenic impurities such as lignin and hemicellulose and only mild chemical treatments are required to ensure its purity [24]. The nanofibrillar porous structure of bacterial cellulose is similar to the extracellular matrix, this making BC one of the most recommended materials for tissue engineering scaffolds [14,25,26]. Previous studies highlighted that BC has the ability to reduce the inflammatory response and increase the rate of tissue regeneration when it is used as wound dressing [25–27]. Due to its excellent mechanical properties, BC was considered a promising material for the development of vascular grafts, dental implants, artificial skin, and blood vessels [25,28–32].

Pure cellulose lacks solubility in common organic solvents. This represents an issue when it comes to processing techniques where a stable polymeric solution is required to prepare the final material (e.g., electrospinning, phase inversion). Cellulose was dissolved so far only in mixtures of highly toxic solvents (e.g., ionic liquids, carbon disulfide, N-methyl-morpholyne-N-oxide, dimethylformamide) [33]; however, for biomedical applications such as tissue engineering, they are not recommended because even small traces of such solvents could cause a substantial biocompatibility decrease. Consequently, cellulose derivatives were developed, their improved dissolution ability in less noxious solvents, or even water, encouraging their use as an alternative to pure cellulose [3]. Moreover, cellulose derivatives maintain the biocompatibility features of pristine cellulose presenting mild or no foreign body reaction during in vivo assays [34].

Graphene (GE) is an allotrope of carbon produced by top down (e.g., mechanical or chemical exfoliation of graphite, chemical synthesis) or bottom up techniques (e.g., chemical vapor deposition, pyrolysis, epitaxial growth) [35]. GE has some unique properties such as high specific surface area, superior electrical and thermal conductivity, and excellent mechanical properties. The free π electrons and reactive sites, generated by the plane carbon-carbon bonds in GE's aromatic structure, ensure it a facile surface functionalization [36,37]. However, the hydrophobicity and strong interactions between sheets hinder the dispersion of GE in aqueous or organic environments [38,39]. This issue was solved with the development of graphene derivatives such as graphene oxide (GO) and reduced graphene oxide (rGO) which possess specific surface groups that allow them to be effectively dispersed in a wide range of solvents or to be incorporated in polymeric matrices (Figure 1) [40,41]. For example, graphene oxide presents hydroxyl functional groups on the upper and bottom surface as well as carboxylic groups on the edges. This chemical structure is characterized by a hydrophilic character that enables GO's dispersion in water and polar solvents and facilitates hydrogen bonding with polymeric matrices [42]. Reduced graphene oxide is characterized by a lower hydrophilicity and oxygen content but an enhanced electrical conductivity. It was showed that the addition of rGO in polymer composites can increase the thermal stability, improve the bioactivity and mechanical properties, and also provide an appropriate medium for electrical stimulation procedures [40,43,44]. A special type of graphene derivative is represented by graphene quantum dots (GQDs). This nano-scaled form of graphene is made up of one or a few GE layers, with lateral dimensions smaller than 10 nm (Figure 1). GQDs photoluminescence and quantum confinement effect recommend them for applications in bioimagistics, biosensors and photocatalysis devices [45].

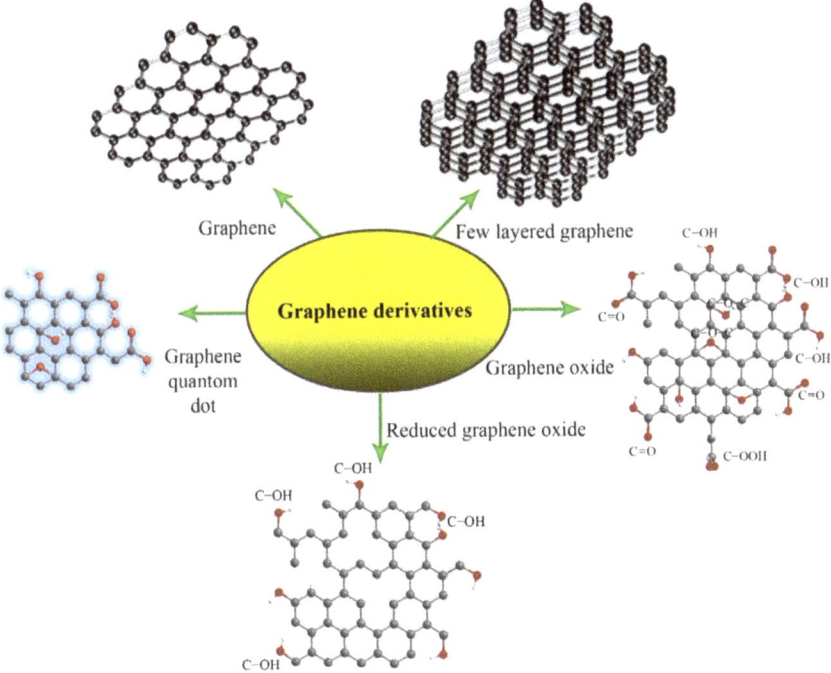

Figure 1. The chemical structures of graphene and its derivatives [46]. Reproduced with copyright permission.

Generally, graphene and its derivatives are considered biocompatible and non-cytotoxic, still, their preparation method highly influences the in vivo and in vitro tests because the residual solvents and reagents used during the synthesis procedures can interact with cells and tissues, thus inducing cytotoxicity and oxidative stress. The hydrazine used to obtain reduced graphene oxide was found to be particularly noxious for human mesenchymal stem cells (hMSCs) [47]. Eco-friendly reduction methods were developed to diminish these residual chemicals-induced adverse effects. For example, Erdal et al. used a microwave-induced hydrothermal reaction and caffeic acid (Caf), a green reducing agent, to produce nanosized reduced graphene oxide (n-rGO), starting from commercial α-cellulose.

Cellulose was treated with an aqueous solution of H_2SO_4 in a microwave device. The material was kept at 180 °C and a pressure of 40 bar, for 2 h, under nitrogen flow. Following this treatment, black solid carbon spheres were obtained. The spheres were further dispersed in concentrated nitric acid (HNO_3), ultrasonicated 30 min at 45 °C and heated at 90 °C for 30 min, under magnetic stirring to obtain nGO. The ultrasonication time is a decisive factor in obtaining materials with uniform and reproducible properties [48]. The green reduction process was performed by placing an aqueous suspension of nGO and Caf in the microwave device, using the same conditions as in the case of carbon spheres synthesis [49]. The resulting n-rGO was incorporated in polycaprolactone (PCL) matrices for the production of bioactive and bioresorbable composites [50], 3D scaffolds with drug delivery ability [51] and macroporous scaffolds with applications in bone tissue engineering [52].

The size and oxidation status of GE are also important for cytotoxicity evaluation. GE and GE derivatives may disrupt cell membranes by direct contact. Moreover, small sized GO has a high potential of being internalized into cells via endocytosis and could cause apoptosis at high concentrations [53]. Surface modification with biocompatible molecules or the insertion in biopolymer matrices were the main solutions proposed to minimize the potential cytotoxic character of GE and GE derivatives. In addition, after incorporation in a polymer matrix, the carbonaceous fillers provide cellular binding sites and, in the case of GO, the oxygenated surface groups increase the hydrophilic character, thus improving cellular adhesion [47].

GE and its derivatives have a demonstrated ability to promote stem cells differentiation processes, particularly adipogenesis and osteogenesis by enhancing the adsorption of differentiation factors and cell adhesion [54]. Graphene-induced osteogenesis was found to be related to the activation of the mechanosensitive integrin-focal adhesion kinase (FAK) axis and also to GE's capacity of promoting the paracrine release of pro-osteogenic molecules in its surroundings, as well as enhancing their delivery to the sites of action [55]. According to recent studies, graphene oxide and reduced graphene oxide are pro-angiogenic. The mechanisms for GO and rGO induced angiogenesis include intracellular formation of reactive oxygen and nitrogen species as well as activation of specific serum antibodies (e.g., phospho-eNOS, phospho-Akt) [56]. The potential of reduced graphene oxide (rGO) to enhance angiogenesis was evaluated by Chakraborty et al. using polyvinyl alcohol/carboxymethyl cellulose (PVA/CMC) scaffolds loaded with different concentrations of rGO nanoparticles. Primary biocompatibility studies included in vitro alamarBlue cytotoxicity assays on three different cell lines—fibroblasts NIH3T3, endothelial-like cells (ECV304) and endothelial cells (EA.hy926). The scaffolds showed no toxicity toward the analyzed cell lines, the cellular viability being similar to the control group. It was concluded that when incorporated inside a scaffold, rGO does not present cytotoxicity even if it is used in concentrations higher than the cytotoxicity threshold in free solution (100 ng/mL). The composite scaffolds were implanted in chick chorioallantoic membrane (CAM) models to study their influence on the neovascularization process. Two days following implantation, the number and wall thickness of the blood vessels were substantially increased, compared to the untreated control. Moreover, angiogenesis and arteriogenesis were enhanced as the rGO concentration in the composites increased, whereas neat PVC/CMC scaffolds showed no bioactivity [57].

As a results of their remarkable properties, cellulose/graphene composites were extensively researched particularly for biomedical applications. Cellulose is a versatile, highly available, biodegradable, and biocompatible material, these characteristics recommending it as a low cost, sustainable alternative to petroleum-based plastics or other types of natural polymers used in tissue engineering. Cellulose/graphene composites designed for the TE field can be divided in two main categories—composites where specific types of cellulose (e.g., cellulose paper [58], bacterial cellulose [59], cellulose derivatives [60]) are employed as polymer matrices in which the carbonaceous fillers are dispersed to improve their mechanical characteristics and biocompatibility, or, composite fillers based on CNCs or CNFs combined with GE or GE derivatives that are used to synergistically reinforce other polymer matrices [21] (e.g., polylactic acid—PLA [61,62], polybutylene succinate—PBS [63], polyacrylamide—PAM [38], polycaprolactone—PCL [64]). Composite membranes with graphene [65,66] were also studied for applications in the hemodialysis field [67,68]. The use of GO for reinforcing cellulose acetate membranes led to an increase of bovine serum albumin (BSA) retention from 80% to over 96% [69,70]. The synergistic effect between GO and carbon nanotubes (CNT) used for the preparation of composite cellulose acetate membranes with potential applications in hemodialysis showed good results in the retention of BSA and hemoglobin [71]. These results are due to both the presence of GO, which has a high surface adsorption ability of the proteins that need to be separated [72,73], and also to the weak chemical interactions that emerge between the delocalized electrons on the surface of graphene and the non-participating electrons from the functional groups of the polymer [74,75].

The enumerated types of cellulose/graphene composites were discussed in detail in the following sections of this review.

2. Cellulose Composites with Graphene for Tissue Engineering Applications

An ideal material for tissue engineering should have an excellent biocompatibility with the cellular components and mimic, as much as possible, the extracellular matrix of the recreated tissue. A porous structure with interconnected porosity is among the essential characteristics of a scaffold in order to allow cellular attachment and migration and facilitate the diffusion of nutrients and metabolites throughout its volume [17]. Moreover, the mechanical characteristics must not be neglected because they should match the ones of the host tissue. Studies showed that both stem and mature cells are sensitive to the stiffness of the substrate on which they are seeded and show different adhesion and morphological characteristics depending on it. The scaffolds should also possess a certain degree of bioactivity and interact with the surrounding environment to actively regulate cellular activity [76]. The recreated tissue type dictates the design and functionality of TE scaffolds. For example, in the case of bone tissue engineering, the restoration of normal biomechanical functions is crucial, therefore, the scaffolds must have similar mechanical properties to the native bone and a degradation behavior matching with the novel bone formation rate [77]. Electrical properties are another important aspect, especially for neural tissue engineering, because the neurons proliferation, migration and communication with other cell types is realized by electrical signaling mechanisms [78]. Furthermore, electrical stimulation performed during cell culture was showed to enhance cell migration, proliferation rate and differentiation, being considered a revolutionary tool for tissue engineering and regenerative medicine [79].

In the next sections of this review, the fabrication methods, physico-chemical features and biological performance of cellulose/graphene composites for tissue engineering will be described. As it will be observed, by carefully choosing the type of cellulose and graphene used as well as the preparation technique, it is possible to obtain scaffolds with tunable characteristics depending on the desired application.

2.1. Cellulose Paper-Graphene Composites

Paper-based scaffolds were first developed starting from surgical grade cotton or bacterial cellulose. The cotton-derived substrates were obtained by a simple and cheap paper making process involving cooking and beating the cotton before pressing it in a British sheet forming device. The resulting paper was afterwards immersed in a gelatin solution to enhance its cellular adhesion ability. According to the FESEM and MTT assay, MG63 cells incubated on all of the developed scaffolds presented a normal morphology but the adhesion and proliferation rates increased on the gelatin-modified ones [80]. Cheng et al. used bacterial cellulose and hydrophobic petroleum jelly-liquid paraffin ink to create cost effective tissue models (~4 cents per single device) by the matrix assisted sacrificial 3D printing technique. The fugitive ink was 3D printed on the wet BC pellicles in a predetermined pattern. The pellicles were air dried and then heated at 70 °C to liquefy the ink and remove it by rinsing with n-hexane and distilled water, thus resulting well-defined microchannels inside the cellulosic matrix. The obtained devices were tested as vascularized breast tumor models by seeding green fluorescence protein (GFP)-labeled human umbilical vein endothelial cells (HUVECs) inside the microchannels and MCF-7 cells on the surrounding cellulosic matrix. The drug response of the tissue model was also evaluated by injecting tamoxifen in the endothelialized microchannels [81].

Graphene-enriched scaffolds improve stem cell viability and osteogenesis by participating in the activation of physiologically relevant mechano-transduction pathways. GE's topographical features and electrical conductivity facilitate cell anchorage and hydroxyapatite formation ability with physio electrical signal transfer, thus enhancing osteogenesis [64,82]. A novel graphene-cellulose (G-C) paper scaffold was developed by Li et al. for applications such as in vitro modeling of human bone development and regeneration, or bone patches and plugs to facilitate in vivo osteogenesis following injuries. Commercial tissues, blotting paper and filter paper were tested as substrates for the fabrication of G-C composites. The papers were laser cut to 1 cm × 1 cm size and aqueous dispersions of GO were deposited on the substrates. The resultant GO-coated papers were dried at 100 °C for 2 min and reduced in 50 mM L-ascorbic acid solution at 80 °C for 3 h (Figure 2 left) [58]. It was found that the G-C papers electrical conductivity and mechanical properties were positively influenced by the rGO and could be tuned by modifying the number of deposited rGO layers. The composites presented an improved biocompatibility, translated by a higher surface live cell density, compared to the uncoated paper. This was attributed to an increase in hydrophilicity caused by rGO addition that favored human adipose derived stem cells (hASDCs) adhesion. Higher levels of alkaline phosphatase (ALP) were observed for the G-C papers-cultured cells, this suggesting that rGO also guided the cellular differentiation into an osteogenic lineage.

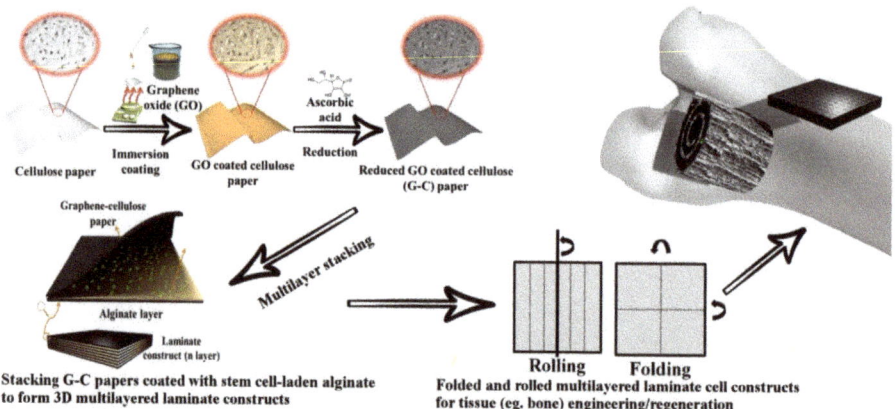

Figure 2. The fabrication stages of G-C paper and origami inspired cell-laden constructs (**left**); Rolling and folding of the 3D structures for applications in bone tissue regeneration (**right**) [58]. Reproduced with copyright permission.

An interesting fact was that the G-C papers could be laminated with alginate and folded or rolled to obtain origami inspired cell-laden constructs as shown in Figure 2 (right). Similar cellulose paper/cell-laden structures were previously obtained using commercial chromatography papers and HS-5 human bone marrow stromal cells suspended in Matrigel [83] or filter papers and MDA-MB-231-GFP breast cancer cells suspended also in Matrigel [84]. In another study, sodium alginate containing 3T3 mouse fibroblasts was used as bioink to create ultrafine patterns on chromatography papers treated with an alginate crosslinking solution ($CaCl_2$) [85]. However, in this studies, cellulose was employed just as a support for the cell-embedded hydrogels, compared to the G-C papers where the modified cellulose itself played an important role in cellular growth and proliferation. The 3D structures were obtained by suspending hASDCs in an alginate solution followed by drop-wise deposition of a small amount of hADSCs-laden alginate on the upper surface of each G-C paper and stacking of multiple such sheets before crosslinking them by immersion CaCl2 (Figure 2 left). Cross sectional SEM images showed that these 3D constructs had a stratified structure, the alginate hydrogel effectively binding the G-C papers. Viable cells were observed at the hydrogel-paper interface immediately post assembly and after a 42 days study period, therefore, it was concluded that the 3D G-C-paper/alginate structures are able to provide a long term cellular support [58].

The developed G-C papers also have the potential to be used for cellular electrical simulation (ES) due to the electrical conductivity provided by the incorporated rGO. This subject was thoroughly analyzed by the same research group. The study was made by integrating the electroactive paper in polystyrene (PS) chambers to obtain electrodes. The electrode assembly process is illustrated in Figure 3. Briefly, the G-C papers were cut into strips and mounted in parallel on a glass substrate. The PS chamber, with an open bottom and removable lid, was attached on top of the glass substrate using silicone and copper tapes were glued on both edges, perpendicular to the G-C strips, to connect them. The electrode was afterwards coupled to an electrical simulator using copper wires [82].

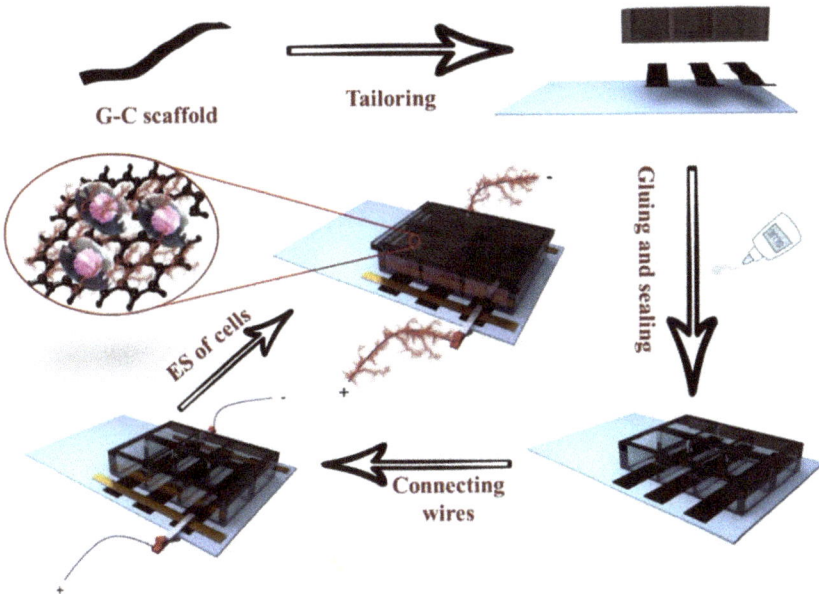

Figure 3. Schematic representation of the assembly and functionality of a G-C-based electrode for cellular electrostimulation [82]. Reproduced with copyright permission.

During electrochemical characterization by cyclic voltammetry and electrochemical impedance spectroscopy, the G-C electrodes showed high stability, lower impedance, and higher charge injection capacity than commercial gold electrodes. hADSCs were cultured on the G-C scaffolds with or without electrical stimulation. At the end of the 28 days study period it was found that electrically simulated cells showed increased proliferation, mineral deposition and ALP expression compared to control samples. According to these results, it was considered that the developed G-C electrodes could represent an alternative to conventional metal electrodes that present the risk of corrosion-related cell compatibility issues [82].

2.2. Bacterial Cellulose-Graphene Composites

Bacterial cellulose/graphene composites (BC/GE) represent an intensively researched subject in tissue engineering. There are two essential requirements for the successful preparation of BC-based composites with GE—the finding of an appropriate synthesis method that maintains the intrinsic nanofibrous structure of BC and the homogenous dispersion of the carbon-based material in the cellulosic matrix [86]. Both graphene and graphene oxide are promising materials for biomedical applications, still, recent studies suggest that the incorporation of GE and GO into 3D nanofibrous scaffolds results in different biological properties, and GO-based scaffolds have a better biocompatibility and bioactivity compared to the ones containing GE [87]. GE and GO-reinforced BC scaffolds were prepared using an accessible membrane-liquid interface culture (MILIC) method. The MILIC technique consisted of the pulverization of a GE or GO-containing culture medium onto BC pellicles obtained from conventional static cultures. A thin layer of GE/BC or GO/BC was formed on the neat BC surface and served as a new substrate for the next hydrogel layer as illustrated in Figure 4. The process was repeated until a desired thickness was reached. After purification and freeze drying, the morphologies, structure, mechanical properties and biocompatibility of BC/GE and BC/GO were compared between them and with pristine BC scaffolds as control.

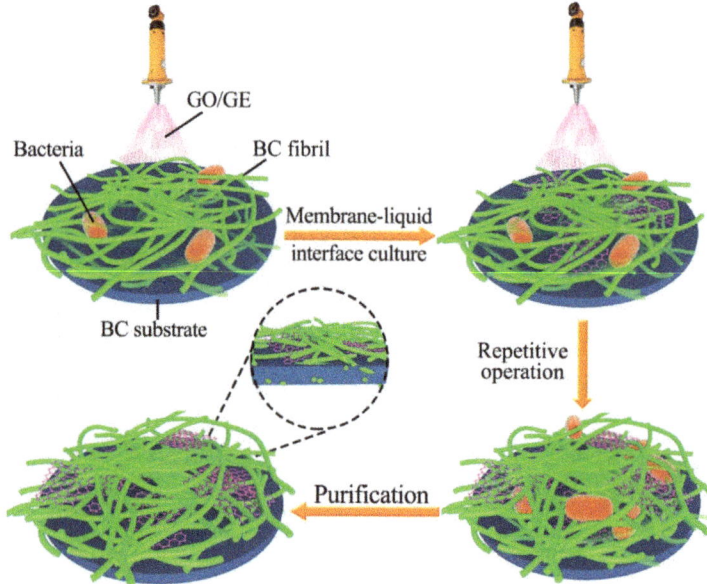

Figure 4. Schematic representation of the steps involved in the MILIC fabrication method used for the GO/BC and GE/BC scaffolds [87]. Reproduced with copyright permission.

Cellular viability assays were conducted using mouse embryonic osteoblasts (MC3T3-E1). As expected, BC/GO scaffolds displayed a better cellular adhesion, spreading, proliferation and osteogenic differentiation compared to neat BC and BC/GE. A possible explanation could be that GO's hydrophilic surface groups provide an improved cytocompatibility [87,88].

Bacterial cellulose/reduced graphene oxide (rGO) films were also studied for applications in biomedical device fabrication, biosensors, and tissue engineering. The composites were obtained using a bacteria-mediated reduction technique. Gluconacetobacter intermedius (BC 41) was cultured in a mixture of culture medium and GO in a static incubator. During the 14 days culture period, BC/GO composites self-assembled in situ and the GO on the surface of the cellulose fibers was biochemically reduced. Cross section SEM images revealed an interconnected structure comprised of stacked rGO sheets linked together by BC nanofibrils. This distinct structure was associated with the favorable mechanical characteristics of the composites, mechanical resistance being a characteristic required to ensure their functionality under the harsh conditions in living organisms. The electrochemical performance of rGO obtained by bacterial reduction was inferior to chemically synthesized rGO; however, it still exhibited a high charge carrying capacity at a given voltage and was considered sufficient to be used for cellular electrical simulation or to collect physiological signals in biosensors. Human marrow mesenchymal stem cells (HMSCs) were used to monitor the cellular response of the BC/rGO films in terms of cellular adhesion and proliferation. Tissue culture polystyrene (TCP) and plain rGO films were set as control groups. After 7 days in culture, HMSCs were present in a higher number on BC/rGO substrates (9.65 × 104) compared to the control groups (6.89 × 104—rGO and 9.15 × 104—TCP). Also, BC/rGO-grown cells displayed better cell attachment and retention when observed using confocal microscopy after 3 days of culture on the analyzed substrates [89].

For enhanced bioactivity, graphene oxide was coated with hydroxyapatite (HA) by a wet chemical precipitation route. Briefly, calcium hydroxide and ortho-phosphoric acid were incorporated in a GO solution (1 mg/mL concentration) by magnetic stirring. The residue was aged for 24 h, washed with distilled water and dried in a hot air oven. The resulting GO-HA complex was incorporated in bacterial cellulose matrices by impregnating the wet BC membrane with the GO-HA suspension in ethanol, under continuous stirring for 24 h. Due to the osteoinductive and osteoconductive properties of hydroxyapatite [90], GO-HA could ensure a higher viability of osteoblasts compared to unmodified GO. The biological characterization was performed on normal (NIH3T3) and osteosarcoma (MG-63) cell lines, and consisted of methyl thiazolyl tetrazolium (MTT) assay and alkaline phosphatase (ALP) activity measuring in culture supernatants. The viability of both MG-63 and NIH3T3 cells in the presence of BC/GO-HA composites was better compared to BC, GO and BC/HA and a higher ALP activity was observed, particularly at a concentration of 50 µg/mL. The results were confirmed by the phase-contrast microscopy images that showed an increased cell density on the BC/GO-HA composites surface compared to the other materials tested [91].

Another area where bacterial cellulose/graphene composites showed promising results is represented by neural tissue engineering. The use of bacterial cellulose/graphene-based scaffolds for cell-based regenerative therapy could solve the main problems associated with this technique, more specifically, the decreased cellular viability, poor integration within the host brain tissue and decreased tendency of implanted stem cells to differentiate toward functional neurons. For an effective reconstruction of injured brain tissues, the used scaffold should reduce inflammation and apoptosis, promote restorative processes, neurite outgrowth as well as axonal elongation [92]. A first attempt to obtain such constructs was made by Si et al. The research group developed BC/GO nanocomposite hydrogels using a facile one-step in situ biosynthesis. A commercial aqueous dispersion of GO nanosheets was added to the BC culture medium, followed by intense stirring for 60 min. The resulting BC/GO pellicles were soaked in deionized water at 90 °C for 2 h and boiled in sodium hydroxide (NaOH) solution for 15 min for purification. The materials were then washed until they reached a neutral pH. SEM images revealed that the GO nanosheets were uniformly dispersed within the BC matrix and the 3D fibrous network and porous structure of BC was kept after the incorporation of the inorganic compounds. The good structural properties of BC/GO composites were attributed to the in situ biosynthesis method that favored strong interactions between the hydroxyl groups on both BC and GO surface. Also, GO maintained its crystal structure the characteristic crystal lattices being visible in TEM images. Tensile testing of BC/GO composites was performed using a universal material testing instrument, under ambient temperature and humidity. The inclusion of GO in the BC structure lead to a notable increase of the tensile strength and Young modulus, as showed by the stress-strain curves. This improvement was considered the result of the strong interfacial interactions and homogenous dispersion of the carbonaceous filler inside the polymeric matrix, which favored an effective load transfer from BC to GO [86].

Further studies on biosynthesized BC/GO scaffolds were conducted by Kim et al. They reported a non-genetic manipulation method of Acetobacter xylinum that resulted in the synthesis of carbon-hybridized BC hydrogels. The bacteria was incubated for 14 days in a culture medium containing GO stabilized with a comb-like amphiphilic polymer (APCLP) for a better dispersion.

The resulting composites were purified with NaOH, washed with distilled water and vacuum dried before characterization. The crystallinity of the hybrid BC/GO scaffolds was lower compared to neat BC while the porosity was higher. These characteristics were associated with the accelerated formation of the BC pellicles, and the perturbation of individual polysaccharide chains crystallization in the culture medium containing APCLP-GO, which lead to a densification of the cellulose nanofibrils, thus increasing the overall porosity. For the biological assessment, a neuronal network was constructed by seeding rat embryonic hippocampal neurons (E18) (Figure 5a) on both sides of the synthesized scaffolds and neat BC pellicles for comparison purpose. As the neurons developed inside the scaffolds (Figure 5b), their terminations interconnected vertically, thus forming a long range neuronal network referred as "minibrain" (Figure 5c,d). Neurons cultured on BC and BC/GO substrates had an accelerated neuronal processes development, compared to the ones cultured on conventional flat glass substrates that presented only early stage dendrites expansion. These differences were attributed to the nanofibrillar ECM-like structure of BC that may simulate cellular development and even guide neurite pathfinding [93].

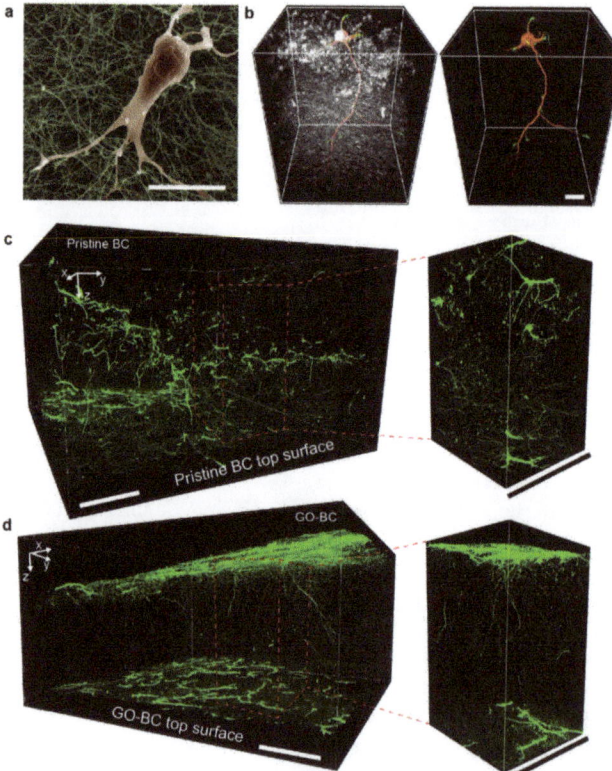

Figure 5. Pseudo-colored SEM image of an E18 neuron seeded on neat BC (**a**) 3D confocal fluorescence images of an E18 neuron development inside neat BC scaffolds (**b**) 3D confocal fluorescence images of Tuj1 red and Phalloidin green-stained neurons cultured on neat BC (**c**) and BC/GO hydrogels (**d**) [93]. Reproduced with copyright permission.

No significant differences in embryonic rat hippocampal neurons development were observed between neat BC and BC/GO. The general conclusion was that GO had little or no influence and the accelerated growth of the neuronal processes (compared to conventional glass substrates) was mainly attributed to the fibrous structure of BC. However, in a different study, neural development was studied using human neural stem cells and the results indicated that GO had an important role in guiding and accelerating the cellular differentiation process toward the neuronal lineage and also enhanced neurites formation, elongation, and branching. The study was conducted by Park et al. that developed 3D hybrid scaffolds based on bacterial cellulose and amphiphilic comb-like polymer (APCLP)-covered graphene oxide flakes, and applied them as brain cortex mimetics in motor cortectomy rat models [92]. The GO flakes, prepared by a chemical exfoliation process (Hummers method), were coated with APCLP for a uniform dispersion, and added in the bacterial culture medium. The obtained membranes were purified with NaOH, washed with distilled water, and dried overnight at 60 °C. Human neural stem cells (F3), isolated from embryonic brains, were seeded within the BC and BC/GO-APCLP scaffolds. The cellular development was observed via phase-contrast microscopy. It was observed that two days after seeding, long neurites started growing from the BC/GO-APCLP-cultured F3 cells whiles the BC-cultured ones showed no substantial changes in morphology. Immunofluorescence assays were further employed for a better understanding of the cellular proliferation and differentiation processes occurring in the scaffolds. Luciferase activity indicated that the number of F3 cells increased constantly until the 8th day of the study but mature neuronal markers (MAP2) and synaptic vesicle proteins (synaptophysin), were present only in BC/GO-APCLP scaffolds. The cell-enriched scaffolds were implanted to motor cortex-ablated rats with mimicked trauma injuries and the cellular behavior using in vivo molecular imaging. According to the bioluminescence signals, representing the number of viable F3 cells, the cells cultured on BC/GO-APCLP and neat BC scaffolds had a higher survival rate (12 days vs. 10 days) compared to the cell-only treated group (conventional cell therapy). After the test period, excised brains were analyzed by immunohistochemistry. Similar to the in vitro tests, most of the cells cultured on BC/GO-APCLP scaffolds showed strong homogenous staining on MAP2 and synaptophysin. Only a few BC-cultured cells presented the same characteristics respectively no cells in the case of the control (cell-only treated) group [92].

2.3. Cellulose Derivatives-Graphene Composites

Electrospinning is a frequently used technique for the production of ECM-mimicking fibrous structures. The electrospinning device uses a high tension source to create an electrical field that draws charged polymer droplets from the tip of a needle to a collector plate, thus resulting micro and nano-scaled fibers [94,95]. In virtue of their native biocompatibility and surface topography that promotes cellular adhesion and influences the conformation of adsorbed adhesion proteins (e.g., fibronectin, vitronectin), electrospun cellulose-based fibers could be worthy candidates for the production of tissue engineering scaffolds [96,97]. Among cellulose derivatives, cellulose acetate (CA), in particular, showed good fiber forming ability in a variety of solvents, flexibility and excellent mechanical strength in fibrous form [98]. Cellulose acetate-based scaffolds have the ability to support osteoblasts growth, phenotype retention and bone formation [99], this encouraging their application as scaffolds for bone tissue engineering.

Liu et al. incorporated GO in CA solutions in acetone/dimethyl formamide (DMF) and electrospun the resulting mixture to obtain hybrid CA/GO nanofibrous mats. The composites presented a uniform and smooth surface and a decreased fiber diameter compared to neat CA. The Raman spectrum of CA/GO showed characteristic peaks at 1300 cm^{-1} (graphene D band) and 1580 cm^{-1} (graphene G band), with increased intensity at higher GO contents. The incorporation of GO into the CA fibers improved their mechanical properties, a higher strain at break and an increased Young's modulus being observed in the stress-strain curves of the composites, compared to neat CA. To investigate the cellular adhesion behavior, human mesenchymal stem cells (hMSCs) were cultured for 1, 2, 4 and 8 h on CA/GO and the results were compared with the control group represented by cells grown on conventional tissue culture polystyrene (TCP) substrates. As expected, hMSCs adhered more efficiently onto the hybrid scaffolds, most likely due to their increased hydrophilicity and high surface area that creates a good environment for cellular retention. The nanofibrous scaffolds were immersed in simulated body fluid (SBF) and the biomineralization process was observed by SEM. The production of calcium phosphate increased proportionally to the SBF incubation time and the concentration of GO in the nanofibers. It is well known that biomineralization in hybrid materials is governed by the availability of functional groups [100]. In this case, the hydroxyl and carboxyl functional groups in facilitated the deposition of Ca^{2+} ions and facilitated the formation of apatite crystals on the nanofibrous mats. HMSCs seeded on GO-CA substrates also showed significantly higher alkaline phosphatase activity than the control group, in differentiation medium [94]. In another study, reduced graphene oxide-cobalt composite nanoparticles were incorporated in CA-based electrospun scaffolds to investigate their potential to enhance human mesenchymal stem cells (hMSCs) osteogenic differentiation under alternative magnetic field (AMF). SEM images showed that the copper nanoparticles were well attached and evenly decorated the surface of the rGO sheets. The cellular biocompatibility of the electrospun mats was confirmed by MTT assay and 4',6-diamidino-2-phenylindole (DAPI) nucleus staining performed on the seeded hMSCs (Figure 6a). Even if cellular adhesion was similar for both hybrid and neat CA fibers, the cell growth orientation on CA/rGO-Co was improved (Figure 6b). Moreover, certain genes associated with novel bone formation (e.g., Runx2, OC, Col 1, OCN) presented an enhanced activity when the hybrid scaffolds were exposed to AMF (Figure 6c) [101].

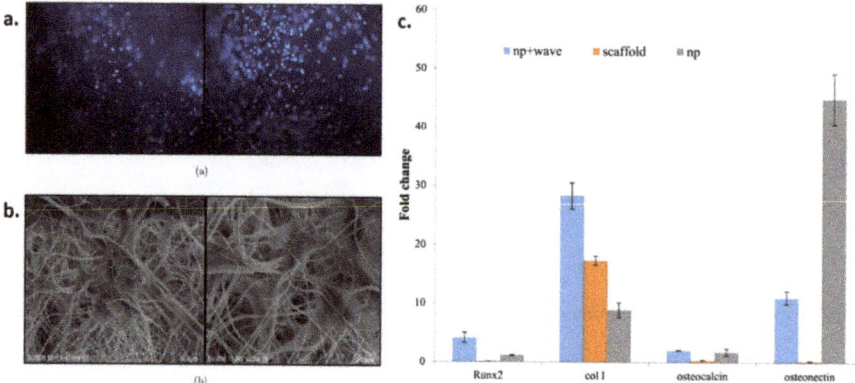

Figure 6. DAPI stained hMSCs 3 days post seeding on CA (left) and CA/rGO-Co (right) scaffolds (**a**) SEM images of the stem cells seeded on CA (left) and CA/rGO-Co (right) nanofibrous scaffolds (**b**) Relative expressions of Run × 2, Col1, OCN and OC genes for hMSCs cultured on CA and CA/rGO-Co, with (np + wave) or without (np) AMF (**c**) [101]. Reproduced with copyright permission.

This study opens up new possibilities for the development of magnetic cellulose-based composites with graphene, besides its demonstrated bioactivity, the carbonaceous structure also being an appropriate support for the immobilization of magnetic nanoparticles.

The development of scaffolds for tumor cell culture is another area of application for nanofibrous CA/GO hybrids. These in vitro models are essential for the evaluation of cytostatics prior to their introduction on the market and also for the study of cancer cells biological features. In a recent study conducted by Wan et al., electrospun CA/GO scaffolds, seeded with human breast cancer cells (MCF-7), were investigated. Cellulose acetate solutions were prepared by dissolving CA in a mixed solvent system of acetone/acetic acid/dichloromethane with a volume ratio of 2/2/1. Subsequently, a GO suspension was added into the CA solution and the system was kept under magnetic stirring for 5 h at room temperature. After vacuum drying, the morphology of the obtained materials was observed using SEM microscopy. It was found that the addition of GO does not significantly influence the scaffold morphology except for the fact that the fibers diameters decreased. The presence of GO and its interactions with the CA matrix were confirmed by the presence of specific D and G bands in the structure of CA/GO, corresponding to ordered sp2 bonded graphitic carbon of GO, and also by the obvious OH peak shift toward lower values in the FTIR spectra of the composites, due to the formation of hydrogen bonds between CA and GO. MCF-7 cell adhesion, viability and proliferation were assessed using live staining with fluoresceindiacetate, rhodamine phalloidin and 4′,6-diamidino-2-phenylindole, followed by observation under fluorescence microscopy. The results revealed a progressive increase in the number of viable cells for both CA and CA/GO scaffolds; however, at the end of the 5 days testing period, there were fewer cells on the neat CA scaffolds compared to CA/GO. These results confirm that GO had a positive influence on the ability of CA scaffolds to sustain MCF-7 cell development [102]. In another study performed by the same research group, electrospun CA/GO microfibrous hybrids were placed on previously synthesized BC pellicles and culture medium was sprayed over them to initiate the in situ BC synthesis inside the CA/GO mats. The layered material was purified by boiling in NaOH solution and washed several times with distilled water until a neutral pH was reached. SEM images revealed the interpenetrated structure of the BC/CA/GO scaffolds. Two different kind of fibers with average diameters of 43.5 nm and 2.2 µm were identified. The purpose of combining nanofibers with microfibers was to obtain a more intricate ECM-like environment, which may notably improve cell-cell and cell-ECM communication. GO was hardly visible in the SEM images because the carbonaceous sheets were embedded in the polymeric matrix, still, the OH peak shift in the FTIR spectrum and the existence of D and G bands at 1345 and 1590 cm^{-1} in the Raman spectra confirmed its presence in the structure of the composites and its interactions with the cellulosic matrix. Biological characterization by CCK-8 and live cell staining procedures indicated that the cancer cell spreading and proliferation was dependent on the GO content, better results being obtained for the GO-incorporated scaffolds compared to neat CA/BC [103].

Phase inversion is a well-established method for the production of cellulose acetate membranes with applications in biomedical engineering and water purification. The process consists of casting the polymer solution on a proper substrate and immersion in a coagulation bath containing a non-solvent to precipitate the membrane. Membranes obtained by this procedure have different structural characteristics than electrospun mats. They are usually asymmetric, consisting of dense layer in top and a porous substructure at the bottom [90]. Ignat et al. used phase inversion to developed cellulose acetate membranes reinforced with graphene oxide and carbon nanotubes (CNT). The polymer was dissolved in N,N′-dimethylformamide under constant stirring and a small amount of NaOH was added to increase the concentration of surface hydroxyl groups for a better polymer-filler compatibility. The carbonaceous fillers were effectively dispersed in the organic matrix by ultrasonication. The membranes were obtained by casting the CA/GO-CNT solution on a glass slide and immersing it in a coagulation bath containing 2-propanol and distilled water. Micro CT analysis showed that CA/GO-CNT membranes exhibited a porous morphology, with open and interconnected porosity, appropriate for cellular migration within the material. The ability of CA/GO-CNT membranes to guide human adipose derived stem cells (hASCs) differentiation toward the adipogenic lineage was evaluated using Oil Red O staining. A higher accumulation of intracellular lipids was observed on the CA/GO-CNT-grown cells, compared to neat CA after 7 and 21 days. Also, a more pronounced Perilipin gene expression was observed in the CA/GO-CNT hybrids compared to the CA reference. Alizarin Red S staining was employed to evaluate the capacity of CA/CNT-GO membranes to induce osteogenesis in hASCs. Neat CA-cultured cells became round, this indicating that osteogenesis was initiated but no further changes were detected during the 21 days test period. In the case of the CA/CNT-GO membranes, the mineralization levels increased proportionally to the content of GO and CNTs. The expression of osteopontin (OPN) was evaluated via qPCR. It was that found that after 21 days, the gene activity was gradually increased on both neat CA and hybrid membranes but in a higher measure for CA-CNT/GO [54]. According to the results of this study, GO and CNT have the ability to selectively improve hASCs differentiation and could be used for the design of novel materials for well-defined tissue engineering applications.

Hydrogels are another category of promising materials for tissue engineering, their soft and highly hydrated structure resembling the ECM of native tissues. Crosslinkers are essential components for the preparation of hydrogels, since the crosslinking process improves the mechanical properties and degradation rate of these materials. The disadvantage of most crosslinkers is their cytotoxicity and poor biodegradability, characteristics that make them inappropriate for biomedical applications [38]. Citric acid-derived graphene quantum dots (GQDs) were employed as safe, biocompatible, crosslinking agents for carboxymethyl cellulose (CMC). GQDs, CMC and glycerol as plasticizing agent were dissolved in distilled water using magnetic stirring. The resulting paste was cast to a polystyrene plate and cured at 60 °C for 24 h to obtain a hydrogel film with thickness of approximately 10 µm. The proposed hydrogel formation mechanism is represented in Figure 7. The carboxylic groups present on GQDs surface dehydrates, thus forming a cyclic anhydride that further reacts with the hydroxyl groups of CMC chains, forming an ester linkage. The reaction continues until the hydrogel is crosslinked via esterification [104].

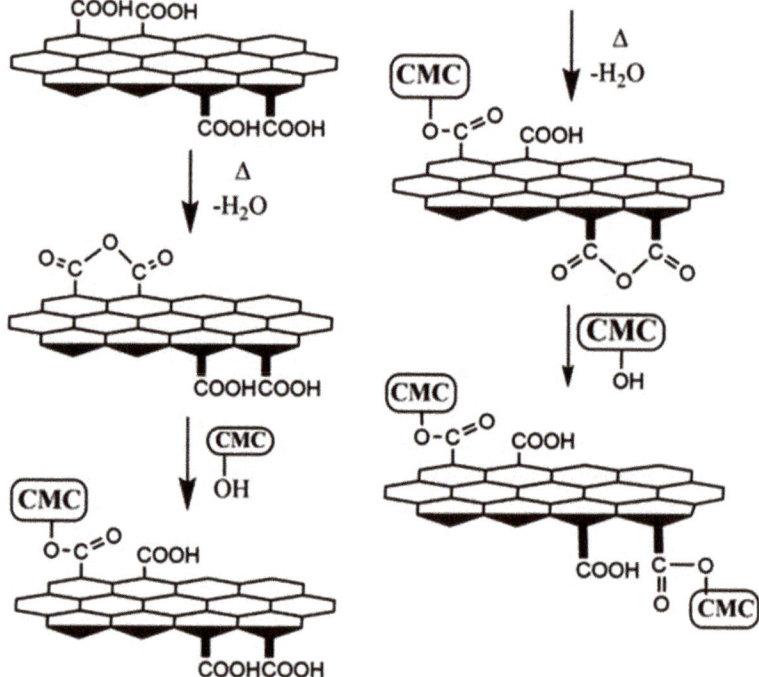

Figure 7. Schematic illustration of the proposed formation mechanism for CMC/GQDs hydrogels [104]. Reproduced with copyright permission.

The drug delivery ability of the CMC/GQDs hydrogels was studied using doxorubicin (DOX) as a model anticancer drug. It was found that the composites released DOX in a pH-dependent manner, most likely due to the pH sensitive swelling of CMC. The hydrogels did not swell in simulated acidic medium (pH 2), thus protecting the loaded drug from degradation at the stomach level, while in the pH range of 4–8.5 corresponding to the duodenum, the highest swelling and drug release were recorded. The swelling and degradation processes also lead to the release of small material fragments. Their potential cytotoxicity on human colon adenocarcinoma HT29 cells was studied by MTT assay and it was found that cellular viability was over 80% even at high concentrations of GQDs (45%) [104,105].

Graphene oxide is a valuable candidate for the development of smart drug delivery systems, its large surface area and oxygenated edge groups ensuring it a high drug loading capacity and potential for further functionalization. Moreover, it was found that due to its amphiphilic character, GO is able to stabilize hydrophobic drugs [106]. Several researchers developed GO-based controlled drug delivery systems for cancer treatment [107–110]. Jiao et al. used carboxymethylcellulose-grafted graphene oxide (CMC/GO) loaded with methotrexate (MTX) to obtain a pH sensitive drug delivery system for colon cancer therapy. CMC was grafted on GO via ethylene diamine by hydrothermal treatment at 90 °C for 10 h. MTX was dissolved in NaOH and predetermined quantities of freeze dried CMC-GO were added under ultrasonication to obtain various drug loading percentages. The drug release rate ranged from 4.76% in simulated buffer solution (SBF) with pH 1 to 67.4% in SBF with pH 7.4. The differences were related to the higher swelling of CMC at basic pH due to the protonation of –COO⁻ groups. The CMC/GO composites presented a negligible cytotoxicity against NIH3T3 cells during MTT assays and the cellular viability was better for the CMC/GO-MTX-treated cells compared to free MTX-treated ones. Metastatic tumor models were created by splenic injection of HT-29 cells to female Balb/c mice. The mice were treated for 5 days with MTX and CMC/GO-MTX administered intra-gastric. Tumor growth was evaluated by Hematoxylin-Eosin staining. A superior tumor inhibition activity was observed for CMC/GO-MTX (83.3%) compared to free MTX (72.2%). Additionally, CMC/GO-MTX-treated mice presented reduced liver metastasis and prolonged survival time [108]. Cancer cells often present an overexpression of the folate receptor, a membrane glycoprotein considered a highly selective tumor marker. The folate receptor binds folic acid with high affinity [111]. Based on this phenomenon, Sahne et al. designed folate-targeting drug delivery systems based on carboxymethylcellulose, polyvinylpyrrolidone (PVP) and spherical graphene oxide nanoparticles synthesized through carbon rehybridization, chemical exfoliation and centrifugation. CMC was modified by thiolation whereas PVP was enriched with amine and thiol-reactive end groups by reversible addition-fragmentation chain transfer (RAFT) polymerization. The functionalized polymers were deposited layer by layer on the surface of GO. Curcumin (Ccm), a polyphenol with antioxidant and antitumoral properties was encapsulated in the CMC layer during the deposition process and monoclonal folic acid antibodies (FA) were grafted on the CCm/CMC-PVP-GO using polyethylene glycol (PEG) as linker. Cytotoxicity studies were performed on MCF7 and Saos 2 cells and 4T1 bearing Balc/c mice were used as tumor models. The Ccm-FA/CMC-PVP-GO presented an inhibition rate of 76 and 81% against Saos 2 and MCF7 tumor cell lines and a 76% antitumor efficiency, expressed by antiangiogenesis, apoptosis and tumor growth inhibition in vivo [109].

2.4. Cellulose Nanocrystals-Graphene Composites

A special type of cellulose-based hydrogels were prepared by Khabibullin et al. using cellulose nanocrystals (CNCs) as building blocks [112]. The characteristics that determined the choice of CNCs for this application are their widespread availability, high mechanical properties, native biocompatibility, and facile surface functionalization. Graphene quantum dots were used as crosslinkers for the CNC-based hydrogels, their addition not only reinforcing the cellulosic structure but also providing it fluorescence properties. The GQDs were dispersed in aqueous solutions of CNCs and the suspensions were vortex mixed for 15 s. The formation of CNC/GQD network was governed by the interactions between the surface hydroxyl and half ester groups on CNCs surface and the carboxyl moieties on GQDs edges (Figure 8b). Hydrophobic interactions could also occur between CNCs hydrophobic faces and GQDs basal planes.

The CNC/GQD hydrogels exhibited a shear thinning behavior during rheological evaluation. At 1% strain, the hybrid suspension formed a hydrogel with G = 80 Pa while at 50% strain, the value of G' decreased to 2 Pa, this signifying gel liquefaction. To exploit this characteristic, the hydrogels performance as injectable material was examined using a 3D printer. The printed hydrogel threads retained their structure and were able to create predetermined patterns. Also, they exhibited variable photoluminescence in the spectral range of 400–680 nm when excited at 365 nm (Figure 8a) [112]. The physico-chemical characteristics of these materials suggest that they could be used as injectable composites for tissue engineering applications, particularly for bioimagistics and biosensing. However, a thorough biocompatibility assessment should also be performed to demonstrate that their use in the biomedical field is risk-free.

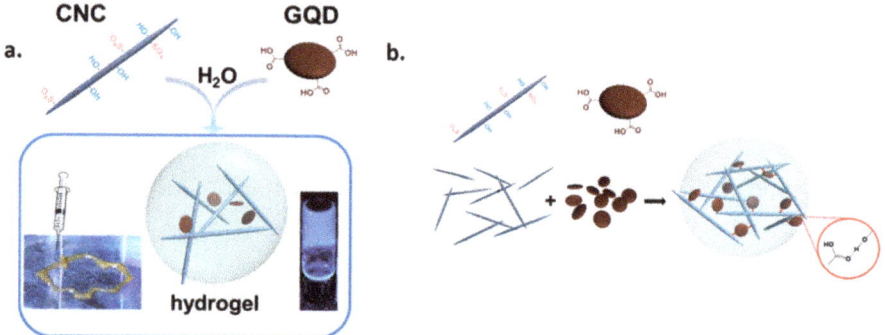

Figure 8. Schematic representation of the CNCs and GQDs hydrogels building blocks and illustration of the shear thinning behavior and photoluminescence properties (**a**) Hydrogen bonds formation between GQDs and CNCs upon mixing (**b**) [112]. Reproduced with copyright permission.

In another study conducted by Kumar et al., GO nanosheets and CNCs were employed as multifunctional crosslinking agents for polyacrylamide-sodium carboxymethyl cellulose (PAA-NaCMC) hydrogels, prepared by in situ free radical polymerization. Due to GO and CNCs hydrophilic nature and ability to provide sites for hydrogen and covalent bonding, both PAA and NaCMC had favorable molecular interactions with GO and formed strong interfacial bonds with CNCs. The synergistic effect of GO and CNC lead to an improvement of the viscoelastic mechanical properties, shape recovery behavior and self-healing ability of the interpenetrating PAA/NaCMC network. Still, further studied must be conducted in order to determinate if these hybrid hydrogels represent appropriate 3D microenvironments for tissue engineering applications [38].

Polylactic acid is a biobased plastic obtained by bacterial fermentation processes. Its biodegradability and biocompatibility recommend it for replacing conventional petroleum-based plastics used in the biomedical field. Still, its mechanical properties need further improvement. CNCs and rGO were used as reinforcing fillers for PLA [61,62]. The polymer was dissolved in chloroform and a chloroform suspension of CNCs and rGO was gradually added into the polymeric solution under vigorous magnetic stirring. The mixture was casted on circular Petri plates and left to dry at room temperature (Figure 9 (left)). The traces of solvent were removed by further drying the PLA films in a laboratory oven, at 40 °C for 3 h. The filler addition increased the tensile strength but also slightly decreased the ductility, thus resulting a lower elongation at break. The biocompatibility of the PLA films was evaluated by seeding NIH3T3 fibroblasts on their surface and observing the changes in cellular morphology 24 h post seeding via fluorescence microscopy (Figure 9 (right)). Minor changes such as small cytoplasmatic lesions were observed but they were not considered a sign of cytotoxicity. An interesting fact was that rGO induced an antibacterial character to the PLA films, as observed during the disk diffusion assay. The samples containing only 0.5 wt % rGO presented an antibacterial activity that was not noticed on neat PLA or PLA/CNC composites (Figure 9 (right)) [61].

This study suggests that rGO is not only a reinforcing agent for biopolymer matrices but can also induce an antimicrobial character, very useful for preventing infections in tissue engineering applications.

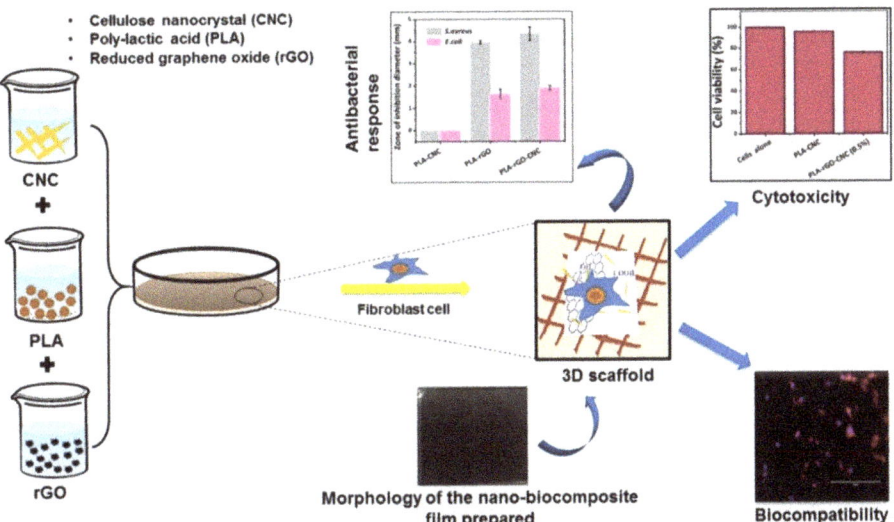

Figure 9. Schematic description of the method used for the preparation of nanocomposite films (**left**); Antibacterial properties, cytotoxicity, biocompatibility, and morphology of the PLA/CNC-rGO composites (**right**) [61]. Reproduced with copyright permission.

Undeniably, cellulose and graphene are highly versatile materials and they can be employed in multiple ways to obtain hybrid composites with superior physico-chemical and biological features compared to cellulose or graphene alone. A summary of the cellulose/graphene composites for tissue engineering applications presented in the previous chapters is found in Table 1.

3. Conclusions and Future Perspectives

This review highlighted some of the recent discoveries regarding cellulose-based composites with graphene and graphene derivatives. As most of the described studies concluded, the combination of cellulose and graphene renders sustainable and cost effective composites with improved physico-chemical features and bioactivity. Cellulose and graphene can be labeled as complementary materials because the incorporation of graphene in cellulose-based matrices minimizes GE's potential cytotoxic character and, in its turn, GE improves the biocompatibility of cellulose by providing cell binding sites and positively interfering in cellular differentiation processes and also its mechanical properties. The versatility of both cellulose and graphene is another important reason cellulose/graphene composites are needed in tissue engineering. Cellulose paper was used to develop electrodes for cellular electrical stimulation or in vivo tissue models for a better understanding of a specific tissue biology and experimental drug testing. Bacterial cellulose and cellulose derivatives were employed in the form of electrospun mats, membranes, and hydrogels as ECM-mimicking scaffolds for enhanced cell adhesion, growth and proliferation whereas cellulose nanocrystals and cellulose nanofibers were incorporated in various biopolymer matrices for a reinforcing effect. Some particularly interesting features for tissue engineering are related to GE and its derivatives ability to guide stem cells differentiation toward specific lineages (osteogenic, adipogenic, angiogenic) and also to the electrical conductivity and fluorescent properties that they provide to cellulose matrices, thus extending the areas of application of cellulose/graphene composites Also, a synergistic reinforcing effect was observed when both cellulose and graphene-based fillers were incorporated in various biopolymer matrices.

Table 1. Components, preparation methods and applications of the cellulose/graphene composites presented in this review.

Cellulose Type	Graphene Type	Composite Preparation Method	Application	Ref.
Cellulose paper (commercial, tissues, blotting paper, filter paper)	rGO	Drop-wise deposition of GO (aqueous dispersion) on the paper substrate and GO reduction with L-ascorbic acid followed by lamination of the G-C papers with alginate Integration of the G-C papers in polystyrene chambers	Multilayered constructs for bone tissue engineering Electrodes for concomitant cell culture and electrical stimulation	[58,59]
BC	GE, GO	Membrane-liquid interface culture (MILIC)	Cell culture scaffolds	[87]
BC	rGO	In situ biosynthesis and bacteria-mediated reduction	Cell culture scaffolds with electrical stimulation potential, biosensors	[89]
BC	GO-HA	Impregnation of the wet BC pellicle with an ethanolic GO-HA suspension	Cell culture scaffolds for bone tissue engineering	[91]
BC	GO	In situ biosynthesis	Cell culture scaffolds for neural tissue engineering	[86]
BC	GO stabilized with APCLP	Non-genetic manipulation of Acetobacter xylinum	E18 neurons culture scaffolds used to construct a 3D neuronal network (minibrain)	[93]
BC	GO covered with APCLP	In situ biosynthesis	Brain cortex mimetics	[92]
CA	GO	Electrospinning	Cell culture scaffolds for bone tissue engineering	[94]
CA	rGO-Co	Electrospinning	Cell culture scaffolds for enhances osteogenic differentiation under alternative magnetic field (AMF)	[101]
CA	GO	Electrospinning	Tumor cell culture scaffolds for	[102]
BC/CA	GO	Electrospinning of CA/GO solution and impregnation with BC culture medium for in situ biosynthesis	Tumor cell culture scaffolds with improved ECM-like features	[103]
CA	GO-CNT	Phase inversion	Membranes for guided hASCs differentiation	[54]
CMC	GQDs	Crosslinking of aqueous CMC suspension by GQDs via esterification	ECM-like scaffolds with pH sensitive drug delivery potential	[104,105]
CMC	GO	Grafting of CMC on GO via hydrothermal treatment	pH sensitive drug delivery systems for colon cancer treatment	[108]
CMC	GO	Layer by layer deposition of CMC and PVP on GO nanoparticles, encapsulation of curcumin in the CMC layer, surface grafting of folic acid antibody using PEG as linker	Folate-targeting drug delivery systems for cancer treatment	[109]
CNCs	GQDs	Crosslinking of CNCs aqueous suspensions by GQDs	Injectable hydrogels with photoluminescence properties	[112]
CNCs	GO	Mixing of CNCs and GO in distilled water	Multifunctional crosslinking agents for PAA/NaCMC hydrogels	[38]
CNCs	rGO	Mixing of CNCs and rGO in chloroform	Reinforcing fillers with antibacterial properties for PLA	[61,62]

The constant research activity in the field of cellulose/graphene composites lead so far to important discoveries in the area of tissue engineering. Even if these materials could represent promising alternatives to currently used medical techniques, the progress toward obtaining clinical products is slow. Future trends consist of the collaboration of clinicians, biomaterial scientists and engineers, with expertise in their own fields, to elaborate marketable and accessible products for therapeutic purposes.

Author Contributions: Conceptualization, M.O. and S.I.V.; resources, M.O. and S.I.V.; data curation, M.O. and S.I.V.; writing—original draft preparation, M.O. and S.I.V.; project administration, S.I.V.; funding acquisition, S.I.V. All authors have read and agreed to the published version of the manuscript.

Funding: This work was supported by a grant of the Romanian National Authority for Scientific Research and Innovation, CNCS-UEFISCDI, project number PN-III-P1-1.2-PCCDI-2017-0407—Intelligent materials for medical applications, sub-project—New generation of hemodialysis composite membranes with derivatized graphene and a grant of the Competitiveness Operational Program 2014–2020, Action 1.1.3: Creating synergies with RDI actions of the EU's HORIZON 2020 framework program and other international RDI programs, MySMIS Code 108792, Acronym project "UPB4H", financed by contract: 250/11.05.2020.

Acknowledgments: M.O. gratefully acknowledge the financial support through project PN-III-P1-1.2-PCCDI-2017-0407—Intelligent materials for medical applications, sub-project—New generation of hemodialysis composite membranes with derivatized graphene. S.I.V. gratefully acknowledge the financial support through the Competitiveness Operational Program 2014-2020, Action 1.1.3: Creating synergies with RDI actions of the EU's HORIZON 2020 framework program and other international RDI programs, MySMIS Code 108792, Acronym project "UPB4H", financed by contract: 250/11.05.2020.

Conflicts of Interest: The authors declare no conflict of interest.

References

1. Profire, L.; Constantin, S.M. Nanomaterials in tissue engineering. In *Polymeric Nanomaterials in Nanotherapeutics*; Vasile, C., Ed.; Elsevier: Amsterdam, The Netherlands, 2019; Chapter 12; pp. 421–436.
2. Berthiaume, F.; Maguire, T.J.; Yarmush, M.L. Tissue engineering and regenerative medicine: History, progress, and challenges. *Ann. Rev. Chem. Biomol. Eng.* **2011**, *2*, 403–430. [CrossRef] [PubMed]
3. Oprea, M.; Voicu, S.I. Recent advances in composites based on cellulose derivatives for biomedical applications. *Carbohydr. Polym.* **2020**, *247*, 116683. [CrossRef] [PubMed]
4. Han, F.; Wang, J.; Ding, L.; Hu, Y.; Li, W.; Yuan, Z.; Guo, Q.; Zhu, C.; Yu, L.; Wang, H.; et al. Tissue engineering and regenerative medicine: Achievements, future, and sustainability in Asia. *Front. Bioeng. Biotechnol.* **2020**, *8*. [CrossRef]
5. Elitok, M.S.; Gunduz, E.; Gurses, H.E.; Gunduz, M. Tissue engineering: Towards development of regenerative and transplant medicine A2-Barh, Debmalya. In *Omics Technologies and Bio-Engineering*; Azevedo, V., Ed.; Academic Press: Cambridge, MA, USA, 2018; Chapter 20; pp. 471–495.
6. Steffens, D.; Braghirolli, D.I.; Maurmann, N.; Pranke, P. Update on the main use of biomaterials and techniques associated with tissue engineering. *Drug Discov. Today* **2018**, *23*, 1474–1488. [CrossRef] [PubMed]
7. Frey, B.M.; Zeisberger, S.M.; Hoerstrup, S.P. Tissue engineering and regenerative medicine-new initiatives for individual treatment offers. *Transfus. Med. Hemother.* **2016**, *43*, 318–319. [CrossRef] [PubMed]
8. Katari, R.; Peloso, A.; Orlando, G. Tissue engineering and regenerative medicine: Semantic considerations for an evolving paradigm. *Front. Bioeng. Biotechnol.* **2014**, *2*, 57. [CrossRef] [PubMed]
9. Tobita, M.; Konomi, K.; Torashima, Y.; Kimura, K.; Taoka, M.; Kaminota, M. Japan's challenges of translational regenerative medicine: Act on the safety of regenerative medicine. *Regen. Ther.* **2016**, *4*, 78–81. [CrossRef]
10. Saul, J.M.; Williams, D.F. Hydrogels in regenerative medicine A2-Modjarrad, Kayvon. In *Handbook of Polymer Applications in Medicine and Medical Devices*; Ebnesajjad, S., Ed.; William Andrew Publishing: Oxford, UK, 2011; Chapter 12; pp. 279–302.
11. Tominac Trcin, M.; Dekaris, I.; Mijović, B.; Bujić, M.; Zdraveva, E.; Dolenec, T.; Pauk-Gulić, M.; Primorac, D.; Crnjac, J.; Špoljarić, B.; et al. Synthetic vs natural scaffolds for human limbal stem cells. *Croat. Med. J.* **2015**, *56*, 246–256. [CrossRef]
12. Goh, S.-K.; Bertera, S.; Olsen, P.; Candiello, J.E.; Halfter, W.; Uechi, G.; Balasubramani, M.; Johnson, S.A.; Sicari, B.M.; Kollar, E.; et al. Perfusion-decellularized pancreas as a natural 3D scaffold for pancreatic tissue and whole organ engineering. *Biomaterials* **2013**, *34*, 6760–6772. [CrossRef]
13. Castells-Sala, C.; Alemany-Ribes, M.; Fernández-Muiños, T.; Recha-Sancho, L.; López-Chicón, P.; Aloy-Reverté, C.; Caballero-Camino, J.; Márquez-Gil, A.; Semino, C.E. Current applications of tissue engineering in biomedicine. *J. Biochips Tiss. Chips* **2013**, *S2*, 1.
14. Stumpf, T.R.; Yang, X.; Zhang, J.; Cao, X. In situ and ex situ modifications of bacterial cellulose for applications in tissue engineering. *Mat. Sci. Eng. C* **2018**, *82*, 372–383. [CrossRef] [PubMed]
15. Naahidi, S.; Jafari, M.; Logan, M.; Wang, Y.; Yuan, Y.; Bae, H.; Dixon, B.; Chen, P. Biocompatibility of hydrogel-based scaffolds for tissue engineering applications. *Biotechnol. Adv.* **2017**, *35*, 530–544. [CrossRef] [PubMed]
16. Mitchell, A.C.; Briquez, P.S.; Hubbell, J.A.; Cochran, J.R. Engineering growth factors for regenerative medicine applications. *Acta Biomater.* **2016**, *30*, 1–12. [CrossRef] [PubMed]

17. Oprea, M.; Panaitescu, D.; Nicolae, C.-A.; Gabor, R.; Frone, A.; Raditoiu, V.; Trusca, R.; Casarica, A. Nanocomposites from functionalized bacterial cellulose and poly(3-hydroxybutyrate-co-3-hydroxyvalerate). *Polym. Degrad. Stab.* **2020**, *179*, 109203. [CrossRef]
18. Hosseinkhani, M.; Mehrabani, D.; Karimfar, M.H.; Bakhtiyari, S.; Manafi, A.; Shirazi, R. Tissue engineered scaffolds in regenerative medicine. *World J. Plast. Surg.* **2014**, *3*, 3–7.
19. Frone, A.N.; Chiulan, I.; Panaitescu, D.M.; Nicolae, C.A.; Ghiurea, M.; Galan, A.-M. Isolation of cellulose nanocrystals from plum seed shells, structural and morphological characterization. *Mater. Lett.* **2017**, *194*, 160–163. [CrossRef]
20. Credou, J.; Berthelot, T. Cellulose: From biocompatible to bioactive material. *J. Mater. Chem. B* **2014**, *2*, 4767–4788. [CrossRef]
21. Trache, D.; Tarchoun, A.F.; Derradji, M.; Hamidon, T.S.; Masruchin, N.; Brosse, N.; Hussin, M.H. Nanocellulose: From Fundamentals to Advanced Applications. *Front. Chem.* **2020**, *8*, 392. [CrossRef]
22. Dong, S.; Hirani, A.A.; Colacino, K.R.; Lee, Y.W.; Roman, M. Cytotoxicity and cellular uptake of cellulose nanocrystals. *Nano LIFE* **2012**, *2*, 1241006. [CrossRef]
23. Souza, S.F.; Mariano, M.; Reis, D.; Lombello, C.B.; Ferreira, M.; Sain, M. Cell interactions and cytotoxic studies of cellulose nanofibers from Curauá natural fibers. *Carbohydr. Polym.* **2018**, *201*, 87–95. [CrossRef]
24. Khan, S.; Ul-Islam, M.; Ikram, M.; Ullah, M.W.; Israr, M.; Subhan, F.; Kim, Y.; Jang, J.H.; Yoon, S.; Park, J.K. Three-dimensionally microporous and highly biocompatible bacterial cellulose–gelatin composite scaffolds for tissue engineering applications. *RSC Adv.* **2016**, *6*, 110840–110849. [CrossRef]
25. Moniri, M.; Boroumand Moghaddam, A.; Azizi, S.; Abdul Rahim, R.; Bin Ariff, A.; Zuhainis Saad, W.; Navaderi, M.; Mohamad, R. Production and status of bacterial cellulose in biomedical engineering. *Nanomaterials* **2017**, *7*, 257. [CrossRef] [PubMed]
26. Torres, F.; Commeaux, S.; Troncoso, O. Biocompatibility of bacterial cellulose based biomaterials. *J. Funct. Biomater.* **2012**, *3*, 864–878. [CrossRef] [PubMed]
27. Liyaskina, E.; Revin, V.; Paramonova, E.; Nazarkina, M.; Pestov, N.; Revina, N.; Kolesnikova, S. Nanomaterials from bacterial cellulose for antimicrobial wound dressing. *J. Phys. Conf. Ser.* **2016**, *784*, 012034. [CrossRef]
28. Klemm, D.; Heublein, B.; Fink, H.P.; Bohn, A. Cellulose: Fascinating biopolymer and sustainable raw material. *Angew. Chem. Int. Ed.* **2005**, *44*, 3358–3393. [CrossRef]
29. Esa, F.; Tasirin, S.M.; Rahman, N.A. Overview of bacterial cellulose production and application. *Agric. Agric. Sci. Proc.* **2014**, *2*, 113–119. [CrossRef]
30. Pandele, A.M.; Comanici, F.E.; Carp, C.A.; Miculescu, F.; Voicu, S.I.; Thakur, V.K.; Serban, B.C. Synthesis and characterization of cellulose acetate-hydroxyapatite micro and nano composites membranes for water purification and biomedical applications. *Vacuum* **2017**, *146*, 599–605. [CrossRef]
31. Pandele, A.M.; Neacsu, P.; Cimpean, A.; Staras, A.I.; Miculescu, F.; Iordache, A.; Voicu, S.I.; Thakur, V.K.; Toader, O.D. Cellulose acetate membranes functionalized with resveratrol by covalent immobilization for improved osseointegration. *Appli. Surf. Sci.* **2018**, *438*, 2–13. [CrossRef]
32. Pandele, A.M.; Constantinescu, A.; Radu, I.C.; Miculescu, F.; Ioan Voicu, S.; Ciocan, L.T.J.M. Synthesis and characterization of pla-micro-structured hydroxyapatite composite films. *Materials* **2020**, *13*, 274. [CrossRef]
33. Kostag, M.; Jedvert, K.; Achtel, C.; Heinze, T.; El Seoud, O.A. Recent advances in solvents for the dissolution, shaping and derivatization of cellulose: Quaternary ammonium electrolytes and their solutions in water and molecular solvents. *Molecules* **2018**, *23*, 511. [CrossRef]
34. Miyamoto, T.; Takahashi, S.; Ito, H.; Inagaki, H.; Noishiki, Y. Tissue biocompatibility of cellulose and its derivatives. *J. Biomed. Mater. Res.* **1989**, *23*, 125–133. [CrossRef]
35. Bhuyan, M.S.A.; Uddin, M.N.; Islam, M.M.; Bipasha, F.A.; Hossain, S.S. Synthesis of graphene. *Int. Nano Lett.* **2016**, *6*, 65–83. [CrossRef]
36. Kenry; Lee, W.C.; Loh, K.P.; Lim, C.T. When stem cells meet graphene: Opportunities and challenges in regenerative medicine. *Biomaterials* **2018**, *155*, 236–250. [CrossRef]
37. Zhang, Q.; Wu, Z.; Li, N.; Pu, Y.; Wang, B.; Zhang, T.; Tao, J. Advanced review of graphene-based nanomaterials in drug delivery systems: Synthesis, modification, toxicity and application. *Mat. Sci. Eng. C* **2017**, *77*, 1363–1375. [CrossRef] [PubMed]
38. Kumar, A.; Rao, K.M.; Han, S.S. Mechanically viscoelastic nanoreinforced hybrid hydrogels composed of polyacrylamide, sodium carboxymethylcellulose, graphene oxide, and cellulose nanocrystals. *Carbohydr. Polym.* **2018**, *193*, 228–238. [CrossRef] [PubMed]

39. Johnson, D.W.; Dobson, B.P.; Coleman, K.S. A manufacturing perspective on graphene dispersions. *Curr. Opin. Coll. Interf. Sci.* **2015**, *20*, 367–382. [CrossRef]
40. Lin, J.; Chen, X.; Huang, P. Graphene-based nanomaterials for bioimaging. *Adv. Drug Deliv. Rev.* **2016**, *105*, 242–254. [CrossRef] [PubMed]
41. Syama, S.; Mohanan, P.V. Safety and biocompatibility of graphene: A new generation nanomaterial for biomedical application. *Int. J. Biol. Macromol.* **2016**, *86*, 546–555. [CrossRef]
42. Azarniya, A.; Eslahi, N.; Mahmoudi, N.; Simchi, A. Effect of graphene oxide nanosheets on the physico-mechanical properties of chitosan/bacterial cellulose nanofibrous composites. *Compos. Part A Appl. Sci. Manuf.* **2016**, *85*, 113–122. [CrossRef]
43. Zheng, F.; Li, R.; He, Q.; Koral, K.; Tao, J.; Fan, L.; Xiang, R.; Ma, J.; Wang, N.; Yin, Y.; et al. The electrostimulation and scar inhibition effect of chitosan/oxidized hydroxyethyl cellulose/reduced graphene oxide/asiaticoside liposome based hydrogel on peripheral nerve regeneration in vitro. *Mat. Sci. Eng. C* **2020**, *109*, 110560. [CrossRef]
44. Foo, M.E.; Gopinath, S.C.B. Feasibility of graphene in biomedical applications. *Biomed. Pharmacother.* **2017**, *94*, 354–361. [CrossRef] [PubMed]
45. Younis, M.R.; He, G.; Lin, J.; Huang, P. Recent advances on graphene quantum dots for bioimaging applications. *Front. Chem.* **2020**, *8*, 424. [CrossRef] [PubMed]
46. Ramezani, M.; Alibolandi, M.; Nejabat, M.; Charbgoo, F.; Taghdisi, S.M.; Abnous, K. Graphene-Based Hybrid Nanomaterials for Biomedical Applications. In *Biomedical Applications of Graphene and 2D Nanomaterials*; Nurunnabi, M., McCarthy, J.R., Eds.; Elsevier: Amsterdam, The Netherlands, 2019; Chapter 6; pp. 119–141.
47. Liao, C.; Li, Y.; Tjong, S.C. Graphene nanomaterials: Synthesis, biocompatibility, and cytotoxicity. *Int. J. Mol. Sci.* **2018**, *19*, 3564. [CrossRef]
48. Voicu, S.; Pandele, M.; Vasile, E.; Rughinis, R.; Crica, L.; Pilan, L.; Ionita, M. The impact of sonication time through polysulfone-graphene oxide composite films properties. *Dig. J. Nanomater. Biostruct.* **2013**, *8*, 1389–1394.
49. Erdal, N.B.; Adolfsson, K.H.; Pettersson, T.; Hakkarainen, M. Green strategy to reduced nanographene oxide through microwave assisted transformation of cellulose. *ACS Sustain. Chem. Eng.* **2018**, *6*, 1246–1255. [CrossRef]
50. Erdal, N.B.; Hakkarainen, M. Construction of bioactive and reinforced bioresorbable nanocomposites by reduced nano-graphene oxide carbon dots. *Biomacromolecules* **2018**, *19*, 1074–1081. [CrossRef] [PubMed]
51. Erdal, N.B.; Yao, J.G.; Hakkarainen, M. Cellulose-derived nanographene oxide surface-functionalized three-dimensional scaffolds with drug delivery capability. *Biomacromolecules* **2019**, *20*, 738–749. [CrossRef]
52. Yadav, A.; Erdal, N.B.; Hakkarainen, M.; Nandan, B.; Srivastava, R.K. Cellulose-derived nanographene oxide reinforced macroporous scaffolds of high internal phase emulsion-templated cross-linked poly(ε-caprolactone). *Biomacromolecules* **2020**, *21*, 589–596. [CrossRef]
53. Zhang, B.; Wei, P.; Zhou, Z.; Wei, T. Interactions of graphene with mammalian cells: Molecular mechanisms and biomedical insights. *Adv. Drug Deliv. Rev.* **2016**, *105*, 145–162. [CrossRef]
54. Ignat, S.R.; Lazăr, A.D.; Şelaru, A.; Samoilă, I.; Vlăsceanu, G.M.; Ioniţă, M.; Radu, E.; Dinescu, S.; Costache, M. Versatile biomaterial platform enriched with graphene oxide and carbon nanotubes for multiple tissue engineering applications. *Int. J. Mol. Sci.* **2019**, *20*, 3868. [CrossRef]
55. Xie, H.; Cao, T.; Franco-Obregón, A.; Rosa, V. Graphene-induced osteogenic differentiation is mediated by the integrin/FAK Axis. *Int. J. Mol. Sci.* **2019**, *20*, 574. [CrossRef] [PubMed]
56. Mukherjee, S.; Sriram, P.; Barui, A.K.; Nethi, S.K.; Veeriah, V.; Chatterjee, S.; Suresh, K.I.; Patra, C.R. Graphene oxides show angiogenic properties. *Adv. Healthc. Mat.* **2015**, *4*, 1722–1732. [CrossRef] [PubMed]
57. Chakraborty, S.; Ponrasu, T.; Chandel, S.; Dixit, M.; Muthuvijayan, V. Reduced graphene oxide-loaded nanocomposite scaffolds for enhancing angiogenesis in tissue engineering applications. *R. Soc. Open Sci.* **2018**, *5*, 172017. [CrossRef] [PubMed]
58. Li, J.; Liu, X.; Tomaskovic-Crook, E.; Crook, J.M.; Wallace, G.G. Smart graphene-cellulose paper for 2D or 3D "origami-inspired" human stem cell support and differentiation. *Coll. Surf. B Biointerf.* **2019**, *176*, 87–95. [CrossRef] [PubMed]
59. Torres, F.G.; Arroyo, J.J.; Troncoso, O.P. Bacterial cellulose nanocomposites: An all-nano type of material. *Mat. Sci. Eng. C* **2019**, *98*, 1277–1293. [CrossRef]

60. Terzopoulou, Z.; Kyzas, G.Z.; Bikiaris, D.N. Recent advances in nanocomposite materials of graphene derivatives with polysaccharides. *Materials* **2015**, *8*, 652–683. [CrossRef]
61. Pal, N.; Dubey, P.; Gopinath, P.; Pal, K. Combined effect of cellulose nanocrystal and reduced graphene oxide into poly-lactic acid matrix nanocomposite as a scaffold and its anti-bacterial activity. *Int. J. Biol. Macromol.* **2017**, *95*, 94–105. [CrossRef]
62. Pal, N.; Banerjee, S.; Roy, P.; Pal, K. Reduced graphene oxide and PEG-grafted TEMPO-oxidized cellulose nanocrystal reinforced poly-lactic acid nanocomposite film for biomedical application. *Mat. Sci. Eng. C* **2019**, *104*, 109956. [CrossRef]
63. Neibolts, N.; Platnieks, O.; Gaidukovs, S.; Barkane, A.; Thakur, V.K.; Filipova, I.; Mihai, G.; Zelca, Z.; Yamaguchi, K.; Enachescu, M. Needle-free electrospinning of nanofibrillated cellulose and graphene nanoplatelets based sustainable poly (butylene succinate) nanofibers. *Mater. Today Chem.* **2020**, *17*, 100301. [CrossRef]
64. Patel, D.K.; Seo, Y.-R.; Dutta, S.D.; Lim, K.-T. Enhanced osteogenesis of mesenchymal stem cells on electrospun cellulose nanocrystals/poly(ε-caprolactone) nanofibers on graphene oxide substrates. *RSC Adv.* **2019**, *9*, 36040–36049. [CrossRef]
65. Pandele, A.M.; Serbanescu, O.S.; Voicu, S.I. Polysulfone composite membranes with carbonaceous structure. synthesis and applications. *Coatings* **2020**, *10*, 609. [CrossRef]
66. Serbanescu, O.; Pandele, A.; Miculescu, F.; Voicu, Ş.I. Synthesis and characterization of cellulose acetate membranes with self-indicating properties by changing the membrane surface color for separation of Gd(III). *Coatings* **2020**, *10*, 468. [CrossRef]
67. Pandele, A.M.; Iovu, H.; Orbeci, C.; Tuncel, C.; Miculescu, F.; Nicolescu, A.; Deleanu, C.; Voicu, S.I. Surface modified cellulose acetate membranes for the reactive retention of tetracycline. *Sep. Purif. Technol.* **2020**, *249*, 117145. [CrossRef]
68. Corobea, M.C.; Muhulet, O.; Miculescu, F.; Antoniac, I.V.; Vuluga, Z.; Florea, D.; Vuluga, D.M.; Butnaru, M.; Ivanov, D.; Voicu, S.I. Novel nanocomposite membranes from cellulose acetate and clay-silica nanowires. *Polym. Adv. Technol.* **2016**, *27*, 1586–1595. [CrossRef]
69. Ionita, M.; Crica, L.E.; Voicu, S.I.; Pandele, A.M.; Iovu, H. Fabrication of cellulose triacetate/graphene oxide porous membrane. *Polym. Adv. Technol.* **2016**, *27*, 350–357. [CrossRef]
70. Ionita, M.; Vasile, E.; Crica, L.E.; Voicu, S.I.; Pandele, A.M.; Dinescu, S.; Predoiu, L.; Galateanu, B.; Hermenean, A.; Costache, M. Synthesis, characterization and in vitro studies of polysulfone/graphene oxide composite membranes. *Compos. Part B Eng.* **2015**, *72*, 108–115. [CrossRef]
71. Ioniță, M.; Crică, L.E.; Voicu, S.I.; Dinescu, S.; Miculescu, F.; Costache, M.; Iovu, H. Synergistic effect of carbon nanotubes and graphene for high performance cellulose acetate membranes in biomedical applications. *Carbohydr. Polym.* **2018**, *183*, 50–61. [CrossRef]
72. Ioniță, M.; Vlăsceanu, G.M.; Watzlawek, A.A.; Voicu, S.I.; Burns, J.S.; Iovu, H. Graphene and functionalized graphene: Extraordinary prospects for nanobiocomposite materials. *Compos. Part B Eng.* **2017**, *121*, 34–57. [CrossRef]
73. Muhulet, A.; Miculescu, F.; Voicu, S.I.; Schütt, F.; Thakur, V.K.; Mishra, Y.K. Fundamentals and scopes of doped carbon nanotubes towards energy and biosensing applications. *Mater. Today Energy* **2018**, *9*, 154–186. [CrossRef]
74. Rusen, E.; Mocanu, A.; Nistor, L.C.; Dinescu, A.; Călinescu, I.; Muștățea, G.; Voicu, Ş.I.; Andronescu, C.; Diacon, A. Design of antimicrobial membrane based on polymer colloids/multiwall carbon nanotubes hybrid material with silver nanoparticles. *ACS Appl. Mater. Interf.* **2014**, *6*, 17384–17393. [CrossRef]
75. Muhulet, A.; Tuncel, C.; Miculescu, F.; Pandele, A.M.; Bobirica, C.; Orbeci, C.; Bobirica, L.; Palla-Papavlu, A.; Voicu, S.I. Synthesis and characterization of polysulfone-TiO_2 decorated MWCNT composite membranes by sonochemical method. *Appl. Phys. A* **2020**, *126*, 233. [CrossRef]
76. Chan, B.; Leong, K. Scaffolding in tissue engineering: General approaches and tissue-specific considerations. *Eur. Spine J.* **2008**, *17*, 467–479. [CrossRef] [PubMed]
77. Fröhlich, M.; Grayson, W.L.; Wan, L.Q.; Marolt, D.; Drobnic, M.; Vunjak-Novakovic, G. Tissue engineered bone grafts: Biological requirements, tissue culture and clinical relevance. *Curr. Stem Cell Res. Ther.* **2008**, *3*, 254–264. [CrossRef]
78. Boni, R.; Ali, A.; Shavandi, A.; Clarkson, A.N. Current and novel polymeric biomaterials for neural tissue engineering. *J. Biomed. Sci.* **2018**, *25*, 90. [CrossRef]

79. Chen, C.; Bai, X.; Ding, Y.; Lee, I.-S. Electrical stimulation as a novel tool for regulating cell behavior in tissue engineering. *Biomater. Res.* **2019**, *23*, 25. [CrossRef] [PubMed]
80. Singh, S.; Dutt, D.; Kaur, P.; Singh, H.; Mishra, N.C. Microfibrous paper scaffold for tissue engineering application. *J. Biomat. Sci. Polym. Ed.* **2020**, *31*, 1091–1106. [CrossRef] [PubMed]
81. Cheng, F.; Cao, X.; Li, H.; Liu, T.; Xie, X.; Huang, D.; Maharjan, S.; Bei, H.P.; Gómez, A.; Li, J.; et al. Generation of cost-effective paper-based tissue models through matrix-assisted sacrificial 3d printing. *Nano Lett.* **2019**, *19*, 3603–3611. [CrossRef]
82. Li, J.; Liu, X.; Crook, J.M.; Wallace, G.G. Electrical stimulation-induced osteogenesis of human adipose derived stem cells using a conductive graphene-cellulose scaffold. *Mat. Sci. Eng. C* **2020**, *107*, 110312. [CrossRef]
83. Derda, R.; Laromaine, A.; Mammoto, A.; Tang, S.; Mammoto, T.; Ingber, D.; Whitesides, G. Paper-supported 3D cell culture for tissue-based bioassays. *Proc. Natl. Acad. Sci. USA* **2009**, *106*, 18457–18462. [CrossRef]
84. Derda, R.; Tang, S.K.; Laromaine, A.; Mosadegh, B.; Hong, E.; Mwangi, M.; Mammoto, A.; Ingber, D.E.; Whitesides, G.M. Multizone paper platform for 3D cell cultures. *PLoS ONE* **2011**, *6*, e18940. [CrossRef]
85. Shahin-Shamsabadi, A.; Selvaganapathy, P.R. ExCeL: Combining extrusion printing on cellulose scaffolds with lamination to create in vitro biological models. *Biofabrication* **2019**, *11*, 035002. [CrossRef] [PubMed]
86. Si, H.; Luo, H.; Xiong, G.; Yang, Z.; Raman, S.R.; Guo, R.; Wan, Y. One-step in situ biosynthesis of graphene oxide–bacterial cellulose nanocomposite hydrogels. *Macromol. Rapid Commun.* **2014**, *35*, 1706–1711. [CrossRef] [PubMed]
87. Luo, H.; Ao, H.; Peng, M.; Yao, F.; Yang, Z.; Wan, Y. Effect of highly dispersed graphene and graphene oxide in 3D nanofibrous bacterial cellulose scaffold on cell responses: A comparative study. *Mater. Chem. Phys.* **2019**, *235*, 121774. [CrossRef]
88. Shang, L.; Qi, Y.; Lu, H.; Pei, H.; Li, Y.; Qu, L.; Wu, Z.; Zhang, W. Graphene and graphene oxide for tissue engineering and regeneration. In *Theranostic Bionanomaterials*; Cui, W., Zhao, X., Eds.; Elsevier: Amsterdam, The Netherlands, 2019; Chapter 7; pp. 165–185.
89. Jin, L.; Zeng, Z.; Kuddannaya, S.; Wu, D.; Zhang, Y.; Wang, Z. Biocompatible, free-standing film composed of bacterial cellulose nanofibers-graphene composite. *ACS Appl. Mater. Interf.* **2016**, *8*, 1011–1018. [CrossRef] [PubMed]
90. Oprea, M.; Voicu, Ş.I. Recent advances in applications of cellulose derivatives-based composite membranes with hydroxyapatite. *Materials* **2020**, *13*, 2481. [CrossRef] [PubMed]
91. Ramani, D.; Sastry, T.P. Bacterial cellulose-reinforced hydroxyapatite functionalized graphene oxide: A potential osteoinductive composite. *Cellulose* **2014**, *21*, 3585–3595. [CrossRef]
92. Park, J.B.; Sung, D.; Park, S.; Min, K.-A.; Kim, K.W.; Choi, Y.; Kim, H.Y.; Lee, E.; Kim, H.S.; Jin, M.S.; et al. 3D graphene-cellulose nanofiber hybrid scaffolds for cortical reconstruction in brain injuries. *2D Mater.* **2019**, *6*, 045043.
93. Kim, D.; Park, S.; Jo, I.; Kim, S.M.; Kang, D.H.; Cho, S.P.; Park, J.B.; Hong, B.H.; Yoon, M.H. Multiscale modulation of nanocrystalline cellulose hydrogel via nanocarbon hybridization for 3D neuronal bilayer formation. *Small* **2017**, *13*, 1700331. [CrossRef]
94. Jun, I.; Han, H.-S.; Edwards, J.R.; Jeon, H. Electrospun fibrous scaffolds for tissue engineering: Viewpoints on architecture and fabrication. *Int. J. Molec. Sci.* **2018**, *19*, 745. [CrossRef]
95. Lannutti, J.; Reneker, D.; Ma, T.; Tomasko, D.; Farson, D. Electrospinning for tissue engineering scaffolds. *Mat. Sci. Eng. C* **2007**, *27*, 504–509. [CrossRef]
96. Lord, M.; Foss, M.; Besenbacher, F. Influence of nanoscale surface topography on protein adsorption and cellular response. *Nano Today* **2010**, *5*, 66–78. [CrossRef]
97. Javed, K.; Krumme, A.; Viirsalu, M.; Krasnou, I.; Plamus, T.; Vassiljeva, V.; Tarasova, E.; Savest, N.; Mere, A.; Mikli, V.; et al. A method for producing conductive graphene biopolymer nanofibrous fabrics by exploitation of an ionic liquid dispersant in electrospinning. *Carbon* **2018**, *140*, 148–156. [CrossRef]
98. Angel, N.; Guo, L.; Yan, F.; Wang, H.; Kong, L. Effect of processing parameters on the electrospinning of cellulose acetate studied by response surface methodology. *J. Agric. Food Res.* **2020**, *2*, 100015. [CrossRef]
99. Liu, X.; Shen, H.; Song, S.; Chen, W.; Zhang, Z. Accelerated biomineralization of graphene oxide-incorporated cellulose acetate nanofibrous scaffolds for mesenchymal stem cell osteogenesis. *Coll. Surf. B Biointerf.* **2017**, *159*, 251–258. [CrossRef]

100. Arakaki, A.; Shimizu, K.; Oda, M.; Sakamoto, T.; Nishimura, T.; Kato, T. Biomineralization-inspired synthesis of functional organic/inorganic hybrid materials: Organic molecular control of self-organization of hybrids. *Org. Biomol. Chem.* **2015**, *13*, 974–989. [CrossRef]
101. Hatamie, S.; Mohamadyar-Toupkanlou, F.; Mirzaei, S.; Ahadian, M.M.; Hosseinzadeh, S.; Soleimani, M.; Sheu, W.-J.; Wei, Z.H.; Hsieh, T.-F.; Chang, W.-C.; et al. Cellulose acetate/magnetic graphene nanofiber in enhanced human mesenchymal stem cells osteogenic differentiation under alternative current magnetic field. *SPIN* **2019**, *09*, 1940011. [CrossRef]
102. Wan, Y.; Lin, Z.; Gan, D.; Cui, T.; Wan, M.; Yao, F.; Zhang, Q.; Luo, H. Effect of graphene oxide incorporation into electrospun cellulose acetate scaffolds on breast cancer cell culture. *Fibers Polym.* **2019**, *20*, 1577–1585. [CrossRef]
103. Wan, Y.; Lin, Z.; Zhang, Q.; Gan, D.; Gama, M.; Tu, J.; Luo, H. Incorporating graphene oxide into biomimetic nano-microfibrous cellulose scaffolds for enhanced breast cancer cell behavior. *Cellulose* **2020**, *27*, 4471–4485. [CrossRef]
104. Rakhshaei, R.; Namazi, H.; Hamishehkar, H.; Rahimi, M. Graphene quantum dot cross-linked carboxymethyl cellulose nanocomposite hydrogel for pH-sensitive oral anticancer drug delivery with potential bioimaging properties. *Int. J. Biol. Macromol.* **2020**, *150*, 1121–1129. [CrossRef]
105. Javanbakht, S.; Namazi, H. Doxorubicin loaded carboxymethyl cellulose/graphene quantum dot nanocomposite hydrogel films as a potential anticancer drug delivery system. *Mat. Sci. Eng. C* **2018**, *87*, 50–59. [CrossRef]
106. Ghawanmeh, A.A.; Ali, G.A.; Algarni, H.; Sarkar, S.M.; Chong, K.F. Graphene oxide-based hydrogels as a nanocarrier for anticancer drug delivery. *Nano Res.* **2019**, *12*, 973–990. [CrossRef]
107. Rao, Z.; Ge, H.; Liu, L.; Zhu, C.; Min, L.; Liu, M.; Fan, L.; Li, D. Carboxymethyl cellulose modified graphene oxide as pH-sensitive drug delivery system. *Int. J. Biol. Macromol.* **2018**, *107*, 1184–1192. [CrossRef] [PubMed]
108. Jiao, Z.; Zhang, B.; Li, C.; Kuang, W.; Zhang, J.; Xiong, Y.; Tan, S.; Cai, X.; Huang, L. Carboxymethyl cellulose-grafted graphene oxide for efficient antitumor drug delivery. *Nanotech. Rev.* **2018**, *7*, 291–301. [CrossRef]
109. Sahne, F.; Mohammadi, M.; Najafpour, G.D. Single-Layer Assembly of Multifunctional carboxymethylcellulose on graphene oxide nanoparticles for improving in vivo curcumin delivery into tumor cells. *ACS Biomat. Sci. Eng.* **2019**, *5*, 2595–2609. [CrossRef]
110. Mianehrow, H.; Moghadam, M.H.M.; Sharif, F.; Mazinani, S.J. Graphene-oxide stabilization in electrolyte solutions using hydroxyethyl cellulose for drug delivery application. *Int. J. Pharm.* **2015**, *484*, 276–282. [CrossRef] [PubMed]
111. Sudimack, J.; Lee, R.J. Targeted drug delivery via the folate receptor. *Adv. Drug. Deliv. Rev.* **2000**, *41*, 147–162. [CrossRef]
112. Khabibullin, A.; Alizadehgiashi, M.; Khuu, N.; Prince, E.; Tebbe, M.; Kumacheva, E.J.L. Injectable shear-thinning fluorescent hydrogel formed by cellulose nanocrystals and graphene quantum dots. *Langmuir* **2017**, *33*, 12344–12350. [CrossRef]

Publisher's Note: MDPI stays neutral with regard to jurisdictional claims in published maps and institutional affiliations.

© 2020 by the authors. Licensee MDPI, Basel, Switzerland. This article is an open access article distributed under the terms and conditions of the Creative Commons Attribution (CC BY) license (http://creativecommons.org/licenses/by/4.0/).

Review

Recent Advances in Applications of Cellulose Derivatives-Based Composite Membranes with Hydroxyapatite

Madalina Oprea [1,2] and Stefan Ioan Voicu [2,3,*]

[1] National Institute for Research and Development in Chemistry and Petrochemistry ICECHIM, Splaiul Independentei 202, 060021 Bucharest, Romania; madalinna.calarasu@gmail.com
[2] Faculty of Applied Chemistry and Materials Science, University Politehnica of Bucharest, Gheorghe Polizu 1-7, 011061 Bucharest, Romania
[3] Advanced Polymer Materials Group, Faculty of Applied Chemistry and Material Science, University Poltehnica of Bucharest, Gheorghe Polizu 1-7, 011061 Bucharest, Romania
* Correspondence: svoicu@gmail.com or stefan.voicu@upb.ro; Tel.: +40-721-165-757

Received: 14 May 2020; Accepted: 28 May 2020; Published: 29 May 2020

Abstract: The development of novel polymeric composites based on cellulose derivatives and hydroxyapatite represents a fascinating and challenging research topic in membranes science and technology. Cellulose-based materials are a viable alternative to synthetic polymers due to their favorable physico-chemical and biological characteristics. They are also an appropriate organic matrix for the incorporation of hydroxyapatite particles, inter and intramolecular hydrogen bonds, as well as electrostatic interactions being formed between the functional groups on the polymeric chains surface and the inorganic filler. The current review presents an overview on the main application fields of cellulose derivatives/hydroxyapatite composite membranes. Considering the versatility of hydroxyapatite particles, the hybrid materials offer favorable prospects for applications in water purification, tissue engineering, drug delivery, and hemodialysis. The preparation technique and the chemical composition have a big influence on the final membrane properties. The well-established membrane fabrication methods such as phase inversion, electrospinning, or gradual electrostatic assembly are discussed, together with the various strategies employed to obtain a homogenous dispersion of the inorganic particles in the polymeric matrix. Finally, the main conclusions and the future directions regarding the preparation and applications of cellulose derivatives/hydroxyapatite composite membranes are presented.

Keywords: membrane; cellulose; hydroxyapatite; water purification; tissue engineering

---

1. Introduction

Among the functional materials currently known, membranes possess a unique characteristic and that is selectivity [1]. Another aspect is represented by the fact that they were the first functional materials known on earth—the membrane of the first unicellular organism [2]. Polymeric membranes were manufactured on a wide scale as filtering materials after the Second World War from the practical necessity of obtaining drinkable water from affected or contaminated natural sources. The first polymer applied industrially to obtain these membranes was cellulose nitrate, an explosive powder used to fabricate bombs and as a consequence, available in large quantities. Membrane technology was constituted in an individual scientific field starting with the research conducted by Loeb and Sourirajan that explained the formation mechanism of asymmetric membranes and the coagulation phenomenon of a polymer from concentrated solution in the presence of a non-solvent [3]. Once the formation mechanism of polymeric membranes was established and understood, more and more materials were developed for different practical applications such as gas separation [4–6], protein concentration [7–9], heavy metals separation [10], and removal of environmental pollutants [11–13]. As time passed, beyond their primary role as filtering materials used to obtain drinking water, membranes were employed in more and more advanced applications, the field with the most advanced performance demands being biomedical engineering. To fulfill the requests of such applications, composite membranes were developed by incorporating nanostructured inorganic fillers in the polymeric matrix thus resulting in a synergistic performance by combining the advantages of organic and inorganic materials [14]. One of the first niche domains targeted the obtainment of hemodialysis membranes [15]. Hemodialysis is the membrane process designed for patients with chronic kidney disease [16] and it is used to substitute kidney functions once every two days. The main chemical species that are separated during this process are urea, uric acid, and excess of creatinine and salts from the human body [17]. Other biomedical applications based on the extracorporeal blood circulation is the artificial lung used especially during open-heart surgeries [18,19] or experimental studies for an artificial liver based on composite membranes functionalized with porcine hepatocytes [20].

Due to the porosity and semi-permeable properties, polymeric membranes are also used in the development of controlled drug delivery devices [21], in an attempt to avoid the toxic effects of high quantities of pharmaceutically active substances on the human body. Another increasingly researched field of application is the one of osseointegration membranes that are placed at the interface between a metallic implant and the bone, with the purpose of favoring the pre-osteoblasts growth and spreading, in order to integrate the implant into the host bone tissue [22–24]. These latter membranes are based especially on composites with hydroxyapatite. Hydroxyapatite is one of the natural bone components, the so called "soft component". It can be used as such, to obtain composites [25,26] or with the addition of other elements, like silver, to obtain hydroxyapatite with antibacterial properties [27]. Hydroxyapatite also has a high synthesis versatility because it can be of animal [28,29] or synthetic origin [30]. The present review desires to offer the reader an overview on the recent progress made in the domain of composite membranes based on cellulose derivatives and hydroxyapatite. Why these two components? They both present the great advantage of originating from natural sources, which is associated with biocompatibility of the resulting composites. More than that, once inserted into the human body, in the case of tissue engineering scaffolds, cellulose presents the remarkable property of bioresorbability, with only glucose molecules resulting after its hydrolysis and degradation. Cellulose derivatives-based membranes for various applications such as water purification, tissue engineering, osseointegration, drug delivery, and hemodialysis will be presented, all having in common hydroxyapatite as a filler agent.

Pure cellulose is frequently turned into cellulose derivatives to overcome the drawbacks related to its poor solubility in common organic solvents [31]. Due to the low cost and widespread availability of cellulose [32], cellulose acetate (CA) and carboxymethylcellulose (CMC) particularly are among the most used cellulose-based matrices for the incorporation of inorganic hydroxyapatite particles in order to obtain hybrid composite membranes with improved physico-chemical and biological characteristics.

Cellulose acetate is a cellulose ester formed by partial or full acetylation of the free hydroxyl groups in the anhydroglucose unit. Depending on the acetyl content that usually ranges between 29% and 48%, mono-, di-, and triacetate can be differentiated [33]. The ability of facile processing by various techniques and its broad range of applications make cellulose acetate the most commonly synthesized cellulose derivative worldwide, the global production of CA from biomass being projected by Global Industry Analysis to be 751.1 thousand metric tons until 2024 [34]. The classic cellulose acetylation process is based on the reaction between wood or cotton pulp with acetic anhydride as the acetylation agent and sulfuric acid as catalyst in an acetic acid reaction media [35]. Currently, this approach is used industrially but more and more research is being conducted on the use of agro-industrial residues and environmentally friendly synthesis routes that involve replacing the sulfuric and acetic acids by eco-friendly reagents [36]. Cellulose acetate is employed for the production of a variety of consumer goods including textiles, photographic films, personal hygiene products, and cigarette filters [37] but its resistance to the action of chemical agents, good thermal stability, flexibility, and mechanical strength [38], coupled with low fouling susceptibility and a hydrophilic nature [39], recommend this polymer especially for the production of membranes, with applications in industrial and biomedical processes (e.g., adsorption, separation, catalysis, biosensing, drug delivery, or tissue regeneration). Cellulose acetate was used in the purification process of contaminated resources of natural gas starting with the mid 1980s, when several companies applied dried CA membranes for CO_2/CH_4 natural gas separations. The hydrophilic nature of cellulose acetate makes the membranes suitable for assisting chemical and biochemical reactions as well as for the removal of polar compounds or specific organic–organic separations using pervaporation. The separation of helium particularly is of great interest in the natural gas purification process, due to its high added value. Asymmetric CA membranes presented an acceptable permeability for He and a good He/N_2 and He/CH_4 selectivity, being considered economically feasible for usage in three stage membrane processes with recycle streams [40]. Carboxymethyl cellulose, one of the most important cellulose ethers, is synthesized by treating alkali cellulose with monochloroacetic acid or sodium monochloro-acetate in an aqueous sodium hydroxide (NaOH) medium [41]. Wood residues, cotton linters, paper sludge, and agricultural waste biomass such as orange peels, corncobs, sugarcane bagasse, rice, or corn husks were used so far as cellulose sources for the preparation of carboxymethylcellulose [42]. As the reaction takes place, the hydroxyl groups in the cellulose backbone are replaced with carboxymethyl groups in the C6 > C3 > C2 order [43]. The chemical structure obtained following etherification is responsible for the unique properties of carboxymethylcellulose, such as water solubility, non-toxicity, biodegradability, transparency, and good film forming ability [44]. Owing to these characteristics, carboxymethylcellulose is already applied in the food, pharmaceutical, and daily-use chemical industries as an emulsifier, thickener, and a flocculating or chelating agent [45,46]. Currently, carboxymethylcellulose based materials are investigated for biomedical applications such as tissue engineering [47] and drug delivery [48] mainly in the shape of hydrogels, membranes, and nanoparticles [49].

Some of the most popular membrane manufacturing techniques include electrospinning, gradual electrostatic assembly, and phase inversion by immersion precipitation or solvent evaporation.

Phase inversion is a popular method for the preparation of cellulose acetate membranes. The first step of this process consists in the dissolution of the polymer in an appropriate solvent, such as hexafluoro-2-propanol [50], formic acid [51], N, N-dimethylformamide [52], acetone [53–55], or a mixed solvent system of the latter two [56], to obtain a homogenous polymeric solution. Afterwards, the obtained solution is cast on a glass plate and submerged in a coagulation bath containing a non-solvent, usually distilled water. The penetration of the solvent into the non-solvent and non-solvent into the polymeric solution cause demixing and polymer precipitation with the formation of a membrane with an asymmetric structure, composed in most cases of a thin film top layer also called "skin", a support porous substructure, and a bottom layer [54,57].

Gradual electrostatic assembly is based on the spontaneous interactions between two oppositely charged polysaccharides, mixed in an aqueous solution. The electrostatic interactions are followed by polymeric chain entanglement and hydrogen bond formation, this resulting in a polyelectrolyte complex membrane. Due to its anionic nature, carboxymethylcellulose was used in combination with cationic polymers such as chitosan (CS) to prepare such composites [58].

Cellulose acetate was among the first electrospun polymers [59]; it is considered suitable for electrospinning because it can be easily dissolved in common organic solvents and maintains high mechanical strength during membrane fabrication [60]. To date, there are no studies that report the effective electrospinning of carboxymethylcellulose without the addition of another polymer; nonetheless, blends of CMC with polyvinylpyrrolidone (PVP) [61], polyvinyl alcohol (PVA) [62], or polyethylene oxide (PEO) [63,64] were successfully electrospun into nanofibrous membranes with applications especially in the biomedical area. The electrospinning experiments usually take place at room temperature under normal atmospheric conditions. The device consists of three major parts: high voltage power supply, feeding nozzle or spinneret, and a grounded collecting plate (metal plate or rotating drum). The polymeric solution is inserted into a capillary tube connected to a feeding nozzle. A high voltage source is used to inject a certain polarity charge into the polymeric solution or melt. When the electric field reaches a critical value the repulsive electrical forces surpass the surface tension forces at the tip of the nozzle and the solution is accelerated towards the opposite polarity collector. The solvent is evaporated and polymeric fibers are formed [65]. Solution viscosity, polymer concentration, and molecular weight are important factors that influence the electrospinning process and an optimal balance between them must be achieved in order to generate uniform fibers. Due to their unique properties, electrospun fibers have been successfully applied in various domains such as environmental engineering, pharmaceutics, optoelectronics, biomedicine, and biotechnology [66].

Lately, an increased research interest was directed towards the utilization of cellulose derivatives/hydroxyapatite composite membranes in environmental or biomedical engineering (e.g., water purification, bone tissue regeneration, wound healing, controlled drug delivery). It was found that there is a good compatibility between cellulose-based matrices and hydroxyapatite, the inorganic particles interacting with the organic component by inter and intramolecular hydrogen bonds, and also by ion-dipole forces formed between the calcium ions of HA and the functional groups on the cellulose derivatives surface [60]. For example, in the case of CA, the positively charged calcium ions bind with the negatively charged carboxylate groups and hydrogen bonds are formed between the hydroxyl groups of cellulose acetate and hydroxyapatite (Figure 1).

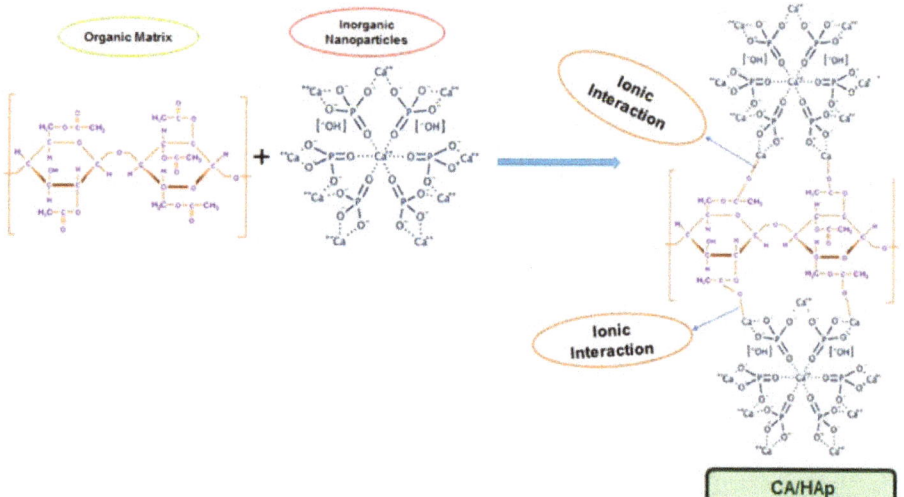

Figure 1. Proposed formation mechanism of cellulose acetate/hydroxyapatite hybrid composites (reproduced with permission from Ref. [60]).

It is demonstrated that hydroxyapatite (HA) particles are good mineral adsorbers and have the capacity to bind divalent heavy metal ions, hence they were researched for the removal of harmful substances in drinking water [53]. Hydroxyapatite particles can also be used as additives for cellulose acetate to improve the morphology and properties of the membranes, this resulting in a better separation performance and higher water flux values [54]. More than that they could act as a catalyst for the precipitation process involved in phase inversion, by increasing the viscosity of the solution [39,57]. The most common synthesis methods for membranes are phase inversion with two possible variants—precipitation in a non-solvent and solvent evaporation. The most versatile is the precipitation of the membrane in a non-solvent. The non-solvent flow through solution polymer film will determine the porosity and morphology depending on several parameters—viscosity, temperature, and miscibility with the polymer solvent. These properties directly influence the speed of membrane formation—the higher the formation speed, the smaller the pore diameter with a high distribution of pores at the surface. A slow process of membrane coagulation will lead to pores with a large diameter and a lower distribution of pores on the surface. Each possibility is preferred depending on the application of the synthesized membrane. The presence of hydroxyapatite in a polymer solution acts in two ways—firstly it will increase the viscosity of the solution and second it will influence the speed of non-solvent through the polymer solution film. The first influence is always higher, so from a more viscous initial solution, membranes with a decreased diameter of pores will be obtained. This can be translated into a more efficient separation. More than that, hydroxyapatite particles itself, in the structure of the membrane, participate in the separation process due to their porosity and ability to retain small chemical species like cations, pesticides, and dyes.

An arising issue is represented by the aggregation tendency of the inorganic particles but various strategies such as ultrasound-assisted mixing [52], surfactants addition [54], or modification of the hydroxyapatite surface [56] are investigated for an improved dispersion. The aggregation tendency of hydroxyapatite particles in aqueous media was successfully overcome by using electrospinning as membrane fabrication technique. Electrospun membranes have some advantages over phase inversion membranes in terms of volume and aspect ratio, specific area, and porosity, but, according to a recent study, HA concentrations higher than 3 wt/v% generated a "beads on a string" morphology in the case of CA solutions (15 wt/v%) in a mixed solvent system of acetone and N,N-dimethylformamide (DMF) [60]. Therefore, the inorganic filler concentration must be carefully chosen in order to obtain smooth nanofibers. Due to its biocompatibility, bioactivity, and osteoconductive properties, hydroxyapatite distinguished itself among materials used for bone tissue regeneration. However, the direct use of hydroxyapatite in such applications was associated with poor mechanical and chemical stability especially in the case of synthetic particles [67]. Therefore hybrid polymer/hydroxyapatite composites were developed, natural polymers such as collagen [68,69], polylactic acid [24,70], chitosan [71,72], and cellulose [73] being preferred, instead of synthetic ones [74], for the preparation of novel composite materials with superior bioactivity compared to pure components.

2. Applications of Cellulose Derivatives/Hydroxyapatite Composite Membranes

2.1. Water Purification

Membrane technology received great attention and was widely applied in the process of separation of toxic compounds from water. Various studies were focused on the improvement of cellulose derivatives/hydroxyapatite composite membranes morpho-structural characteristics in order to ensure an increased water flow without affecting their filtration ability. For example, Pandele et al. prepared cellulose acetate/hydroxyapatite membranes by immersion precipitation and used ultrasound assisted mixing to achieve a homogenous filler dispersion throughout the organic phase. The membranes presented a dense surface with embedded micro-structured hydroxyapatite crystals and conical shaped pores on the bottom side (Figure 2) [52]. In this case, scanning electron microscopy presents surface aspect changes and the emergence of hydroxyapatite crystals in the composite membranes, both on the active and on the porous surface of the membrane. On the active surface, a slight decrease of the average diameter of the pores and shape in the case of composite membranes was observed compared to membranes made of neat cellulose acetate (explained by weak chemical interaction between the acetate groups of the polymer and the phosphate groups of HA). In such cases these interactions can lead to internal defects in the membrane structure. A difference in pore diameter on active surface and porous surface of the membranes can also be easily observed. This can be explained by the precipitation of the polymer film into a non-solvent which generate an asymmetric structure of the membrane with a conical shape of pores in cross section of the membrane [2,12]. Water flows increased by increasing the concentration of hydroxyapatite, ranging from 8.29 L/m^{2}h for the CA membrane to 20.96, 23.25, or

26.73 L/m²h for the composite membranes with 1, 2, and 4 wt. % HA due to the uniform addition of hydroxyapatite particles within the membrane structure.

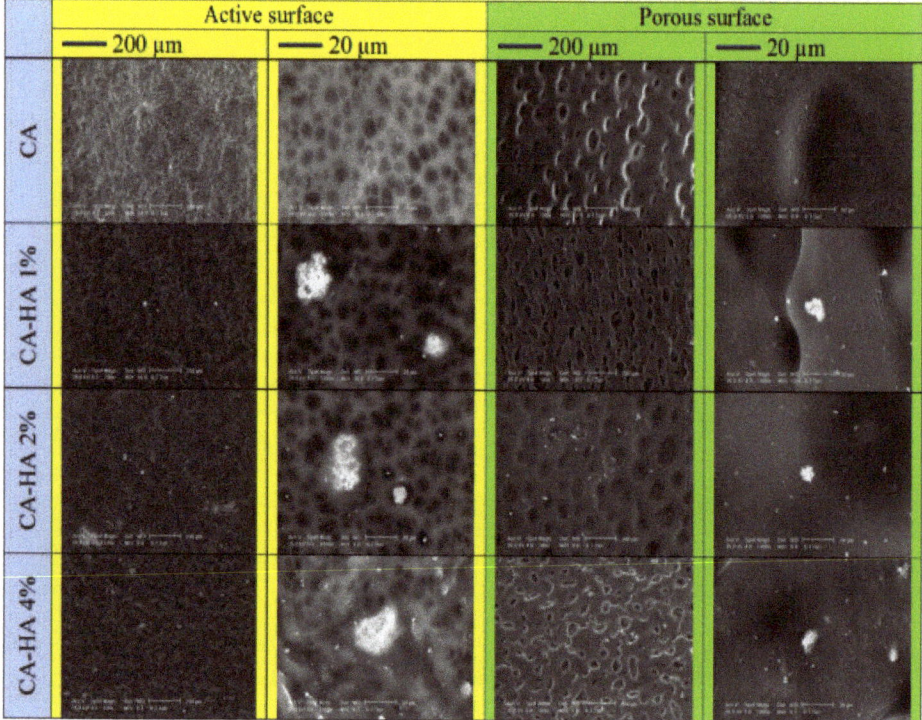

Figure 2. SEM images of the cellulose acetate/hydroxyapatite membranes prepared by immersion precipitation (reproduced with permission from Ref. [52]).

Despite the sonication treatment, high amounts of filler cannot be dispersed uniformly throughout the membrane and hydroxyapatite clusters with high concentrations are formed. Another dispersion improvement strategy was reported by Ohland et al. and it is represented by the functionalization of the hydroxyapatite particles in a methanol plasma atmosphere, prior to their ultrasound-assisted incorporation in the polymeric solution. X-Ray Photoelectron Spectroscopy (XPS) analysis of the functionalized hydroxyapatite revealed an increase in the oxygen atomic percentage and also in the oxygen/carbon ratio, this indicating that the methanol plasma precursor successfully provided hydroxyl groups that were further transferred to the particles surface, this increasing their hydrophilicity and compatibility with the polymeric matrix. The composite membranes prepared using phase inversion with partial solvent evaporation before precipitation presented an anisotropic morphology (Figure 3a). The plasma-treated hydroxyapatite particles were well dispersed, aggregates being observed only at 5 wt. % HA concentrations and higher (Figure 3b). The incorporation of the functionalized particles in the cellulose acetate matrix improved the membrane affinity and lead to an increased water and salt flux in forward and reverse osmosis permeation tests (Figure 3c,d) [56]. In another study conducted by the same research group, the plasma functionalized hydroxyapatite particles were incorporated in both layers of a dual-structured membrane composed of a porous cellulose acetate support obtained by phase inversion and a selective polyamide layer synthesized over the porous support by interfacial polymerization. The addition of the functionalized hydroxyapatite particles in the cellulose acetate support improved the hydrophilicity and reduced internal concentration polarization, which further increased water flux in forward osmosis without affecting the reverse salt flux. More than that, the incorporation of functionalized hydroxyapatite in the selective polyamide layer reorganized the polymeric chains and improved its affinity with water, this resulting in lower diffusion resistance and improvement of water permeability in reverse and forward osmosis [75].

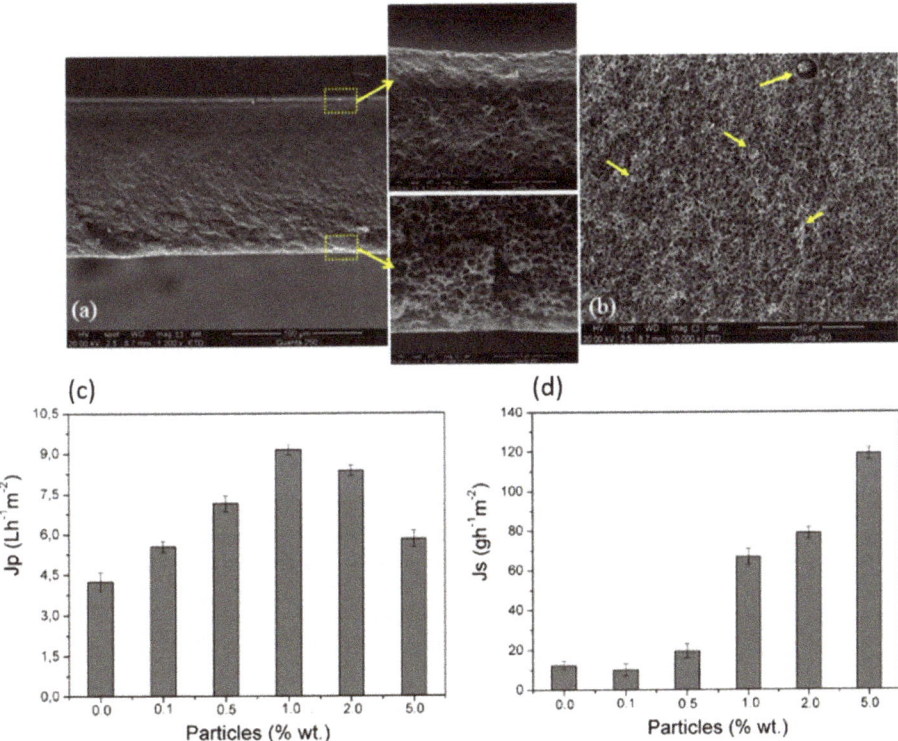

Figure 3. SEM images of the anisotropic cellulose acetate/hydroxyapatite membranes with details of top and bottom layers (**a**); particle agglomerates observed in the cross-section of the membrane containing 5 wt. % hydroxyapatite (HA) (**b**); water (**c**); and salt flux (**d**) of the composite membranes obtained during permeation tests (reproduced with permission from Ref. [56]).

High performance ultrafiltration membranes based on cellulose acetate and nano-hydroxyapatite were prepared using the surfactant assisted phase inversion method to prevent particle agglomeration. The nano-powder was first dispersed in 12-hydroxystearic acid, an amphiphilic surfactant, and then mixed homogenously with the polymeric solution, acetone being used as solvent in both cases. The filler loading percentage was varied between 10 and 30 wt. %, the best results being obtained for the membrane containing 30 wt. % hydroxyapatite. The higher water flux values and the increase in salt rejection capability of the composite membranes were attributed to the formation of hydrophilic nano-pores in the top layer and in the pore walls of the middle layer, coupled with the negative surface charge. These characteristics are thought to also be involved in the modification of the fouling behavior [54]. Overall, the addition of an amphiphilic surfactant during the membrane preparation process was shown to be an effective way to prevent the hydroxyapatite particle aggregation and allowed the effective incorporation of a high filler percentage in the organic matrix. The membranes obtained by this method presented superior performances in terms of water flux values (34.96 L/m^2 h bar) and fouling resistance compared to previously reported studies [52,56,75].

Azzaoui et al. used polyethylene glycol (PEG) 1000 as surfactant, during the hydroxyapatite synthesis stage, to obtain highly dispersed particles that were further incorporated into cellulose acetate [53] or hydroxyethyl cellulose acetate (HECA) [76] matrices. The hybrid membranes were prepared using phase inversion by controlled evaporation, a procedure also known as solvent casting. Morphology studies showed a good compatibility between both of the two organic matrices (CA and HECA) and hydroxyapatite particles. The thermal stability of the composites increased proportionally with the loading of HA but was slightly lower compared to the neat polymers. For example, the decomposition temperature of HECA/HA/PEG ranged between 269–353 °C while neat HECA decomposed between 280–375 °C, respectively, CA/HA/PEG composites decomposed between 355–375 °C while neat CA membranes presented a maximum decomposition temperature of ~376.88 °C. Nuclear Magnetic Resonance (NMR) spectra revealed the presence of hydrogen bonding between carbon–oxygen groups in HECA and the hydroxyl groups on the inorganic particles surface. The CA/HA/PEG membranes were further tested for their ability to separate toxic compounds from water. They presented recovery rates of adsorbed bisphenol-A, a toxic and carcinogenic organic compound from polluted water, comparable with the commercial Florisil®® adsorbent for chromatography, notably in the case of the 30/60/10 CA/HA/PEG composite (Figure 4). The study concluded that thin CA/HA/PEG composite membranes coupled with gas chromatography-mass spectrometry (GC–MS) analysis could be successfully used for the retention and detection of potential toxic compounds in water such as bisphenol-A [53].

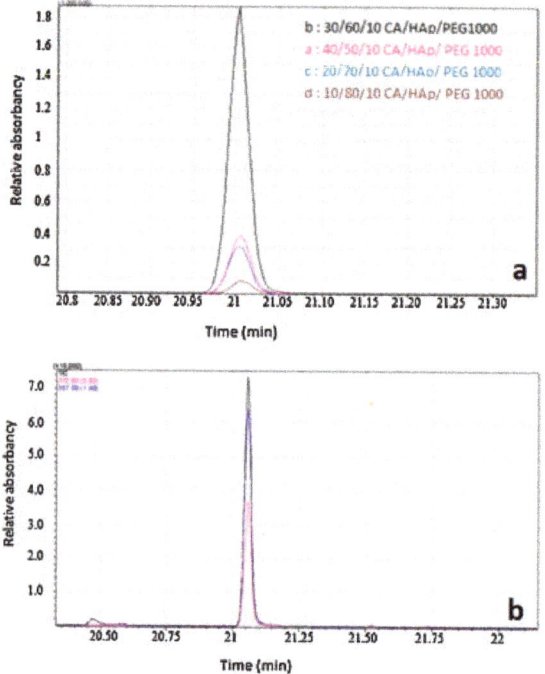

Figure 4. Chromatograms of bisphenol-A desorbed from cellulose acetate/hydroxyapatite/poly(ethylene glycol) (CA/HA/PEG) composite membranes (**a**) and commercial Florisil (**b**) (reproduced with permission from Ref. [53]).

Other examples of toxic compounds in water include heavy metals, pesticides, polychlorinated biphenyls, and inorganic and organic poisons [77]. Nanofibrillated cellulose acetate/hydroxyapatite membranes were tested for the adsorption of Fe (III) and Pb (II) ions in wastewater. The experimental adsorption procedures, conducted using solutions of 10 ppm metal ion concentration, showed that CA/HA nanofibers recorded a 100% removal efficiency for Pb (II) and 95.5% for Fe (III) while neat CA nanofibers recorded only a 19.3% and 22.6% removal of Pb (II) and Fe (III) ions respectively. The adsorption potential was highly increased by the addition of hydroxyapatite due to the chemical and physical affinity between the positively charged metal ions and the functional groups on the filler particles surface. Porosity also played an important role in the improvement of adsorption rates, the heavy metal ions hydroxide precipitates being retained not only on the surface but also in the depth of channels and pores of the membranes [60]. Similar results were obtained in the case of cellulose pulp/carboxymethyl cellulose/nano-hydroxyapatite composite paper sheets tested for the adsorption of iron, lead, and cadmium ions in polluted water [78].

Carboxymethylcellulose was also used in combination with hydroxyapatite to prepare biodegradable nanocomposite membranes for the removal of bisphenol-A from water. The membranes were fabricated by two different methods, the conventional solution casting and the double decomposition procedure. Double decomposition is a common method used for the synthesis of hydroxyapatite. It consists of the decomposition of both calcium and phosphate sources in an aqueous solution to form hydroxyapatite precipitates [79]. In this case $Ca(NO_3)_2 \cdot 4H_2O$ and $(NH_4)_2HPO_4$ were precipitated in an aqueous CMC solution and lysine, an essential amino acid with antibacterial activity due to its amine group [80], was used as a plasticizer. The membrane fabricated by double decomposition had superior characteristics in terms of filler dispersion and flexibility. Due to the improved dispersion of the inorganic particles in the polymeric matrix, the double decomposition membrane kept its transparency after the addition of HA, while the casted one became opaque. The bisphenol-A extraction efficiency was evaluated using gas chromatography coupled with mass spectrometry. The highest recovery rate was recorded for the membranes obtained by double decomposition containing a mass ration of 20:70:10 CMC/HA/lysine. These membranes also presented good antifungal and antimicrobial properties against *Candida albicanis*, Gram-positive *Bacillus subtillus* and Gram-negative *Escherichia coli*, this indicating their multi-functionality and potential to be used for both water purification and biomedical applications [44].

2.2. Bone Tissue Engineering

The use of cellulose derivatives/hydroxyapatite composite membranes as artificial scaffolds for in vitro regeneration of bone tissue was reported in several studies. Osteoblasts are anchorage dependent cells whose attachment on scaffolds depends on the surface area and porosity. Electrospinning cellulose acetate results in biocompatible membranes, with appropriate structural characteristics for cellular attachment and migration. The thin fibers serve as attachment sites for cells while the interconnected porosity allows the cellular migration inside the membrane. Studies showed that the fibrous cellulose acetate membranes are very flexible and the pores can dynamically expand to accommodate growing cells (Figure 5a). The bioactivity is ensured by the hydroxyapatite particles that are involved in both initial adsorption of adhesion proteins such as fibronectin and vitronectin, that will further serve as anchorage sites for osteoblasts (Figure 5b,c) and in cellular differentiation and promotion of calcium phosphates deposition [81].

Surface modifications via oxygen plasma or alkaline treatments were reported for the improvement of the membranes biological activity by mechanisms such as migration of the hydroxyapatite filler towards the fiber surface or the increase in hydrophilicity and deacetylation of cellulose acetate [82]. According to a study conducted by Tao et al., electrospun cellulose acetate/silk fibroin/hydroxyapatite membranes are an efficient carrier for the sustained release of bone morphogenetic protein-2 (BMP-2). Silk fibroin contains ligands that promote cell adhesion, migration, and proliferation and can also support the differentiation of mesenchymal stem cells into osteoblasts. BMP-2 is a growth factor that plays an important role during the development and regeneration of bone tissue. Coaxial electrospinning was employed for the membranes preparation, the fibers consisting in a cellulose acetate core and a silk fibroin shell in which nano-hydroxyapatite and BMP-2 were included. The presence of hydroxyapatite and the controlled release of BMP-2 simulated the proliferation and osteogenic differentiation of bone marrow mesenchymal stem cells seeded on the fibrous membranes and also the cell—mediated calcium accumulation. The fibrous composites were implanted directly on a bone defect site to study their influence on the tissue regeneration in vivo. It was observed that the presence of the material significantly increased the rate of novel bone tissue formation in rat cranial defects after 12 weeks of implantation [83].

Figure 5. Differences between osteoblast cells grown on neat cellulose acetate fibers (**a**) and composite cellulose acetate/nano-hydoxyapatite fibers at low (**b**) and high (**c**) magnification (reproduced with permission from Ref. [81]).

Natural bone is an anisotropic material composed of cells, a collagen matrix, and calcium phosphates in the form of hydroxyapatite. Inspired by the constituents of the natural bone, hybrid membranes based on cellulose acetate, hydroxyapatite, and cellulose microfibers were studied as materials for bone tissue engineering. Cellulose microfibers extracted from raw cotton were selected as reinforcing agents for the cellulose acetate matrix due to their resemblance with collagen. The microfibers were coated with hydroxyapatite by immersion in simulated body fluid (SBF) prior to their dispersion in the cellulose acetate solution using cetrimonium bromide (CTAB) as surfactant. The thin membranes obtained after solvent evaporation presented a porous morphology, favorable for osseointegration and were flexible enough to be rolled and stacked, if required, to increase their mechanical strength and anisotropy [84]. Hydroxyapatite deposited from SBF was also used for the coating of cellulose acetate/polyvinylpyrrolidone (PVP) electrospun membranes. The immiscibility of the two polymers coupled with the low surface tension of PVP lead to a phase separation during the electrospinning process, this causing the formation of a dual fiber structure composed of an inner CA core and an outer PVP discontinuous shell. As the PVP content increased, the uniform cylindrical fibers morphology was gradually lost, flat fibers with multiple surface grooves being obtained. However, the composite fibers provided more favorable conditions for biomineralization, the grooves and cavities facilitating the hydroxyapatite particles deposition and crystals growth. Additionally, the thermal stability was improved due to the formation of hydrogen bonds between CA and PVP [85]. Yamaguchi et al. also reported a successful method for the fabrication of hydroxyapatite coated cellulose-based membranes by biomineralization in SBF. Cellulose acetate nanofibers were deacetylated, then treated with sodium hydroxide and subsequently carboxymethylated using a sodium carbonate/chloroacetic acid mixture to obtain sodium carboxymethylcellulose (NaCMC) nanofibers. It was noticed that the NaOH concentration, biomineralization time, and bicarbonate ion (HCO_3^-) density had an important influence on the hydroxyapatite crystals formation and growth. At higher NaOH and HCO_3^- concentrations, a large number of fine hydroxyapatite crystals were formed on the nanofibers probably due to the increased substitution degree of the carboxyl groups and the intermolecular interactions between HA particles and HCO_3^- ions. A longer biomineralization time lead to crystal growth and formation of aggregates in the hollow spots between the nanofibers (Figure 6). A mineralization time of 6 h was found to be optimal for the proliferation of MC3T3-E1 cells, the fine attachment and homogenous dispersion of hydroxyapatite on the NaCMC nanofibers favoring cellular anchorage and development [86].

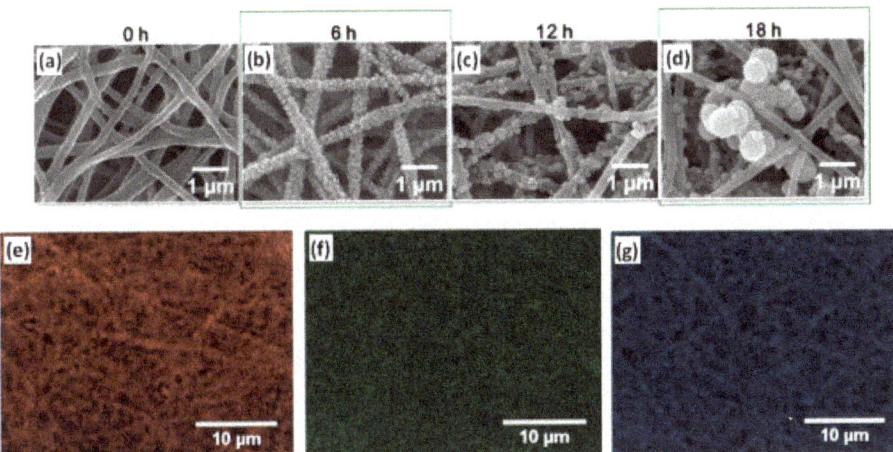

Figure 6. SEM images of hydroxyapatite coated sodium carboxymethylcellulose nanofibers mineralized for 0 h (**a**), 6 h (**b**), 12 h (**c**), 18 h (**d**), and elemental mapping images of the NaCMC/HA nanofibers mineralized for 6 h: carbon map (**e**), calcium map (**f**), and phosphorus map (**g**) (reproduced with permission from Ref. [86]).

Other cellulose based matrices reported in literature for biomimetic mineralization include composites of ethyl cellulose (EC) and hydroxypropyl cellulose (HPC) with polyacrylic acid (PAA). The composites were soaked in salt solutions containing small amounts of PAA to prevent mineral precipitation and thermostated at 30 °C for 5 days, the solutions being frequently changed to maintain an optimal pH value (8.5–9). The mineralization rate was proportional to the content of PAA that served as an ion adsorbent. XRD and EDS analysis revealed that the minerals were incorporated in the polymeric matrix in the form of amorphous calcium phosphate, hydroxyapatite, and PAA-Ca^{2+} complexes. The thermal and mechanical properties were remarkably improved after mineralization, especially in the case of EC/PAA composites that presented a higher percentage of deposited inorganic compounds [87].

Recent studies showed that carboxymethylcellulose stimulates adhesion, spreading, and migration of mouse fibroblasts in vitro and also has the potential to induce osteogenic differentiation. These findings support its use in the fabrication of materials for bone tissue regeneration [88]. Carboxymethylcellulose blends with polyvinyl pyrrolidone (PVP) and sodium tripolyphosphate as crosslinking agents were reported for the preparation of electrospun nanofibrous membranes with applications in bone tissue engineering. For an improved bioactivity, hydroxyapatite doped with different quantities of zinc and manganese was incorporated in the CMC/PVP solutions prior to the electrospinning process. According to previous studies, the slow release of zinc and manganese ions at the defect site has the ability to enhance novel bone formation and growth around implants and also improves cellular adhesion. The cellular proliferation ability of human osteoblasts (HOS) seeded on the composite membranes was evaluated by 3-[4-5-dimethylthiazole-2-yl]-2,5-diphenyltetrazolium bromide (MTT) assays at different time periods (1, 3, and 7 days). The cellular proliferation and attachment increased proportionally to the culture time, this indicating the non-toxicity of the fibers (Figure 7). The antimicrobial activity and hemocompatibility were also evaluated. The highest zone of microorganism's growth inhibition being observed for the membrane containing 60 wt. % hydroxyapatite doped with 0.1 M Zn–Mn (PC1-60). This membrane was also the most hemocompatible, with hemolysis values lower than 3% when compared to the other formulations [61].

Polyethylene oxide (PEO) was also found to be an effective additive for the improvement of CMCs electrospinnability, the emergence of hydrogen bonds between the macromolecules of CMC and PEO, combined with the polymeric chains entanglement, resulting in a crosslinked structure which allows the spinning of nanofibers. A detailed study on the influence of the electrospinning parameters on the formation of nanofibers from aqueous mixtures of carboxymethylcellulose/polyethylene oxide/nano-hydroxyapatite was performed by Gasparic et al. It was found that a blend of 7 wt. % CMC and 5 wt. % PEO (CMC:PEO 1:1 wt./wt.), electrospun using a voltage of 60–65 kV and a distance of 150 mm between electrodes, at a relative humidity lower than 50% at 20 °C, produced high quality nanofibrous membranes. The addition of hydroxyapatite nanoparticles broadened the size distribution of the fibers but their diameter remained on the nanoscale. Additionally, the prepared membranes were hydrophobized by impregnation with alkenyl succinic anhydride, to extend their range of applications by making them insoluble in water. Cellular viability results were similar to the ones obtained for commercial collagen/apatite scaffolds. Osteoblasts grown on the hydrophobized membranes presented similar viability but reduced cellular attachment compared to the commercial substrates [64]. Another strategy to improve the properties of CMC was reported by Qi et al. and it consisted in the protonation of CMC nonwoven sheets. To prepare the nonwoven sheets, pre-purified cotton linters were dissolved in cuprammonium and the obtained solution was ejected from the spinning nozzle of a net conveyor. The resultant cellulose nonwoven sheets were then carboxymethylated using a mixture of sodium chloroacetate and sodium hydroxide [49]. For protonation, the sheets were immersed for 1 h in a mixture of nitric acid and methanol. The protonated sheets were then loaded with calcium phosphates by an alternate soaking method in aqueous solutions of $CaCl_2$/Tris-HCl and NaH_2PO_4 to obtain flexible and easy to handle scaffolds for bone regeneration. XRD spectra showed that the deposited calcium phosphate was a mixed phase of brushite (dicalcium phosphate dihydrate) and hydroxyapatite. The highly soluble brushite could act as a calcium resource and promote mineralization during osteogenesis while hydroxyapatite contributes to osteoconductivity. Further analysis indicated that a higher protonation degree determined a more effective loading due to the suppression of the fibers swelling during the soaking process, this leaving the fiber gaps open for calcium phosphates deposition. Also, the sheet dissolution time can be adjusted according to the desired application by modifying the protonation degree [88].

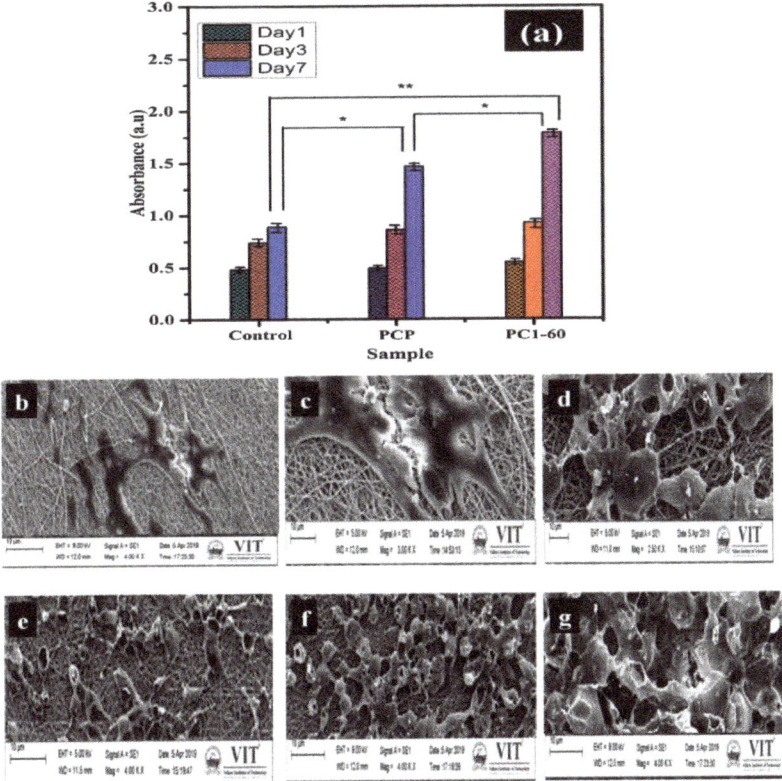

Figure 7. MTT assay results for HOS cells seeded on CA/PVP (PCP) and CA/PVP/HA 0.1M Zn–Mg (PC1-60) membranes on the 1st, 3rd, and 7th days (**a**); microscopy images showing cellular growth (**b–d**) and attachment (**e–g**) on PCP and PC1-60 membranes (reproduced with permission from Ref. [61]). * Significant difference from all in day 3 ($p < 0.5$). ** Significant difference from all in day 7 ($p < 0.5$).

Guided bone regeneration is a surgical procedure that uses barrier membranes, with or without bone substitution materials, to avoid the interfering of non-osteogenic components in tissue regeneration, prior to the installment of metallic implants. The barrier membrane plays a vital role in the guided bone regeneration process, therefore, the optimization of membrane materials, both in terms of barrier and bioactive properties, represents an important research topic [89]. Composites based on cellulose acetate and hydroxyapatite represent a suitable option for guided bone regeneration as long as an optimal synthesis route is developed to ensure best membrane performances in this field. In a recent study conducted by Dascalu et al., the influence of tree key synthesis parameters—ultrasonic dispersion time, hydroxyapatite particles size, and powder concentration—on the final cellulose acetate/bovine-bone derived hydroxyapatite membrane characteristics was investigated. Two sorts of hydroxyapatite powders (20 µm vs. 40 µm particle size) were added in different concentrations (20%, 30%, and 40%) in cellulose acetate solutions and sonicated for 1 or 4 min. The biocomposite membranes were obtained by immersion precipitation, using distilled water as non-solvent. The variation of the synthesis parameters influenced both surface and bulk characteristics of the membranes. It was determined that a uniform dispersion and a homogenous surface and volume aspect is favored by smaller hydroxyapatite particles sizes and a longer sonication time even at high concentrations (Figure 8). In vitro biocompatibility assays were also performed using mouse pre-osteoblasts (MC3T3-E1) and the results were favorable, all the analyzed membranes eliciting a good cellular response in terms of adhesion and viability [90].

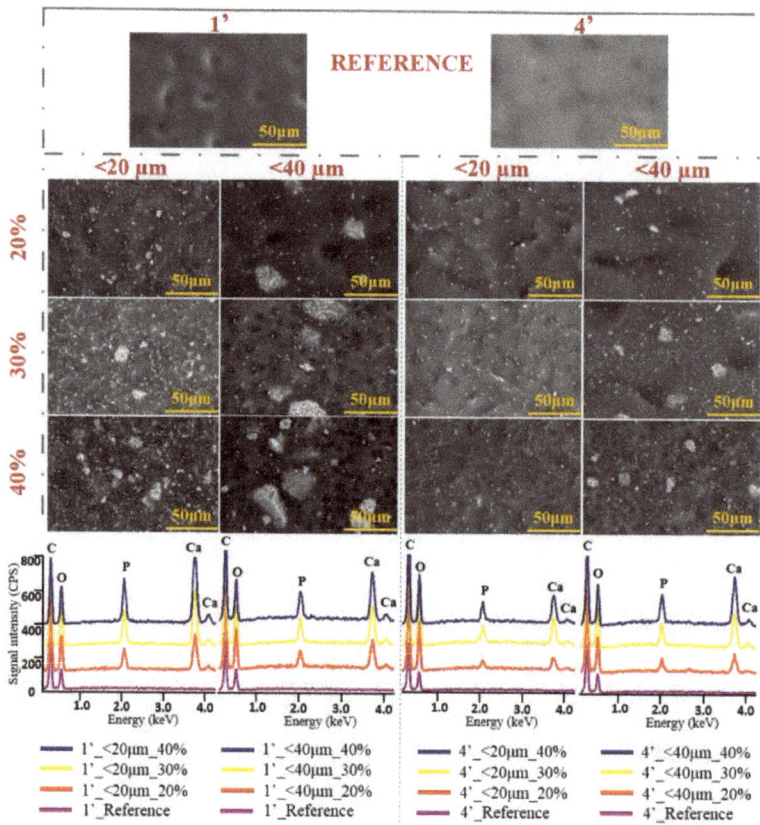

Figure 8. The influence of the three key synthesis parameters—ultrasonic dispersion time, hydroxyapatite particles size, and powder concentration—on the morpho-compositional features of the composite cellulose acetate/hydroxyapatite membranes (reproduced with permission from Ref. [91]).

Carboxymethylcellulose/chitosan blends were also studied in combination with nano-hydroxyapatite (n-HA) for the preparation of membranes with applications in guided bone regeneration. Different techniques were employed for the preparation of such membranes but gradual electrostatic assembly was found to be the most effective for obtaining a uniform, homogenous structure. Jiang et al. studied three types of blending methods to determine which one is optimal for the electrostatic self-assembly of sodium carboxymethyl cellulose (NaCMC) and chitosan (CS) without phase separation or aggregation of the nano-hydroxyapatite particles or polymeric components. Layered casting was used in early studies, more specifically, a certain amount of nano-hydroxyapatite was homogenized into an aqueous NaCMC solution and the mixture was poured in a Petri plate. Afterwards, a chitosan solution in acetic acid was casted slowly on the NaCMC/n-HA wet membrane and the tri-composite material was dried at room temperature and crosslinked with $CaCl_2$. However, SEM analysis showed that the chitosan solution did not penetrate into the NaCMC/n-HA liquid membrane and the electrostatic interactions were limited only at the interface between the two phases [91]. To avoid the formation of this double-layered structure, layered casting was replaced with solution blending. The process consisted of the preparation of two different solutions, CS in acetic acid and NaCMC/n-HA slurry in distilled water, which were further mixed together under continuous stirring. The ternary blending mixture was dried at room temperature, immersed in NaOH

5% solution for 24 h, and then dried again. It was noticed that a membrane cannot be formed using this technique, the organic components having the tendency of agglomeration into macroscopic particles or clusters due to the immediate electrostatic interactions that appear between the cationic groups of CS and anionic groups of NaCMC when the two separate solutions are mixed together. Finally, it was established that the optimal blending method was the incorporation of n-HA slurry and chitosan powder into the aqueous NaCMC solution followed by dropwise addition of acetic acid under intense stirring for the gradual dissolution of the CS particles. The extended chitosan chains connect to the surrounding NaCMC matrix through electrostatic interactions and intermolecular hydrogen bonds forming a polymer composite reinforced with n-HA particles that also form hydrogen bonds with the organic matrix. This homogenous structure provided enhanced hydrostability to the membrane, preventing its progressive collapse into pieces, as observed in the case of the layered casted and solution blended membranes when soaked into water. The nano-hydroxyapatite content influenced the tensile strength and elongation rate, which presented an increasing trend with the n-HA percentage up to 60 wt. %, a sharp decrease being recorded at higher values [58]. This optimized blending technique was used for the preparation of NaCMC/CS/n-HA composites for bone tissue regeneration. The membrane containing 60 wt. % n-HA displayed the highest osteoblast viability and osteocalcin expression during the cellular compatibility tests. For an improved integration with the anisotropic natural bone tissue, the membranes were treated by mechanical perforation with a pore size of 300 μm (1 mm pore-to-pore spacing) and curled in a concentric manner to obtain spiral-cylindrical scaffolds with an osteon-like structure (Figure 9). After rolling, the membranes presented a tensile strength comparable to that of cancellous bone (4.91 MPa) and were therefore considered appropriate to support new bone formation at the site of implantation. The materials were implanted in radial defects models of New Zealand white rabbits and radiopacity was observed in the whole implant area after 12 weeks due to the new bone tissue formation throughout the entire scaffold and its remodeling into cortical bone. Histological analysis of the cross-sections collected from the middle part of the scaffold revealed that the newly formed bone tissue grew along the spiral wall and bone marrow penetrated the entire cylinder forming a medullar cavity in the center, this indicating a successful osseointegration and functional reconstruction facilitated by the spiral-cylindrical architecture. The study concluded that the NaCMC/CS/n-HA spiral-cylindrical scaffolds have a high potential for the treatment of large bone defects and could also be effective in the healing of critical-sized segmental bone defects [92].

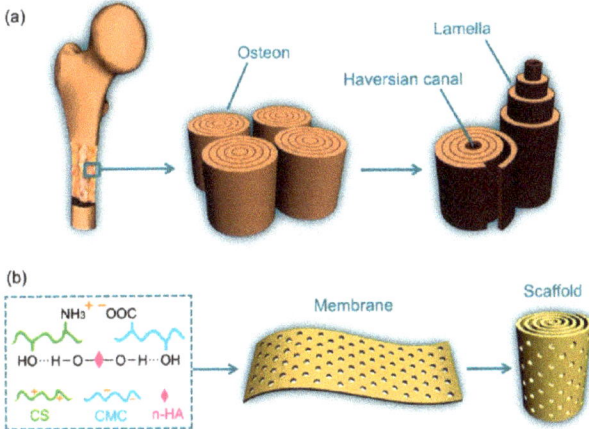

Figure 9. Structural comparison between natural bone (**a**) and the CS/NaCMC/n-HA spiral-cylindrical scaffold (**b**) (reproduced with permission from Ref. [93]).

2.3. Wound Healing

Multiple studies emphasized the important role of calcium in the normal homeostasis of the skin and keratinocyte proliferation and differentiation [93]. Nano-hydroxyapatite was used as a calcium source and incorporated in cellulose acetate/gelatin solutions. The composite membranes were prepared by electrospinning and their potential as wound healing mats was evaluated via a full-thickness excision wound model in healthy adult male Wistar rats. The results showed that the concentration of nanoparticles was in direct correlation with porosity, hydrophilicity, water vapor transmission rate, and cellular proliferation. The synthesized dressings had higher wound closure percentages compared to the sterile gauze used as control and also improved collagen synthesis, re-epithelialization, neovascularization and the overall cosmetic appearance of the wounds (Figure 10). Best results were obtained for the cellulose acetate/gelatin membranes with a hydroxyapatite content of 25 mg/10 cc of polymer solution, a higher inorganic particles percentage leading to hyperplasia of the epidermal layer and foreign body reactions [50].

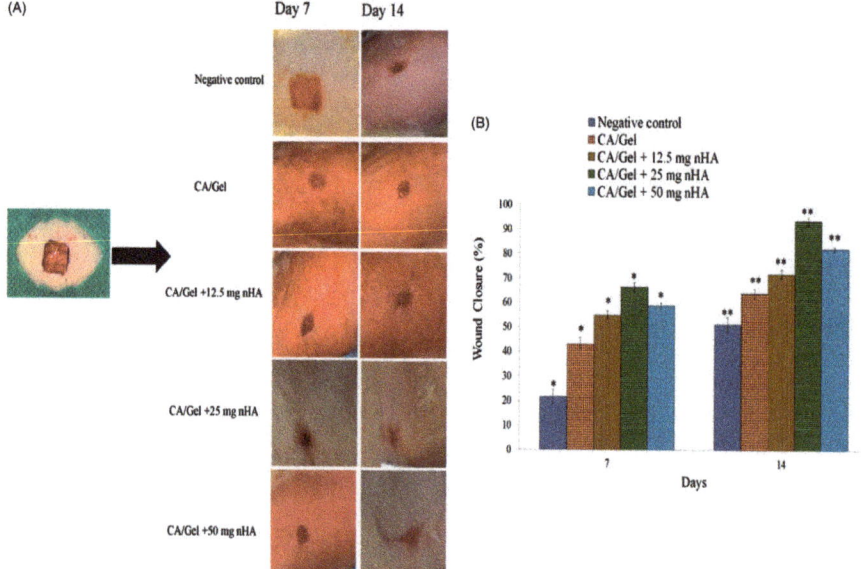

Figure 10. Cosmetic appearance of the wounds treated with the cellulose acetate/gelatin/nano-hydroxapatite dressings 7 and 14 days post-wounding (**A**) and histograms comparing the wound closure percentages at the end of the 7th and 14th day post-wounding (**B**) (reproduced with permission from Ref. [50]). * Significant difference from all in day 7 ($p < 0.5$). ** Significant difference from all in day 14 ($p < 0.5$).

2.4. Controlled Drug Delivery

Drug delivery systems are gaining momentum in the biomedical field due to their advantages such as controlled release of the therapeutic substance, which ensure the maintenance of its level within a desired range and the need for fewer administrations associated with an increased patient comfort [94]. Nystatin (Nys), a polyene antibiotic obtained from *Streptomyces noursei* bacterial strain, was used as model drug to evaluate the controlled drug delivery ability of cellulose acetate/hydroxyapatite membranes. The composites were obtained by adding calculated amounts of hydroxyapatite and nystatin in the cellulose acetate solution (CA/Nys mass ratio ~3.34), followed by casting on Petri plates and solvent evaporation [55]. The release rate of nystatin was higher when hydroxyapatite was present in the membranes. The release enhancement was associated to the increased membrane porosity generated by the addition of HA. The connection between membrane porosity and the drug release rate was also observed by Wang et al. that prepared CA membranes for the transdermal delivery of scopolamine using PEG 1000 as pore-forming agent. The scopolamine release was found to be proportional to the PEG content, a higher PEG percentage generating a more porous structure and higher release rates respectively [95]. Another important aspect that influenced the Nys release was related to the retention of drug molecules by the hydroxyapatite crystals formed on the membrane surface, thus favoring the drug dispersion and also increasing its exposure to the dissolution media during the release tests. Hydroxyapatite and nystatin were also observed in the cross-section SEM images as white spots spread over the internal surface of the membrane, therefore the particles were not retained entirely on the surface and a part of them were incorporated in the polymeric matrix. The CA membranes achieved a sustained Nys release for up to 50 h, the CA/HA ones releasing the drug faster. According to literature, they can be characterized as an effective controlled drug delivery device because the drug release occurred over a period of time greater than 24 h [96]. Chitosan (CS) [97] and polycaprolactone (PCL) [98] were also investigated as controlled delivery devices for nystatin in the form of films and capsules. However, the CS films maintained a gradual nystatin release for only 5 h, reaching a plateau phase in approximately 1–2 h while the PCL nanocapsules completely released the drug after 9 h. Consequently, CA/HA-based delivery systems are a viable option for applications where a prolonged drug delivery is required.

2.5. Hemodialysis

Hemodialysis is the process of removal of waste and extra water from blood using an external filter called a dialyzer, which contains a semipermeable membrane [99]. Cellulose acetate was among the first polymers used for the production of hemodialysis membranes, the renewability, biodegradability, and biocompatible nature as well as low cost and good toughness recommending it for this application [100]. Due to the binding sites on the particles surface, hydroxyapatite can effectively retain biomolecules. This aspect is very important because in hemodialysis the toxins should be removed and useful biomolecules like proteins and carbohydrates should be retained. Hayder et al. investigated the effect of hydroxyapatite on the morphology, water flux, urea clearance, and glucose and bovine serum albumin (BSA) retention of cellulose acetate/polyethylene glycol membranes prepared by phase inversion. After the addition of hydroxyapatite, macro-pores were formed in the compact cellulose acetate matrix due to the leech out of filler particles. The porosity and increased hydrophilicity of the composite membranes determined a significant increase in water and urea permeation, the calculated permeability of the composites membranes being 0.075 L/h m bar for water and 0.035 L/h m bar for urea compared to 0.005 and 0.0037 L/h m bar in the case of pure CA. However, only low quantities of glucose and bovine serum albumin passed through the membranes [51]. These results, together with the reported hemocompatibility of cellulose acetate [101], show promising prospects for the use of hydroxyapatite as functional filler in hemodialysis membranes.

3. Conclusions and Future Perspectives

The present review was focused on the systematical presentation of the preparation methods and functional applications of cellulose derivatives/hydroxyapatite composite membranes. The manufacturing techniques varied from phase inversion by immersion precipitation or solvent evaporation to electrospinning and gradual electrostatic assembly, the optimal method being selected depending on the type of cellulose derivative used as an organic matrix. For example, cellulose acetate was found to be appropriate especially for electrospinning and phase inversion while the anionic nature of carboxymethylcellulose recommended it for gradual electrostatic assembly in combination with oppositely charged polymers. The reviewed studies showed that the addition of hydroxyapatite can impart desired characteristics to the surface and structure of the membrane thus improving its performance. However, an important issue is represented by the aggregation tendency of the hydroxyapatite particles especially when phase inversion is used for membrane preparation. Various strategies were employed to obtain a homogenous filler dispersion, the addition of surfactants, plasma modification of the inorganic particles, and ultrasound assisted mixing showing promising results. Regarding the application fields of the hybrid cellulose derivatives/hydroxyapatite membranes, water purification, hemodialysis, and tissue engineering are among the domains intensively researched.

Future perspectives are mainly related to the biomedical applications of composite membranes with hydroxyapatite, particularly osseointegration combined with the controlled release of therapeutic agents. The synthesis of hydroxyapatite with controlled particle dimensions to favor an optimal osseointegration represents one of the challenges of the near future. Too large dimensions hinder the development of pre-osteoblasts, while too small ones are characterized by an inefficient activity of these particles. Another future direction is represented by the obtainement of hydroxyapatite from natural sources, other than bone (e.g., snail shells or seashells) as well as its wide scale synthesis from anorganic precursors. A domain that will know an explosive development will be the one of controlled release of pharmaceutically active substances from hydroxyapatite particles, antibiotics and cytostatics being the main classes of drugs that will be studied for such applications. In the dental field, composite membranes are already used to favor osseointegration. The capacity of the membrane to release antibiotics, locally, from particles of hydroxyapatite will enhance both quality and efficiency of the medical act and will also improve the patient's life. The utilization of composite membranes with hydroxyapatite for such applications will provide multiple simultaneous advantages—the membrane will favor implant osseointegration in the case of major bone defects generated by the extraction of the tumoral mass and the targeted release of cytostatics will be associated with an augmented treatment efficiency and with the minimization of the toxic impact of these drugs on the human body.

Principle practical gaps are related to the control of porosity for both components—hydroxyapatite and the membrane. New methods for hydroxyapatite production will be accompanied by processing steps like controlled milling or sintering in order to provide the desired porosity for the required application. In the domain of membranes the porosity can be easily controlled, but a major challenge will be the synthesis of membranes with macroporosity and nanoparticles in their structure. Most probably, the easiest way to address this problem will be the chemical bonding between hydroxyapatite and the cellulose derivative, which will remain a future challenge.

Author Contributions: Conceptualization, M.O. and S.I.V.; resources, M.O. and S.I.V.; data curation, M.O. and S.I.V.; writing—original draft preparation, M.O. and S.I.V.; project administration, S.I.V.; funding acquisition, S.I.V. All authors have read and agreed to the published version of the manuscript.

Funding: This work was supported by a grant of the Romanian National Authority for Scientific Research and Innovation, CNCS-UEFISCDI, project number PN-III-P1-1.2-PCCDI-2017-0407—Intelligent materials for medical applications, sub-project—New generation of hemodialysis composite membranes with derivatized graphene.

Acknowledgments: The authors gratefully acknowledge the financial support through project PN-III-P1-1.2-PCCDI-2017-0407—Intelligent materials for medical applications, sub-project—New generation of hemodialysis composite membranes with derivatized graphene.

Conflicts of Interest: The authors declare no conflict of interest.

References

1. Ulbricht, M. Advanced functional polymer membranes. *Polymer* **2006**, *47*, 2217–2262. [CrossRef]
2. Lehn, J.-M. *Supramolecular Chemistry: Concepts and Perspectives*; Wiley: Hoboken, NJ, USA, 1995.
3. Loeb, S. *The Loeb-Sourirajan Membrane: How It Came About*; American Chemical Society (ACS): Washington, DC, USA, 1981; Volume 153, pp. 1–9.
4. Tiwari, R.R.; Jin, J.; Freeman, B.; Paul, D.R. Physical aging, CO_2 sorption and plasticization in thin films of polymer with intrinsic microporosity (PIM-1). *J. Membr. Sci.* **2017**, *537*, 362–371. [CrossRef]
5. Zhou, H.; Tao, F.; Liu, Q.; Zong, C.; Yang, W.; Cao, X.; Jin, W.; Xu, N. Microporous Polyamide Membranes for Molecular Sieving of Nitrogen from Volatile Organic Compounds. *Angew. Chem. Int. Ed.* **2017**, *56*, 5755–5759. [CrossRef]
6. Zhou, H.; Jin, W. Membranes with Intrinsic Micro-Porosity: Structure, Solubility, and Applications. *Membranes* **2018**, *9*, 3. [CrossRef]

7. Voicu, Ş.I.; Dobrica, A.; Sava, S.; Ivan, A.; Naftanaila, L. Cationic surfactants-controlled geometry and dimensions of polymeric membrane pores. *J. Optoelectron. Adv. Mater.* **2012**, *14*, 923–928.
8. Ioniță, M.; Crica, L.E.; Voicu, S.I.; Dinescu, S.; Miculescu, F.; Costache, M.; Iovu, H. Synergistic effect of carbon nanotubes and graphene for high performance cellulose acetate membranes in biomedical applications. *Carbohydr. Polym.* **2018**, *183*, 50–61. [CrossRef]
9. Voicu, Ş.I.; Pandele, A.; Tanasă, E.; Rughinis, R.; Crica, L.; Pilan, L.; Ionita, M. The impact of sonication time through polysulfone-graphene oxide composite films properties. *Dig. J. Nanomater. Biostruct.* **2013**, *8*, 1389–1394.
10. Serbanescu, O.; Pandele, A.; Miculescu, F.; Voicu, S.I. Synthesis and Characterization of Cellulose Acetate Membranes with Self-Indicating Properties by Changing the Membrane Surface Color for Separation of Gd(III). *Coatings* **2020**, *10*, 468. [CrossRef]
11. Raicopol, M.D.; Andronescu, C.; Voicu, S.I.; Vasile, E.; Pandele, A.M. Cellulose acetate/layered double hydroxide adsorptive membranes for efficient removal of pharmaceutical environmental contaminants. *Carbohydr. Polym.* **2019**, *214*, 204–212. [CrossRef]
12. Thakur, V.; Voicu, S.I. Recent advances in cellulose and chitosan based membranes for water purification: A concise review. *Carbohydr. Polym.* **2016**, *146*, 148–165. [CrossRef]
13. Satulu, V.; Mitu, B.; Pandele, A.; Voicu, S.; Kravets, L.; Dinescu, G. Composite polyethylene terephthalate track membranes with thin teflon-like layers: Preparation and surface properties. *Appl. Surf. Sci.* **2019**, *476*, 452–459. [CrossRef]
14. Castro-Muñoz, R.; Agrawal, K.V.; Coronas, J. Ultrathin permselective membranes: The latent way for efficient gas separation. *RSC Adv.* **2020**, *10*, 12653–12670. [CrossRef]
15. Stamatialis, D.; Papenburg, B.J.; Gironès, M.; Saiful, S.; Bettahalli, S.N.; Schmitmeier, S.; Wessling, M. Medical applications of membranes: Drug delivery, artificial organs and tissue engineering. *J. Membr. Sci.* **2008**, *308*, 1–34. [CrossRef]
16. Corobea, M.; Muhulet, O.; Miculescu, F.; Antoniac, I.V.; Vuluga, Z.; Florea, D.; Vuluga, D.M.; Butnaru, M.; Ivanov, D.; Voicu, S.I.; et al. Novel nanocomposite membranes from cellulose acetate and clay-silica nanowires. *Polym. Adv. Technol.* **2016**, *27*, 1586–1595. [CrossRef]
17. Falkenhagen, D.; Strobl, W.; Hartmann, J.; Schrefl, A.; Linsberger, I.; Kellner, K.-H.; Aussenegg, F.; Leitner, A. Patient safety technology for microadsorbent systems in extracorporeal blood purification. *Artif. Organs* **2002**, *26*, 84–90. [CrossRef]
18. Flendrig, L.M.; La Soe, J.W.; Jörning, G.G.; Steenbeek, A.; Karlsen, O.T.; Bovée, W.M.; Ladiges, N.C.; Velde, A.A.T.; Chamuleau, R.A. In vitro evaluation of a novel bioreactor based on an integral oxygenator and a spirally wound nonwoven polyester matrix for hepatocyte culture as small aggregates. *J. Hepatol.* **1997**, *26*, 1379–1392. [CrossRef]
19. Shih, C.; Lee, K.-R.; Lai, J. 60Co γ-ray irradiation modified poly(4-methyl-pentene) membrane for oxygenator. *Eur. Polym. J.* **1994**, *30*, 629–634. [CrossRef]
20. Sauer, I.M.; Neuhaus, P.; Gerlach, J.C. Concept for modular extracorporeal liver support for the treatment of acute hepatic failure. *Metab. Brain Dis.* **2002**, *17*, 477–484. [CrossRef]
21. Leoni, L.; Boiarski, A.; Desai, T.A. Characterization of Nanoporous Membranes for Immunoisolation: Diffusion Properties and Tissue Effects. *Biomed. Microdevices* **2002**, *4*, 131–139. [CrossRef]
22. Corobea, M.S.; Albu-Kaya, M.; Ion, R.; Cimpean, A.; Miculescu, F.; Antoniac, I.; Raditoiu, V.; Sirbu, I.; Stoenescu, M.; Voicu, S.I.; et al. Modification of titanium surface with collagen and doxycycline as a new approach in dental implants. *J. Adhes. Sci. Technol.* **2015**, *29*, 1–14. [CrossRef]
23. Voicu, S.I.; Condruz, R.M.; Mitran, V.; Cimpean, A.; Miculescu, F.; Andronescu, C.; Miculescu, M.; Thakur, V. Sericin Covalent Immobilization onto Cellulose Acetate Membrane for Biomedical Applications. *ACS Sustain. Chem. Eng.* **2016**, *4*, 1765–1774. [CrossRef]
24. Pandele, A.M.; Constantinescu, A.E.; Radu, I.; Miculescu, F.; Voicu, S.I.; Ciocan, L.T. Synthesis and Characterization of PLA-Micro-structured Hydroxyapatite Composite Films. *Materials* **2020**, *13*, 274. [CrossRef]
25. Miculescu, F.; Maidaniuc, A.; Voicu, S.I.; Thakur, V.; Stan, G.; Ciocan, L.T. Progress in Hydroxyapatite–Starch Based Sustainable Biomaterials for Biomedical Bone Substitution Applications. *ACS Sustain. Chem. Eng.* **2017**, *5*, 8491–8512. [CrossRef]

26. Miculescu, F.; Maidaniuc, A.; Miculescu, M.; Batalu, N.D.; Ciocoiu, R.C.; Voicu, S.I.; Stan, G.; Thakur, V. Synthesis and Characterization of Jellified Composites from Bovine Bone-Derived Hydroxyapatite and Starch as Precursors for Robocasting. *ACS Omega* **2018**, *3*, 1338–1349. [CrossRef]
27. Maidaniuc, A.; Miculescu, M.; Voicu, S.I.; Ciocan, L.T.; Niculescu, M.; Corobea, M.; Rada, M.E.; Miculescu, F. Effect of micron sized silver particles concentration on the adhesion induced by sintering and antibacterial properties of hydroxyapatite microcomposites. *J. Adhes. Sci. Technol.* **2016**, *30*, 1829–1841. [CrossRef]
28. Miculescu, F.; Mocanu, A.C.; Stan, G.; Miculescu, M.; Maidaniuc, A.; Cimpean, A.; Mitran, V.; Voicu, S.I.; Machedon-Pisu, T.; Ciocan, L.T. Influence of the modulated two-step synthesis of biogenic hydroxyapatite on biomimetic products' surface. *Appl. Surf. Sci.* **2018**, *438*, 147–157. [CrossRef]
29. Maidaniuc, A.; Miculescu, F.; Andronescu, C.; Miculescu, M.; Matei, E.; Pencea, I.; Csaki, I.; Machedon-Pisu, T.; Ciocan, L.T.; Voicu, S.I.; et al. Induced wettability and surface-volume correlation of composition for bovine bone derived hydroxyapatite particles. *Appl. Surf. Sci.* **2018**, *438*, 158–166. [CrossRef]
30. Miculescu, F.; Mocanu, A.-C.; Dascălu, C.A.; Maidaniuc, A.; Batalu, D.; Berbecaru, A.C.; Voicu, S.I.; Miculescu, M.; Thakur, V.; Ciocan, L.T. Facile synthesis and characterization of hydroxyapatite particles for high value nanocomposites and biomaterials. *Vacuum* **2017**, *146*, 614–622. [CrossRef]
31. Sirviö, J.A.; Heiskanen, J.P. Room-temperature dissolution and chemical modification of cellulose in aqueous tetraethylammonium hydroxide–carbamide solutions. *Cellulose* **2019**, *27*, 1933–1950. [CrossRef]
32. Khiari, R.; Belgacem, M.N. Potential for using multiscale Posidonia oceanica waste. In *Lignocellulosic Fibre and Biomass-Based Composite Materials*; Elsevier BV: Amsterdam, The Netherlands, 2017; pp. 447–471.
33. Alfassi, G.; Rein, D.M.; Shpigelman, A.; Cohen, Y.; Rein, D.M. Partially Acetylated Cellulose Dissolved in Aqueous Solution: Physical Properties and Enzymatic Hydrolysis. *Polymers* **2019**, *11*, 1734. [CrossRef]
34. Global Industry Analysis. Cellulose Acetate (MCP-2035). Available online: https://www.strategyr.com/market-report-cellulose-acetate-forecasts-global-industry-analysts-inc.asp (accessed on 5 May 2020).
35. Xu, J.; Wu, Z.; Wu, Q.; Kuang, Y. Acetylated cellulose nanocrystals with high-crystallinity obtained by one-step reaction from the traditional acetylation of cellulose. *Carbohydr. Polym.* **2020**, *229*, 115553. [CrossRef] [PubMed]
36. Araújo, D.; Castro, M.C.R.; Figueiredo, A.; Vilarinho, M.; Machado, A. Green synthesis of cellulose acetate from corncob: Physicochemical properties and assessment of environmental impacts. *J. Clean. Prod.* **2020**, *260*, 120865. [CrossRef]
37. Puls, J.; Wilson, S.A.; Hölter, D. Degradation of Cellulose Acetate-Based Materials: A Review. *J. Polym. Environ.* **2010**, *19*, 152–165. [CrossRef]
38. Wsoo, M.A.; Shahir, S.; Bohari, S.P.M.; Nayan, N.H.M.; Razak, S.I.A. A review on the properties of electrospun cellulose acetate and its application in drug delivery systems: A new perspective. *Carbohydr. Res.* **2020**, *491*, 107978. [CrossRef]
39. Dobos, A.M.; Filimon, A.; Bargan, A.; Zaltariov, M.-F. New approaches for the development of cellulose acetate/tetraethyl orthosilicate composite membranes: Rheological and microstructural analysis. *J. Mol. Liq.* **2020**, *309*, 113129. [CrossRef]
40. Martin-Gil, V.; Ahmad, M.; Castro-Muñoz, R.; Fila, V. Economic Framework of Membrane Technologies for Natural Gas Applications. *Sep. Purif. Rev.* **2018**, *48*, 298–324. [CrossRef]
41. Altunina, L.; Tikhonova, L.; Yarmukhametova, E. Method for Deriving Carboxymethyl Cellulose. *Eurasian Chem. J.* **2016**, *3*, 49. [CrossRef]
42. Golbaghi, L.; Khamforoush, M.; Hatami, T. Carboxymethyl cellulose production from sugarcane bagasse with steam explosion pulping: Experimental, modeling, and optimization. *Carbohydr. Polym.* **2017**, *174*, 780–788. [CrossRef]
43. Shui, T.; Feng, S.; Chen, G.; Li, A.; Yuan, Z.; Shui, H.; Kuboki, T.; Xu, C.C. Synthesis of sodium carboxymethyl cellulose using bleached crude cellulose fractionated from cornstalk. *Biomass Bioenergy* **2017**, *105*, 51–58. [CrossRef]
44. Azzaoui, K.; Mejdoubi, E.; Lamhamdi, A.; Jodeh, S.; Hamed, O.; Berrabah, M.; Jerdioui, S.; Salghi, R.; Akartasse, N.; Errich, A.; et al. Preparation and characterization of biodegradable nanocomposites derived from carboxymethyl cellulose and hydroxyapatite. *Carbohydr. Polym.* **2017**, *167*, 59–69. [CrossRef]
45. Karataş, M.; Arslan, N. Flow behaviours of cellulose and carboxymethyl cellulose from grapefruit peel. *Food Hydrocoll.* **2016**, *58*, 235–245. [CrossRef]

46. Chen, H. 5-Lignocellulose biorefinery product engineering. In *Lignocellulose Biorefinery Engineering*; Chen, H., Ed.; Woodhead Publishing: Cambridge, UK, 2015; pp. 125–165. [CrossRef]
47. Chen, Y.; Cui, G.; Dan, N.; Huang, Y.; Bai, Z.; Yang, C.; Dan, W. Preparation and characterization of dopamine–sodium carboxymethyl cellulose hydrogel. *SN Appl. Sci.* **2019**, *1*, 609. [CrossRef]
48. Singh, V.; Joshi, S.; Malviya, T. Carboxymethyl cellulose-rosin gum hybrid nanoparticles: An efficient drug carrier. *Int. J. Boil. Macromol.* **2018**, *112*, 390–398. [CrossRef]
49. Ohta, S.; Nishiyama, T.; Sakoda, M.; Machioka, K.; Fuke, M.; Ichimura, S.; Inagaki, F.; Shimizu, A.; Hasegawa, K.; Kokudo, N.; et al. Development of carboxymethyl cellulose nonwoven sheet as a novel hemostatic agent. *J. Biosci. Bioeng.* **2015**, *119*, 718–723. [CrossRef]
50. Samadian, H.; Salehi, M.; Farzamfar, S.; Vaez, A.; Ehterami, A.; Sahrapeyma, H.; Goodarzi, A.; Ghorbani, S. In vitro and in vivo evaluation of electrospun cellulose acetate/gelatin/hydroxyapatite nanocomposite mats for wound dressing applications. *Artif. Cells Nanomed. Biotechnol.* **2018**, *46*, 964–974. [CrossRef]
51. Hayder, A.; Hussain, A.; Khan, A.N.; Waheed, H. Fabrication and characterization of cellulose acetate/hydroxyapatite composite membranes for the solute separations in Hemodialysis. *Polym. Bull.* **2017**, *75*, 1197–1210. [CrossRef]
52. Pandele, A.M.; Comanici, F.; Carp, C.; Miculescu, M.; Voicu, S.; Thakur, V.; Serban, B. Synthesis and characterization of cellulose acetate-hydroxyapatite micro and nano composites membranes for water purification and biomedical applications. *Vacuum* **2017**, *146*, 599–605. [CrossRef]
53. Azzaoui, K.; Lamhamdi, A.; Mejdoubi, E.M.; Berrabah, M.; Hammouti, B.; Elidrissi, A.; Fouda, M.M.; Al-Deyab, S.S.; Lamhamdi, A. Synthesis and characterization of composite based on cellulose acetate and hydroxyapatite application to the absorption of harmful substances. *Carbohydr. Polym.* **2014**, *111*, 41–46. [CrossRef]
54. Ciobanu, G.; Ciobanu, O. High-performance ultrafiltration mixed-matrix membranes based on cellulose acetate and nanohydroxyapatite. *Desalin. Water Treat.* **2015**, *57*, 1–9. [CrossRef]
55. Ciobanu, G.; Ana-Maria, B.; Luca, C. Nystatin-loaded Cellulose Acetate/Hydroxyapatite Biocomposites. *Revista de Chimie* **2013**, *64*, 1426–1429.
56. Ohland, A.L.; Salim, V.M.M.; Borges, C.P. Plasma functionalized hydroxyapatite incorporated in membranes for improved performance of osmotic processes. *Desalination* **2019**, *452*, 87–93. [CrossRef]
57. Zare, S.; Kargari, A. Membrane properties in membrane distillation. In *Emerging Technologies for Sustainable Desalination Handbook*; Elsevier BV: Amsterdam, The Netherlands, 2018; pp. 107–156.
58. Jiang, H.; Zuo, Y.; Cheng, L.; Wang, H.; Gu, A.; Li, Y. A homogenous CS/NaCMC/n-HA polyelectrolyte complex membrane prepared by gradual electrostatic assembling. *J. Mater. Sci. Mater. Electron.* **2010**, *22*, 289–297. [CrossRef]
59. Anton, F. Process and Apparatus for Preparing Artificial Threads. U.S. Patent US1975504A, 2 October 1934.
60. Hamad, A.A.; Hassouna, M.S.; Shalaby, T.I.; Elkady, M.F.; Elkawi, M.A.A.; Hamad, H.A. Electrospun cellulose acetate nanofiber incorporated with hydroxyapatite for removal of heavy metals. *Int. J. Boil. Macromol.* **2020**, *151*, 1299–1313. [CrossRef]
61. Kandasamy, S.; Narayanan, V.; Sumathi, S. Zinc and manganese substituted hydroxyapatite/CMC/PVP electrospun composite for bone repair applications. *Int. J. Boil. Macromol.* **2019**, *145*, 1018–1030. [CrossRef]
62. El-Newehy, M.; El-Naggar, M.E.; Alotaiby, S.; El-Hamshary, H.; Moydeen, M.; Al-Deyab, S. Preparation of biocompatible system based on electrospun CMC/PVA nanofibers as controlled release carrier of diclofenac sodium. *J. Macromol. Sci. Part A* **2016**, *53*, 566–573. [CrossRef]
63. Shi, D.; Wang, F.; Lan, T.; Zhang, Y.; Shao, Z. Convenient fabrication of carboxymethyl cellulose electrospun nanofibers functionalized with silver nanoparticles. *Cellulose* **2016**, *23*, 1899–1909. [CrossRef]
64. Gašparič, P.; Kurecic, M.; Kargl, R.J.; Maver, U.; Gradišnik, L.; Hribernik, S.; Kleinschek, K.S.; Smole, M.S. Nanofibrous polysaccharide hydroxyapatite composites with biocompatibility against human osteoblasts. *Carbohydr. Polym.* **2017**, *177*, 388–396. [CrossRef]
65. Raghavan, P.; Nageswaran, S.; Thakur, V.; Ahn, J.-H. Electrospinning of Cellulose: Process and Applications. In *Nanocellulose Polymer Nanocomposites*; Wiley: Hoboken, NJ, USA, 2014; pp. 311–340.
66. Sill, T.J.; Von Recum, H.A. Electrospinning: Applications in drug delivery and tissue engineering. *Biomaterials* **2008**, *29*, 1989–2006. [CrossRef]

67. Fragal, E.H.; Cellet, T.S.; Fragal, V.H.; Companhoni, M.V.; Nakamura, T.U.; Muniz, E.C.; Silva, R.; Rubira, A.F. Hybrid materials for bone tissue engineering from biomimetic growth of hydroxiapatite on cellulose nanowhiskers. *Carbohydr. Polym.* **2016**, *152*, 734–746. [CrossRef]
68. Zhang, Z.; Ma, Z.; Zhang, Y.; Chen, F.; Zhou, Y.; An, Q. Dehydrothermally crosslinked collagen/hydroxyapatite composite for enhanced in vivo bone repair. *Colloids Surf. B Biointerfaces* **2018**, *163*, 394–401. [CrossRef]
69. Montalbano, G.; Molino, G.; Fiorilli, S.; Vitale-Brovarone, C. Synthesis and incorporation of rod-like nano-hydroxyapatite into type I collagen matrix: A hybrid formulation for 3D printing of bone scaffolds. *J. Eur. Ceram. Soc.* **2020**. [CrossRef]
70. Ma, B.; Han, J.; Zhang, S.; Liu, F.; Wang, S.; Duan, J.; Sang, Y.; Jiang, H.; Li, N.; Ge, S.; et al. Hydroxyapatite nanobelt/polylactic acid Janus membrane with osteoinduction/barrier dual functions for precise bone defect repair. *Acta Biomater.* **2018**, *71*, 108–117. [CrossRef]
71. Domínguez, J.H.L.; Jiménez, H.T.; Cocoletzi, H.H.; Hernández, M.G.; Banda, J.A.M.; Nygren, H. Development and in vivo response of hydroxyapatite/whitlockite from chicken bones as bone substitute using a chitosan membrane for guided bone regeneration. *Ceram. Int.* **2018**, *44*, 22583–22591. [CrossRef]
72. Nazeer, M.A.; Yilgor, E.; Yilgor, E. Intercalated chitosan/hydroxyapatite nanocomposites: Promising materials for bone tissue engineering applications. *Carbohydr. Polym.* **2017**, *175*, 38–46. [CrossRef]
73. Torgbo, S.; Sukyai, P. Fabrication of microporous bacterial cellulose embedded with magnetite and hydroxyapatite nanocomposite scaffold for bone tissue engineering. *Mater. Chem. Phys.* **2019**, *237*, 121868. [CrossRef]
74. Swetha, M.; Sahithi, K.; Moorthi, A.; Srinivasan, N.; Ramasamy, K.; Selvamurugan, N. Biocomposites containing natural polymers and hydroxyapatite for bone tissue engineering. *Int. J. Boil. Macromol.* **2010**, *47*, 1–4. [CrossRef]
75. Ohland, A.L.; Salim, V.M.M.; Borges, C.P. Nanocomposite membranes for osmotic processes: Incorporation of functionalized hydroxyapatite in porous substrate and in selective layer. *Desalination* **2019**, *463*, 23–31. [CrossRef]
76. Azzaoui, K.; Mejdoubi, E.; Lamhamdi, A.; Zaoui, S.; Berrabah, M.; Elidrissi, A.; Hammouti, B.; Fouda, M.M.; Al-Deyab, S.S. Structure and properties of hydroxyapatite/hydroxyethyl cellulose acetate composite films. *Carbohydr. Polym.* **2015**, *115*, 170–176. [CrossRef]
77. Hellawell, J.M. Toxic substances in rivers and streams. *Environ. Pollut.* **1988**, *50*, 61–85. [CrossRef]
78. El-Aziz, M.E.A.; Saber, E.; El-Khateeb, M. Preparation and characterization of CMC/HA-NPs/pulp nanocom-posites for the removal of heavy metal ions. *KGK Rubberpoint* **2019**, *72*, 36–41.
79. Minh, D.P.; Rio, S.; Sharrock, P.; Sebei, H.; Lyczko, N.; Tran, N.D.; Raii, M.; Nzihou, A. Hydroxyapatite starting from calcium carbonate and orthophosphoric acid: Synthesis, characterization, and applications. *J. Mater. Sci.* **2014**, *49*, 4261–4269. [CrossRef]
80. Cutrona, K.J.; Kaufman, B.A.; Figueroa, D.; Elmore, D.E. Role of arginine and lysine in the antimicrobial mechanism of histone-derived antimicrobial peptides. *FEBS Lett.* **2015**, *589*, 3915–3920. [CrossRef]
81. Gouma, P.; Xue, R.; Goldbeck, C.; Perrotta, P.; Balázsi, C. Nano-hydroxyapatite—Cellulose acetate composites for growing of bone cells. *Mater. Sci. Eng. C* **2012**, *32*, 607–612. [CrossRef]
82. Kwak, D.H.; Lee, E.J.; Kim, D.J. Bioactivity of cellulose acetate/hydroxyapatite nanoparticle composite fiber by an electro-spinning process. *J. Nanosci. Nanotechnol.* **2014**, *14*, 8464–8471. [CrossRef]
83. Tao, C.; Zhang, Y.; Li, B.; Chen, L. Hierarchical micro/submicrometer-scale structured scaffolds prepared via coaxial electrospinning for bone regeneration. *J. Mater. Chem. B* **2017**, *5*, 9219–9228. [CrossRef]
84. Nisar, F.; Bin Khalid, U.; Akram, M.A.; Javed, S.; Mujahid, M. Fabrication of Cellulose Acetate/Cellulose-HA Composite Films for Bone Fixation. *Key Eng. Mater.* **2018**, *778*, 325–330. [CrossRef]
85. Hou, J.; Wang, Y.; Xue, H.; Dou, Y. Biomimetic Growth of Hydroxyapatite on Electrospun CA/PVP Core-Shell Nanofiber Membranes. *Polymers* **2018**, *10*, 1032. [CrossRef]
86. Yamaguchi, K.; Prabakaran, M.; Ke, M.; Gang, X.; Chung, I.-M.; Um, I.C.; Gopiraman, M.; Kim, I.S. Highly dispersed nanoscale hydroxyapatite on cellulose nanofibers for bone regeneration. *Mater. Lett.* **2016**, *168*, 56–61. [CrossRef]
87. Ogiwara, T.; Katsumura, A.; Sugimura, K.; Teramoto, Y.; Nishio, Y. Calcium Phosphate Mineralization in Cellulose Derivative/Poly(acrylic acid) Composites Having a Chiral Nematic Mesomorphic Structure. *Biomacromolecules* **2015**, *16*, 3959–3969. [CrossRef]

88. Qi, P.; Ohba, S.; Hara, Y.; Fuke, M.; Ogawa, T.; Ohta, S.; Ito, T. Fabrication of calcium phosphate-loaded carboxymethyl cellulose non-woven sheets for bone regeneration. *Carbohydr. Polym.* **2018**, *189*, 322–330. [CrossRef]
89. Elgali, I.; Omar, O.; Dahlin, C.; Thomsen, P. Guided bone regeneration: Materials and biological mechanisms revisited. *Eur. J. Oral Sci.* **2017**, *125*, 315–337. [CrossRef] [PubMed]
90. Dascălu, C.-A.; Maidaniuc, A.; Pandele, A.M.; Voicu, S.I.; Machedon-Pisu, T.; Stan, G.E.; Cîmpean, A.; Mitran, V.; Antoniac, I.V.; Miculescu, F. Synthesis and characterization of biocompatible polymer-ceramic film structures as favorable interface in guided bone regeneration. *Appl. Surf. Sci.* **2019**, *494*, 335–352. [CrossRef]
91. Liuyun, J.; Yubao, L.; Chengdong, X. A novel composite membrane of chitosan-carboxymethyl cellulose polyelectrolyte complex membrane filled with nano-hydroxyapatite I. Preparation and properties. *J. Mater. Sci. Mater. Electron.* **2009**, *20*, 1645–1652. [CrossRef] [PubMed]
92. Jiang, H.; Zuo, Y.; Zou, Q.; Wang, H.; Du, J.; Li, Y.; Yang, X. Biomimetic Spiral-Cylindrical Scaffold Based on Hybrid Chitosan/Cellulose/Nano-Hydroxyapatite Membrane for Bone Regeneration. *ACS Appl. Mater. Interfaces* **2013**, *5*, 12036–12044. [CrossRef]
93. Lansdown, A.B.G. Calcium: A potential central regulator in wound healing in the skin. *Wound Repair Regen.* **2002**, *10*, 271–285. [CrossRef]
94. Yun, Y.H.; Lee, B.K.; Park, K. Controlled Drug Delivery: Historical perspective for the next generation. *J. Control. Release* **2015**, *219*, 2–7. [CrossRef]
95. Wang, F.-J.; Yang, Y.Y.; Zhang, X.-Z.; Zhu, X.; Chung, T.-S.; Moochhala, S.; Chung, T.-S. Cellulose acetate membranes for transdermal delivery of scopolamine base. *Mater. Sci. Eng. C* **2002**, *20*, 93–100. [CrossRef]
96. Ruggiero, R.; Carvalho, V.D.A.; Da Silva, L.G.; Magalhães, D.; Ferreira, J.A.; De Menezes, H.H.M.; De Melo, P.G.; Naves, M.M. Study of in vitro degradation of cellulose acetate membranes modified and incorporated with tetracycline for use as an adjuvant in periodontal reconstitution. *Ind. Crop. Prod.* **2015**, *72*, 2–6. [CrossRef]
97. Aksungur, P.; Sungur, A.; Ünal, S.; Iskit, A.B.; Squier, C.A.; Şenel, S. Chitosan delivery systems for the treatment of oral mucositis: In vitro and in vivo studies. *J. Control. Release* **2004**, *98*, 269–279. [CrossRef]
98. Abousamra, M.M.; Basha, M.; Awad, G.E.; Mansy, S.S. A promising nystatin nanocapsular hydrogel as an antifungal polymeric carrier for the treatment of topical candidiasis. *J. Drug Deliv. Sci. Technol.* **2019**, *49*, 365–374. [CrossRef]
99. Vadakedath, S.; Kandi, V. Dialysis: A Review of the Mechanisms Underlying Complications in the Management of Chronic Renal Failure. *Cureus* **2017**, *9*, e1603. [CrossRef] [PubMed]
100. Dumitriu, C.; Voicu, S.I.; Muhulet, A.; Nechifor, G.; Popescu, S.; Ungureanu, C.; Carja, A.; Miculescu, F.; Andronescu, E.; Pirvu, C. Production and characterization of cellulose acetate—Titanium dioxide nanotubes membrane fraxiparinized through polydopamine for clinical applications. *Carbohydr. Polym.* **2018**, *181*, 215–223. [CrossRef]
101. Sunohara, T.; Masuda, T.; Kawanishi, H.; Takemoto, Y. Fundamental Characteristics of the Newly Developed ATA™ Membrane Dialyzer. *Contrib. Nephrol.* **2016**, *189*, 215–221. [CrossRef] [PubMed]

© 2020 by the authors. Licensee MDPI, Basel, Switzerland. This article is an open access article distributed under the terms and conditions of the Creative Commons Attribution (CC BY) license (http://creativecommons.org/licenses/by/4.0/).

MDPI
St. Alban-Anlage 66
4052 Basel
Switzerland
Tel. +41 61 683 77 34
Fax +41 61 302 89 18
www.mdpi.com

Materials Editorial Office
E-mail: materials@mdpi.com
www.mdpi.com/journal/materials

www.ingramcontent.com/pod-product-compliance
Lightning Source LLC
LaVergne TN
LVHW070653100526
838202LV00013B/956